Springer Complexity

Springer Complexity is an interdisciplinary program publishing the best research and academic-level teaching on both fundamental and applied aspects of complex systems—cutting across all traditional disciplines of the natural and life sciences, engineering, economics, medicine, neuroscience, social and computer science.

Complex Systems are systems that comprise many interacting parts with the ability to generate a new quality of macroscopic collective behavior the manifestations of which are the spontaneous formation of distinctive temporal, spatial or functional structures. Models of such systems can be successfully mapped onto quite diverse "real-life" situations like the climate, the coherent emission of light from lasers, chemical reaction-diffusion systems, biological cellular networks, the dynamics of stock markets and of the internet, earthquake statistics and prediction, freeway traffic, the human brain, or the formation of opinions in social systems, to name just some of the popular applications.

Although their scope and methodologies overlap somewhat, one can distinguish the following main concepts and tools: self-organization, nonlinear dynamics, synergetics, turbulence, dynamical systems, catastrophes, instabilities, stochastic processes, chaos, graphs and networks, cellular automata, adaptive systems, genetic algorithms and computational intelligence.

The three major book publication platforms of the Springer Complexity program are the monograph series "Understanding Complex Systems" focusing on the various applications of complexity, the "Springer Series in Synergetics", which is devoted to the quantitative theoretical and methodological foundations, and the "Springer Briefs in Complexity" which are concise and topical working reports, case studies, surveys, essays and lecture notes of relevance to the field. In addition to the books in these two core series, the program also incorporates individual titles ranging from textbooks to major reference works.

Indexed by SCOPUS, INSPEC, zbMATH, SCImago.

Series Editors

Understanding Complex Systems

Founding Editor: Scott Kelso

Henryk Fukś

Solvable Cellular Automata

Methods and Applications

Springer

Henryk Fukś
Department of Mathematics and Statistics
Brock University
St. Catharines, ON, Canada

ISSN 1860-0832 ISSN 1860-0840 (electronic)
Understanding Complex Systems
ISBN 978-3-031-38702-9 ISBN 978-3-031-38700-5 (eBook)
https://doi.org/10.1007/978-3-031-38700-5

This Springer imprint is published by the registered company Springer Nature Switzerland AG
The registered company address is: Gewerbestrasse 11, 6330 Cham, Switzerland

To my wife and children.

Preface

Mathematica considerat necessitatem complexionis quae est explicatio simplicitatis.

Theodoricus Chartrensis

In the theory of dynamical systems and their applications, the possibility of an explicit determination of the orbit of a dynamical system in a functional form is a valuable property of systems and models. In statistical physics, "exact solvability" is also a desirable property of models, and the usual meaning of this term is that the solutions can be expressed explicitly in terms of some known functions.

Cellular automata are often considered to be fully discrete analogs of partial differential equations. Unfortunately, in contrast to partial differential equations, relatively little is known about the solvability of cellular automata. The goal of this book is to fill this gap by considering two types of initial value problems for cellular automata, the deterministic one and the probabilistic one.

The deterministic problem can be stated as follows: given the initial configuration x, determine the state of the cell i after n iterations of a specific cellular automaton by expressing this state as an explicit function of x and n. The probabilistic problem, on the other hand, is to determine the probability of the occurrence of a given block of symbols after n iterations of a cellular automaton rule, assuming that the initial configuration is drawn from a known distribution—typically the Bernoulli distribution.

The main focus of the book is elementary cellular automata, that is, those with two states and nearest-neighbor interaction. In Chap. 1, we introduce the basic concepts of cellular automata theory and develop the notation used in the rest of the book. The deterministic initial value problem is introduced in the Chap. 2, where we also presented its solutions for selected simple rules. In Chaps. 2–6, various techniques for solving the deterministic problem are introduced, using elementary CA rules of increasing complexity as examples.

Chapter 7 introduces the concept of probability measure in the context of cellular automata and gives a detailed discussion of block probabilities and their mutual dependencies. The probabilistic initial value problem is then introduced in Chap. 8, where many examples of its solution are given, again using elementary cellular automata with increasing complexity as examples. Probabilistic cellular automata are introduced in the subsequent chapter, and probabilistic solutions for several representative cases are constructed.

Chapter 10 presents various applications of the solutions of the initial value problems obtained elsewhere in the book. Solutions are used to elucidate theoretical concepts such as asymptotic emulation or equicontinuity and to analyze several different models such as road traffic model, models of phase transitions and density classification.

Chapter 11 is devoted to the method of approximating cellular automata orbits, known as the local structure theory. Examples of its application to selected elementary rules as well as some probabilistic rules are given, and some of its modifications are discussed. The last chapter presents an overview of solvable and non-solvable elementary rules and addresses possible directions of generalization of results presented in the book.

The Appendix is intended to serve as a reference section. All solution formulae obtained by the author for elementary cellular automata are gathered there. Many of these are derived elsewhere in the book, but all are obtainable using methods introduced in the book. Table 12.1 gives a quick overview of rules which are solvable in both deterministic and probabilistic sense.

In terms of the background required from the reader, the principle followed by the author was to use only as minimal mathematical apparatus as required. General mathematical culture one typically acquires at the undergraduate level while studying mathematics, physics or computer science should be more than sufficient.

St. Catharines, ON, Canada Henryk Fukś
June 2023

Contents

Notation and Symbols

AE	Almost equicontinuous
AED	Almost equicontinuous direction
CA	Cellular automaton
ED	Equicontinuous direction
FSM	Finite state machine
PCA	Probabilistic cellular automaton
RFP	Ruelle–Frobenius–Perron equations
$\mathcal{A}$	Alphabet or symbol set
N	Number of symbols in $\mathcal{A}$
$\mathcal{A}^{\mathbb{Z}}$	Space of all bisequences over $\mathcal{A}$
$\mathcal{A}^n$	Set of n-blocks over $\mathcal{A}$
F, G, H	Upper case letters represent global functions of cellular automata
F_k	Global function of CA with Wolfram number k
F^n	n-th iterate of F
f, g, h	Lowercase letters represent local functions corresponding to F, G, H
$\mathbf{f}, \mathbf{g}, \mathbf{h}$	Lowercase bold letters represent block mapping corresponding to F, G, H
a, b, c	Variables representing symbols of the alphabet
$\mathbf{a}, \mathbf{b}, \mathbf{c}$	Blocks of symbols (words)
x, y, z	Bisequences of symbols
$\mathbf{a}_i$	i-th element of the block $\mathbf{a}$
$[\mathbf{a}]_i$	Cylinder set generated by $\mathbf{a}$ and anchored at i
x_i	i-th element of the bisequence x
σ	Shift transformation
$\mathbf{f}^{-1}(\mathbf{a})$	Set of preimages of $\mathbf{a}$
$\mathbf{f}^{-n}(\mathbf{a})$	Set of n-step preimages of $\mathbf{a}$
$\psi_a(p)$	Density polynomial corresponding to symbol a
$\psi_b(p_1, \cdots, p_n)$	Density polynomial corresponding to block $\mathbf{b}$
$\psi_b(p_1, \cdots, p_n)$	Density polynomial corresponding to set of strings A

$\mu,\ \nu$	Measures over $\mathcal{A}^{\mathbb{Z}}$
$\mathfrak{M}(\mathcal{A}^{\mathbb{Z}})$	Set of shift-invariant probability measures generated by cylinder sets
$\mathfrak{M}^{(k)}(\mathcal{A}^{\mathbb{Z}})$	Set of Markov measures of order k
$\Xi^{(k)}$	Scramble operator of order k
ν_p	Bernoulli measure over $\{0,\ 1\}^{\mathbb{Z}}$ such that probability of 1 is p
$\nu_{1/2}$	Symmetric (uniform) Bernoulli measure over $\{0,\ 1\}^{\mathbb{Z}}$
$p_n(\mathbf{a})$	Probability of occurrence of **a** after n iterations of CA
$w(a\|\mathbf{b})$	Local transition function of PCA
$[P]$	Iverson bracket, equal to 1 if P is true, otherwise 0
$\bar{x}$	Equals to $1 - x$ for $x \in \{0,\ 1\}$
$\mathrm{ev}(n)$	Equals to 1 if n is even, otherwise 0
$I_k(n)$	Equals to 1 if $n \leq k$, otherwise 0

List of Figures

List of Tables

Chapter 1
Deterministic Cellular Automata

1.1 Basic Definitions

In order to define cellular automata, we will first define the space in which they operate. Let $N > 1$ and let $\mathcal{A} = \{0, 1, \ldots, N-1\}$ be called an *alphabet*, or a *symbol set*, and let $\mathcal{A}^{\mathbb{Z}}$ be a space of all bisequences over $\mathcal{A}$, to be called *configuration space*. By a bisequence we mean an infinite sequence of symbols with indices extending from $-\infty$ to ∞, so that $x \in \mathcal{A}^{\mathbb{Z}}$ can be written as $x = \{x_i\}_{i=-\infty}^{\infty}$, where $x_i \in \mathcal{A}$.

The space of bisequences can be easily "metrized", meaning that we can introduce the notion of distance (metric) in $\mathcal{A}^{\mathbb{Z}}$. Many choices of such metrics have been proposed, but one of the most frequently used is the *Cantor metric* defined as $d(x, y) = 2^{-k}$, where $k = \min\{|i| : x_i \neq y_i\}$. Given two bisequences x and y, k is the index closest to the origin such that the bisequences x and y differ in kth position.

$\mathcal{A}^{\mathbb{Z}}$ with the metric d is called a *Cantor space*, that is, compact, totally disconnected and perfect metric space. We will be interested in functions F that map the space $\mathcal{A}^{\mathbb{Z}}$ to itself, also called *transformations*, $F : \mathcal{A}^{\mathbb{Z}} \to \mathcal{A}^{\mathbb{Z}}$. The set of all such functions is too big to develop a meaningful theory, therefore we will impose some additional restrictions on them. As it is the case in mathematical analysis and topology, the functions which exhibit no abrupt changes in values will be of particular interest to us. More precisely, we will consider transformations $F : \mathcal{A}^{\mathbb{Z}} \to \mathcal{A}^{\mathbb{Z}}$ for which arbitrarily small changes in values can be assured by making the changes of arguments sufficiently small. Such functions are called *continuous*. More formally, a function $F : \mathcal{A}^{\mathbb{Z}} \to \mathcal{A}^{\mathbb{Z}}$ is continuous if for every point $x \in \mathcal{A}^{\mathbb{Z}}$ and every $\epsilon > 0$ there exists $\delta > 0$ such that for all $y \in \mathcal{A}^{\mathbb{Z}}$ we have $d(x, y) < \delta \implies d(f(x), f(y)) < \epsilon$.

Example 1.1 As the first example of a continuous transformation on $\mathcal{A}^{\mathbb{Z}}$ we will consider the shift transformation $\sigma : \mathcal{A}^{\mathbb{Z}} \to \mathcal{A}^{\mathbb{Z}}$ defined as $[\sigma(x)]_i = x_{i+1}$. To prove its continuity, let us take arbitrary $x \in \mathcal{A}^{\mathbb{Z}}$ and any $\epsilon > 0$. If $d(\sigma(x), \sigma(y)) < \epsilon$ then $\sigma(x)$ and $\sigma(y)$ differ in a position with index k or $-k$ such that $k > \log_2 \epsilon$. Since σ shifts bisequences to the left, this means that x and y differ in a position with index $k+1$ or $-k+1$, thus we must ensure that x and y differ in a position with absolute value of the index at least $k+1$. Obviously $\log_2 \epsilon + 1 = \log_2(2k)$, therefore

H. Fukś, *Solvable Cellular Automata*, Understanding Complex Systems,
https://doi.org/10.1007/978-3-031-38700-5_1

choosing $\delta = \epsilon/2$ will satisfy $d(x, y) < \delta \implies d(\sigma(x), \sigma(y)) < \epsilon$, as required by the definition of continuity.

Example 1.2 Consider now the transformation $R : \mathcal{A}^{\mathbb{Z}} \to \mathcal{A}^{\mathbb{Z}}$ which reflects the bisequence with respect to the origin, $[R(x)]_i = x_{-i}$. If we want $d(\sigma(x), \sigma(y)) < \epsilon$ then taking $d(x, y) < \epsilon$ will ensure this, because the spatial reflection does not change the distance between bisequences. Therefore, $\delta = \epsilon$ satisfies the required condition $d(x, y) < \delta \implies d(\sigma(x), \sigma(y)) < \epsilon$, demonstrating that R is continuous.

Example 1.3 Let $V : \mathcal{A}^{\mathbb{Z}} \to \mathcal{A}^{\mathbb{Z}}$ be defined by $[V(x)]_i = x_{\lfloor 1/i \rfloor}$ for $i \neq 0$ and $[V(x)]_0 = x_0$, where $\lfloor x \rfloor = \max\{m \in \mathbb{Z} : m \leq x\}$. We will call it the *index inverting transformation*. Consider $x = \ldots 00000, \ldots$, a configuration consisting of all zeros and let $\epsilon = 1/4$. Obviously $V(x) = x$ for this configuration. If we want continuity, we want to find δ ensuring that $d(V(x), V(y)) < 1/4$ for any y for which $d(x, y) < \delta$, and this means ensuring $d(x, V(y)) < 1/4$, i.e., $[V(y)]_i$ must differ from 0 in a position with index of absolute value 5 or more. There is no way to make this happen by choosing y sufficiently close to $x = \ldots 00000, \ldots$, because being close to $x = \ldots 00000, \ldots$ means that the first no-zero symbol in y appears sufficiently far from the origin. If the first no-zero symbol in y appears in position with index n, then it will be in position with index $\lfloor 1/n \rfloor$ in $V(y)$, and $\lfloor 1/n \rfloor < 5$ for any n. The index inverting transformation is, therefore, discontinuous.

Example 1.4 Let $\mathcal{A} = \{0, 1\}$. A block of n consecutive zeros preceded by 1 and followed by 1 will be called a *cluster* of length n. Consider now a transformation $C : \mathcal{A}^{\mathbb{Z}} \to \mathcal{A}^{\mathbb{Z}}$ which replaces all zeros in clusters of odd length by 1's, and leaves everything else unchanged. This transformation is not continuous, and to see it, let us choose $x = \ldots 00000, \ldots$, a configuration consisting of all zeros and let $\epsilon = 1/4$. Obviously $C(x) = x$ for this configuration. Now fix $k > 0$ and take y such that $y_i = 1$ if $|i| = k$ and $y_i = 0$ otherwise. This configuration has a single odd cluster of zeros of length $2k - 1$ centered at the origin. Transformation C will fill this cluster with 1's and $C(y)$ will then consist of block of 1's of length $2k + 1$ centered at the origin and preceded and followed by all zeros, or in other words, $[C(y)]_i = 1$ if $|i| \leq k$ and $[C(y)]_i = 0$ otherwise. Configurations $C(x)$ and $C(y)$ differ already at $i = 0$ position, thus $d(C(x), C(y)) = 2^{-1} = 1$. We can select k as large as we want to make $d(x, y)$ arbitrarily small, but $d(C(x), C(y))$ will always be equal to 1, thus there is no way to fulfil the condition $d(C(x), C(y)) < 1/4$.

If the transformation $F : \mathcal{A}^{\mathbb{Z}} \to \mathcal{A}^{\mathbb{Z}}$ satisfies $\sigma(T(x)) = T(\sigma(x))$ for any $x \in \mathcal{A}^{\mathbb{Z}}$, we will say that F *commutes with shift* or is *shift-commuting*. It turns out that shift-commuting continuous transformations have a very special property, as stated in the following theorem.

Theorem 1.1 (Curtis–Hedlund–Lyndon) *A shift-commuting transformation* $F : \mathcal{A}^{\mathbb{Z}} \to \mathcal{A}^{\mathbb{Z}}$ *is continuous if and only if there exist integer* $r \geq 0$ *and function* $f : \mathcal{A}^n \to \mathcal{A}$ *such that*

$$[F(x)]_i = f(x_{i-r}, x_{i-r+1}, \dots x_{i+r}) \tag{1.1}$$

for all $x \in \mathcal{A}^{\mathbb{Z}}$ *and* $i \in \mathbb{Z}$.

The proof of the above theorem (stated in a slightly different form) appeared in the 1968 paper of Hedlund [1], where he gave credit to two other contributors, Curtis and Lyndon. More modern version of the proof can be found in [2]. Our version follows [3]. Before we attempt the proof, we need the following lemma.

Lemma 1.1 *Let* U *and* V *be compact, disjoint subsets of* $\mathcal{A}^{\mathbb{Z}}$. *Then there exists* $\delta > 0$ *such that* $d(x, y) \geq \delta$ *whenever* $x \in U$ *and* $y \in V$.

Proof We will prove this by contradiction. Assume that there is no positive lower bound on $d(x, y)$. Then we can find sequences $(u_n) \subset U$ and $(v_n) \subset V$ such that $\lim_{n\to\infty} d(u_n, v_n) = 0$. Compactness of U implies that there exist a convergent subsequence of u_n, to be denoted by u_{n_k}, converging to some point $u \in U$. By the triangle inequality we have

$$d(u, v_{n_k}) \leq d(u, u_{n_k}) + d(u_{n_k}, v_{n_k}).$$

Since the right hand side of the above inequality tends to 0 as $k \to \infty$, we must also have $v_{n_k} \to u$ as $k \to \infty$. This means that u is an accumulation point of V, and therefore, since V is compact, it must belong to V. This is impossible because U and V are disjoint. □

Proof *(Curtis–Hedlund–Lyndon theorem)* ($\Leftarrow$). Assume that Eq. (1.1) holds. Let $x \in \mathcal{A}^{\mathbb{Z}}$ and $F(x) = y$. Now choose $\epsilon > 0$ and $k \in \mathbb{N}$ such that $2^{-(k+1)} < \epsilon$. Select $\delta < 2^{-(r+k+1)}$.

Let $u \in \mathcal{A}^{\mathbb{Z}}$ with $d(x, u) < \delta$. Then $x_i = u_i$ for $i = 0, \pm 1, \pm 2 \dots \pm (k + r)$. We will now apply F to both x and u. $F(x)$ and $F(u)$ will be identical on positions with index $i = 0, \pm 1, \pm 2, \dots \pm k$. Therefore, $d(F(x), F(y)) \leq 2^{-(k+1)} < \epsilon$, proving continuity of F.

To demonstrate that F is shift-commuting we take again arbitrary $x \in \mathcal{A}^{\mathbb{Z}}$ and $y = F(x)$. We then have

$$[\sigma(F(x))]_i = [\sigma(y)]_i = y_{i+1} = f(x_{i+1-r}, \dots, x_{i+1+n}).$$

On the other hand,

$$[F(\sigma(x))]_i = f([\sigma(x)]_{i-r}, \dots, [\sigma(x)]_{i+r}) = f(x_{i+1-r}, \dots, x_{i+1+n}),$$

which is the same result as for $[\sigma(F(x))]_i$, showing that F indeed commutes with σ.

($\Rightarrow$). Suppose that F is continuous and shift-commuting. For a symbol $a \in \mathcal{A}$, define $C(a)$ to be the set of all configurations in $\mathcal{A}^{\mathbb{Z}}$ which have the symbol a in the central position (at the origin), that is, $C(a) = \{x \in \mathcal{A}^{\mathbb{Z}} : x_0 = a\}$. Obviously $\mathcal{A}^{\mathbb{Z}} = \bigcup_{a\in\mathcal{A}} C(a)$, and the sets $C(a)$ are pairwise disjoint. They are all closed sets, and since $\mathcal{A}^{\mathbb{Z}}$ is compact, as closed subsets of a compact space they are also compact.

Let $E(a)$ be the preimage of $C(a)$, $E(a) = F^{-1}(C(a))$. We will now use the basic topological property of compact spaces, namely that continuous preimages of compact subsets in such spaces are also compact. This guarantees compactness of $E(b)$ sets. Furthermore, by Lemma 1.1, there exist $\delta > 0$ such that bisequences in different sets $E(a)$ are at least δ apart.

Let us now choose n satisfying $2^{-(n+1)} < \delta$. Suppose we have two bisequences $x, y \in \mathcal{A}^{\mathbb{Z}}$ such that they agree on the central block extending from $-n$ to n, i.e., $x_i = y_i$ for $i \in \{-n, -n+1, \ldots n\}$. Then $d(x, y) \leq 2^{-(n+1)} < \delta$. We know that bisequences belonging to different sets $E(b)$ must be at least δ apart, yet x and y are closer to each other then δ, therefore they must belong to the same set $E(a)$ (sets $E(a)$ cover the entire space $\mathcal{A}^{\mathbb{Z}}$, thus not belonging to any of them is not an option). Images of both x and y, therefore, belong to the same set $C(a)$, and $[F(x)]_0 = [F(x)]_0$. This means that $[F(x)]_0$ depends only on $x_{-n}, x_{-n+1}, \ldots, x_n$, and if we denote this functional dependence by f, we have

$$[F(x)]_0 = f(x_{-n}, x_{-n+1}, \ldots, x_n).$$

In the final step we need to use the fact that F commutes with shift. Since $x_1 = [\sigma(x)]_0$, by induction $x_i = [\sigma^i(x)]_0$. This yields

$$[F(x)]_i = [\sigma^i F(x)]_0 = [F\sigma^i(x)]_0 = f([\sigma^i(x)]_{-n}, \ldots, [\sigma^i(x)]_n) \tag{1.2}$$

$$= f(x_{i-n}, x_{i-n+1}, \ldots, x_{i+n}), \tag{1.3}$$

satisfying Eq. (1.1). □

Continuous transformations of $\mathcal{A}^{\mathbb{Z}}$ commuting with shift will be the main subject of this book. They are usually called *cellular automata*, although in the past they were also known as sliding block codes [3], homogeneous structures, tessellation structures, etc.

In what follows we will provide some basic terminology related to cellular automata needed for our purposes. A broader exposition of the theory of cellular automata can be found in the monograph by Hadeler and Müller [4]. Readers interested in the history of the subject will find some useful material in the book of Ilachinski [5].

It will be convenient to define the cellular automaton in terms of the function f found in the proof of Theorem 1.1, as follows.

Definition 1.1 Given an alphabet $\mathcal{A}$, a positive integer r and a function $f : \mathcal{A}^{2r+1} \to \mathcal{A}$, the transformation $F : \mathcal{A}^{\mathbb{Z}} \to \mathcal{A}^{\mathbb{Z}}$ defined by

$$[F(x)]_i = f(x_{i-r}, x_{i-r+1}, \ldots x_{i+r}) \tag{1.4}$$

will be called *cellular automaton*. Alternatively, we will refer to F as the *the global function* or *global rule* of the cellular automaton. Function f will be called the *local function*, *local rule* or simply the *cellular automaton rule*, and r will be called *radius of the rule*.

Among the examples we have introduced earlier, σ is a cellular automaton (CA) with local function $f(x_0, x_1, x_2) = x_2$. The spatial reflection R is not a cellular automaton because although it is continuous, it does not commute with σ. The transformation C of Example 1.4 commutes with σ but is not continuous, thus it is not a cellular automaton either.

1.2 Elementary Rules and Their Nomenclature

Binary rules (with $\mathcal{A} = \{0, 1\}$) of radius 1 are called *elementary rules*, and they are usually identified by their Wolfram number $W(f)$, defined as [6]

$$W(f) = \sum_{x_1,x_2,x_3=0}^{1} f(x_1, x_2, x_3) 2^{(2^2 x_1 + 2^1 x_2 + 2^0 x_3)}. \tag{1.5}$$

This seemingly strange formula is actually quite easy to interpret. Consider the elementary rule with the local function defined as

$$f(x_1, x_2, x_3) = \begin{cases} x_2 & \text{if } x_1 = 0, \\ x_3 & \text{if } x_1 = 1. \end{cases} \tag{1.6}$$

This local function can also be specified by listing all possible combinations of arguments in reversed lexicographical order, namely

$$\begin{aligned} &f(111) = 1,\ f(110) = 0,\ f(101) = 1,\ f(100) = 0, \\ &f(011) = 1,\ f(010) = 1,\ f(001) = 0,\ f(000) = 0. \end{aligned}$$

Note that we omitted commas between symbols, writing 000 instead of 0, 0, 0. This convention will be frequently used in this book to make formulae more compact. If we now write only the outputs of f in the same order, we obtain 10101100. This string can be treated as a number written in base 2, and converting it to base 10 we obtain

$$1 \cdot 2^7 + 0 \cdot 2^6 + 1 \cdot 2^5 + 0 \cdot 2^4 + 1 \cdot 2^3 + 1 \cdot 2^2 + 0 \cdot 2^1 + 0 \cdot 2^0 = 172,$$

therefore $W(f) = 172$. Instead of saying "CA rule with Wolfram number 172" we will often simply say "rule 172".

Example 1.5 Knowing the Wolfram number of an elementary CA rule we can construct its local function. To do this, we convert the decimal rule number to base 2, adding trailing zeros if needed to make the total number of binary digits equal to 8. For example, if $W(f) = 18$, then we have $18_{10} = 00010010_2$. The binary digits represent outputs of f in the same order as in the previous example, meaning that

$$f(111) = 0,\ f(110) = 0,\ f(101) = 0,\ f(100) = 1,$$
$$f(011) = 0,\ f(010) = 0,\ f(001) = 1,\ f(000) = 0.$$

Rule 18, therefore, can be defined as

$$f(x_1, x_2, x_3) = \begin{cases} 1 \text{ if } x_1x_2x_3 = 100 \text{ or } 001, \\ 0 \text{ otherwise.} \end{cases} \tag{1.7}$$

Extension of Wolfram numbers to alphabets with higher number of symbols and higher neighbourhood radii is straightforward. For rule with N states and n inputs, where $n = 2r + 1$, we have

$$W(f) = \sum_{x_1, x_2, \ldots, x_n \in \mathcal{A}} f(x_1, x_2, \ldots, x_n) N^{\sum_{i=1}^{n} x_i N^{n-i}}. \tag{1.8}$$

Since for $N > 2$ or $n > 3$ the number $W(f)$ becomes very large, it is not often used in such situations. In this book we will only use it for elementary rules.

We should add at this point that for elementary CA there are other rule numbering schemes. One of them is the *Fatès transition code*, introduced in [7]. For a given ECA with a local function f, define *transition function* t as

$$t(x_1, x_2, x_3) = |f(x_1, x_2, x_3) - x_2|.$$

Obviously $t(x_1, x_2, x_3) \neq 0$ only if $f(x_1, x_2, x_3) \neq x_2$. If for some (x_1, x_2, x_3) the transition function is equal to 1, we will say that the (x_1, x_2, x_3) transition is *active*. The active transitions corresponding to (x_1, x_2, x_3) are coded by uppercase letters A to H, as follows,

000	001	010	011	100	101	110	111
A	B	C	D	E	F	G	H.

The Fatès transition code is then a list of active transitions expressed by the above letters.

Example 1.6 For ECA with Wolfram number 232, the local function is given by

$$f(111) = 1,\ f(110) = 1,\ f(101) = 1,\ f(100) = 0,$$
$$f(011) = 1,\ f(010) = 0,\ f(001) = 0,\ f(000) = 0.$$

This can be written as

$$f(x_1, x_2, x_3) = \begin{cases} 1 - x_2 & \text{if } x_1x_2x_3 = 010 \text{ or } 101, \\ x_2 & \text{otherwise,} \end{cases} \tag{1.9}$$

meaning that the only active transitions are $(0, 1, 0)$ and $(1, 0, 1)$, corresponding to letters C and F. Rule 232, therefore, has the Fatès transition code CF.

1.3 Blocks and Block Mappings

A block or *word* of length $n \in \mathbb{N}$ is an ordered set $\mathbf{a} = a_1 a_2 \ldots a_n$, where $a_i \in \mathcal{A}$. We will often refer to such blocks as *n-blocks*. The length of the block will be denoted by $|\mathbf{a}|$. The set of all n-blocks will be denoted by $\mathcal{A}^n$, and $\mathcal{A}^\star = \bigcup_{i=1}^{\infty} \mathcal{A}^i$.

Definition 1.2 Given a cellular automaton with local function f of radius r we define the corresponding *block mapping* $\mathbf{f} : \mathcal{A}^\star \mapsto \mathcal{A}^\star$ as follows. Let $\mathbf{a} = a_1 \ldots a_n \in \mathcal{A}^n$ where $n \geq 2r + 1$. Then $\mathbf{f}(\mathbf{a})$ is a block of length $n - 2r$ given by

$$\mathbf{f}(\mathbf{a}) = \{f(a_i, a_{i+1}, \ldots, a_{i+2r})\}_{i=1}^{n-2r}. \tag{1.10}$$

If $|\mathbf{a}| < 2r + 1$, then we define $\mathbf{f}(\mathbf{a}) = \emptyset$.

The block mapping simply applies the local function f to $\mathbf{a}$ and returns a shorter block, as illustrated in Fig. 1.1. Note that if $\mathbf{a} \in \mathcal{A}^{2r+1}$ then $f(a_1, a_2, a_3) = \mathbf{f}(\mathbf{a})$. Each application of a block mapping decreases the length of the block by $2r$. Therefore, if $|\mathbf{a}| = n$, then $|\mathbf{f}^k(\mathbf{a})| = n - 2kr$.

The block mapping is not generally invertible meaning that $\mathbf{f}^{-1}$ is not a single-valued function. We will, however, use $\mathbf{f}^{-1}(\mathbf{b})$ to denote the set of all blocks $\mathbf{a} \in \mathcal{A}^\star$ such that $\mathbf{f}(\mathbf{a}) = \mathbf{b}$. Elements of $\mathbf{f}^{-1}(\mathbf{b})$ will be called *block preimages* of $\mathbf{b}$ under the rule f. Similarly, the set of n-step preimages of the block $\mathbf{b}$ under the rule f will be defined as $\mathbf{f}^{-n}(\mathbf{b}) = \{\mathbf{a} \in \mathcal{A}^\star : \mathbf{f}^n(\mathbf{a}) = \mathbf{b}\}$. Note that the notion of block preimages has been extensively studied by many authors (usually in somewhat different context), including [8–11].

Example 1.7 Let us take as an example again the elementary rule 172 and $\mathbf{a} = 1001010$. We can compute $\mathbf{f}(\mathbf{a})$ by applying f to all consecutive triples of symbols, that is, $\mathbf{f}(a) = f(100)f(001)f(010)f(101)f(010) = 00111$. If we apply $\mathbf{f}$ again to 00111, we will obtain $\mathbf{f}^2(\mathbf{a}) = 011$, and yet another application of $\mathbf{f}$ yields $\mathbf{f}^3(\mathbf{a}) = 1$. It is sometimes convenient to list consecutive images of $\mathbf{a}$ in a column with symmetric indentation, as follows:

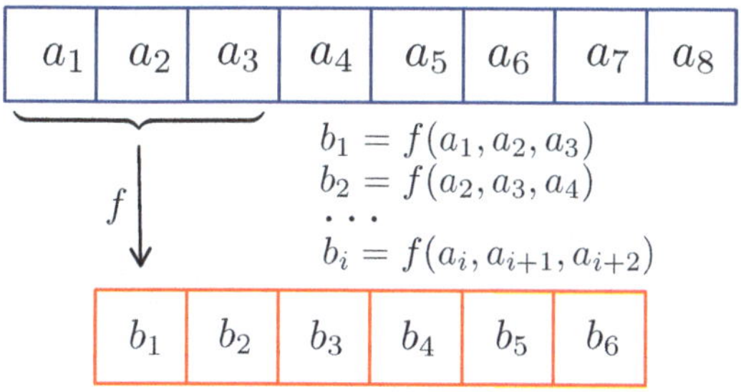

Fig. 1.1 Example of application of the block mapping $\mathbf{f}$ corresponding to a rule f of radius 1 applied to a block $\mathbf{a} = a_1 a_2 \ldots a_8$, so that $\mathbf{f}(\mathbf{a}) = \mathbf{b} = b_1 b_2 \ldots b_6$

```
1001010
 00111
  011
   1
```

The above shows, starting from the top, blocks $\mathbf{a}$, $\mathbf{f}(\mathbf{a})$, $\mathbf{f}^2(\mathbf{a})$ and $\mathbf{f}^3(\mathbf{a})$.

As mentioned, for a given $\mathbf{a}$, there is often more than one block $\mathbf{b}$ such that $\mathbf{f}(\mathbf{b}) = \mathbf{a}$. For example, for rule 172, there are exactly three 4-blocks such that application of $\mathbf{f}$ to them returns 01, namely $\mathbf{f}(0010) = \mathbf{f}(0011) = \mathbf{f}(1101) = 01$. We can, therefore, write $\mathbf{f}^{-1}(01) = \{0010, 0011, 1101\}$. Similarly, we can write

$$\mathbf{f}^{-2}(101) = \{0011101, 0101101, 0111101, 1011101, 1101101, 1111101\}, \quad (1.11)$$

because all the above blocks (and only these blocks) have the property that after applying $\mathbf{f}$ twice to any of them, one obtains 101. It is often useful to visualize blocks and their preimages in a graph in which nodes represent blocks. Two blocks are connected by an edge if one is the image of the other. Such graph must necessarily be a tree, thus we will call it a *preimage tree*. Figure 1.2 shows fragment of a preimage tree for rule 172 rooted at the block we have just discussed.

Example 1.8 The set of block preimages can be empty. For example, consider elementary rule 14, for which

$$f_{14}(x_1, x_2, x_3) = x_3 + x_2 - x_2x_3 - x_1x_3 - x_1x_2 + x_1x_2x_3.$$

One can easily verify that $\mathbf{f}^{-1}(111) = \emptyset$. This can be done by "brute force", by computing images of all 32 blocks of length 5 and checking that none of them is 111. We will say that block 14 belongs to the *garden of Eden* because it has no predecessors (preimages). When a block belongs to the garden of Eden, other blocks

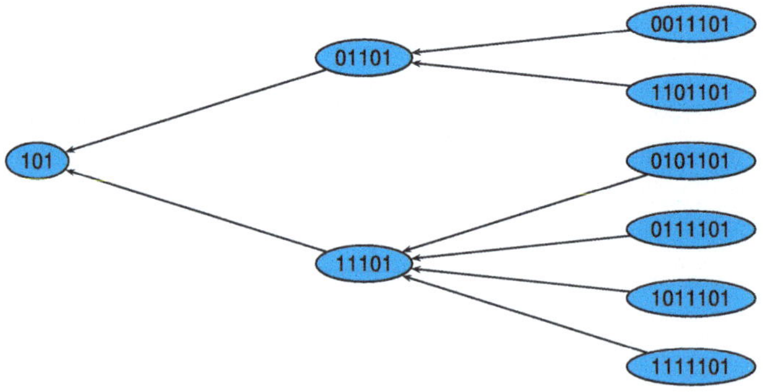

Fig. 1.2 Fragment of a preimage tree for rule 172 rooted at 101

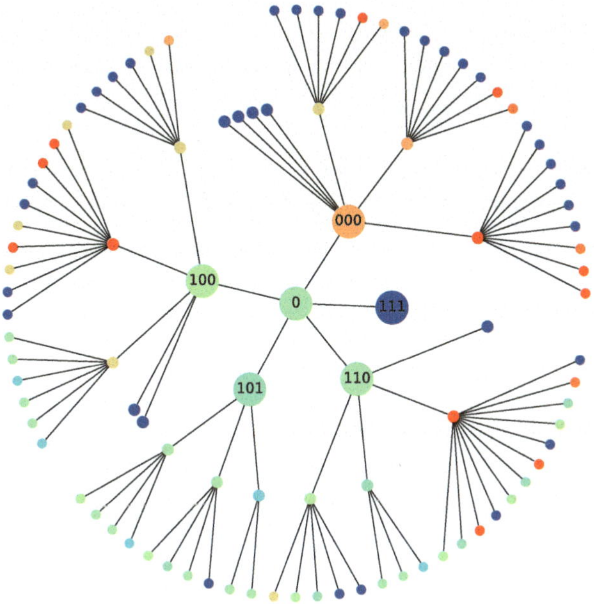

Fig. 1.3 First three levels of the preimage tree for rule 14 rooted at 0

containing it must also belong there. For example, for rule 14, blocks 00111 or 01110 have no preimages either. This is illustrated in Fig. 1.3 which shows the preimage tree for rule 14 rooted at 0. Degrees of nodes are shown in colors ranging from blue to red, with blue representing low degree and red high degree. Dark blue, therefore, indicates that a given node belongs to the garden of Eden. We can see that this is the case for block 111, but there are many other nodes which are dark blue (for clarity we are showing labels of only up to the first level of the tree, other blocks are just marked as circles). For instance, the single blue node on the right of 110 represents 00111. Two blue nodes below 100 are 01110 and 01111. Many other garden of Eden blocks are in the top level of the tree.

Note that if $|\mathbf{a}| = 2r + 1$, then $|\mathbf{f}(\mathbf{a})| = 1$, that is, $\mathbf{f}$ returns a single symbol,

$$\mathbf{f}(\mathbf{a}) = f(a_1, a_2, \ldots, a_{2r+1}).$$

For this reason, if $|\mathbf{a}| = 2r + 1$, we will sometimes write $f(\mathbf{a})$ in place of $\mathbf{f}(\mathbf{a})$ or conversely, slightly abusing our notation. This can be useful to construct local functions corresponding to powers of F. For example, let $G = F^2$. Obviously G is a cellular automaton, and we have

$$\begin{aligned}[G(x)]_i &= [F(F(x))]_i = f([F(x)]_{i-r}, \ldots, [F(x)]_{i+r}) \\ &= f(f(x_{i-2r}, \ldots, x_i), \ldots, f(x_i, \ldots, x_{i+2r})) \\ &= \mathbf{f}^2(x_{i-2r}x_{i-2r+1} \ldots x_{i+2r}). \end{aligned} \tag{1.12}$$

This means that the local function of G is given by $g(\mathbf{a}) = \mathbf{f}^2(\mathbf{a})$, where $\mathbf{a} = a_1 a_1 \ldots a_{2r+1}$. Generalization to higher powers of F is straightforward.

Proposition 1.1 *Let F be a cellular automaton with the corresponding local rule f of radius r. Then F^n is a cellular automaton with its local rule given by the block mapping $\mathbf{f}^n$ restricted to blocks of length $2rn + 1$.*

1.4 Orbits of Cellular Automata

We will be interested in repeated applications of the global rule F to a given bisequence in $\mathcal{A}^{\mathbb{Z}}$.

Definition 1.3 Let $F : \mathcal{A}^{\mathbb{Z}} \to \mathcal{A}^{\mathbb{Z}}$ be a cellular automaton and $x \in \mathcal{A}^{\mathbb{Z}}$. The *orbit of x under the action of F* is the set

$$\{F^n(x)\}_{n=0}^{\infty}.$$

The usual interpretation of n in the above is to treat it as the time variable in the dynamical system starting from x and evolving in discrete time steps by repeated applications of F. The state of the system at time n is, therefore, $F^n(x)$.

Since x is an infinite-dimensional vector, visualisation of its orbit is rather difficult. What is usually done, therefore, is visualisation of orbits of periodic bisequences, that is, bisequences such that $x_{i+L} = x_i$ for any i, where L is the period of the bisequence. In the CA literature this is usually referred to as *imposition of periodic boundary conditions on x*. It is easy to show that if x is L-periodic, then $F^n(x)$ is L-periodic also, for any $n \in \mathbb{N}$. For this reason we can visualize the orbit of L-periodic bisequence by showing only L symbols of $F^n(x)$, e.g. only $[F^n(x)]_0, [F^n(x)]_1, \ldots, [F^n(x)]_{L-1}$. The usual way of doing this is by representing the single period of $F^n(x)$ as vertical array of "cells" with different colors corresponding to different symbols of the alphabet $\mathcal{A}$. We can then plot a sequence of such arrays arranged in rows starting from x, then $F(x)$ below it, then $F^2(x)$, $F^3(x)$, etc. The resulting diagram will be called the *spatiotemporal diagram* of F.

Example 1.9 Let $\mathcal{A} = \{0, 1\}$ with white color representing 0 and blue representing 1. Figure 1.4 shows spatiotemporal diagrams of rules 172 and 18, defined, respectively, in Eqs. (1.6) and (1.7). These diagrams show orbits of an initial configuration (in the top row) with period 50 constructed by randomly assigning to $x_0, x_1, \ldots, x_{49}$ symbols 0 and 1 with the same probability 0.5 for each symbol, independently for each $i \in \{0, 1, \ldots, 49\}$. Only the first 50 rows are shown including the initial configuration, so that the bottom row corresponds to $F^{49}(x)$. The initial configuration x is identical

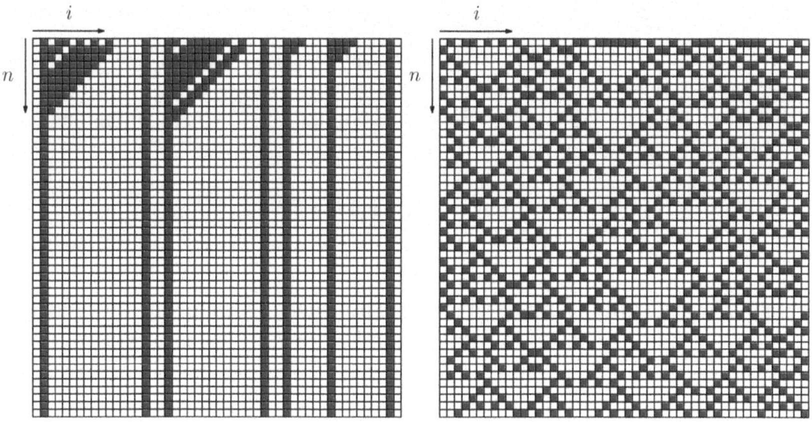

Fig. 1.4 Spatiotemporal diagrams of rules 172 (left) and 18 (right)

in both cases, yet as we can see, the orbits are drastically different. The orbit of x under the action of rule 172 stops changing after 10 time steps and settles down to a configuration consisting of isolated 1'a in the background of all zeros. The orbit for rule 18 continues changing and develops characteristic triangles of varying size distributed throughout the entire pattern.

1.5 Minimal Rules

In Example 1.2, we introduced elementary cellular automata rules (ECA), namely rules with $\mathcal{A} = \{0, 1\}$ and radius $r = 1$. These rules exhibit properties which are somewhat representative to all cellular automata, therefore we will study their orbits in detail.

The first thing to notice is that local rules of ECA are Boolean functions with 3 arguments, $f : \{0, 1\}^3 \to \{0, 1\}$. Since there are 3 arguments and each can have two possible values, the total number of possible inputs of f is $2^3 = 8$. Moreover, each input can produce 2 possible outputs 0 or 1, thus the total number of functions $f : \{0, 1\}^3 \to \{0, 1\}$ is $2^{2^3} = 256$. There are, therefore, 256 elementary rules with corresponding Wolfram numbers ranging from 0 to 255.

The number of ECA is small enough to attempt to consider each one of them separately, but it is possible to exploit their symmetries and reduce their number to a more manageable set of representative cases. The first obvious symmetry is the spatial reflection R which interchanges the roles of the left and the right. Define the action of R on functions and bisequences as follows,

$$\begin{aligned} [R(x)]_i &= x_{-i} \text{ for any } x \in \mathcal{A}^{\mathbb{Z}} \text{ and } i \in \mathbb{Z}, \\ Rf(x_1, x_2, \ldots, x_n) &= f(x_n, x_{n-1}, \ldots, x_1) \text{ for any } f : \{0, 1\}^n \to \{0, 1\}. \end{aligned} \tag{1.13}$$

Another obvious symmetry is the Boolean conjugation C which interchanges the roles of 0 and 1, defined as

$$\begin{aligned} [Cx]_i &= 1 - x_i \text{ for any } x \in \mathcal{A}^{\mathbb{Z}} \text{ and } i \in \mathbb{Z}, \\ Cf(x_1, x_2, \ldots, x_n) &= 1 - f(1 - x_1, 1 - x_2, \ldots, 1 - x_n) \text{ for any } f : \{0, 1\}^n \to \{0, 1\}. \end{aligned} \tag{1.14}$$

Denote by I the identity operator which does not change bisequences nor functions, $Ix = x$, $If = f$. Since R and C commute and $R^2 = I$, $C^2 = I$, symmetries R and C generate abelian group $\{I, R, C, RC\}$ of order 4. This means that given the local function f, there are only four varieties of local functions which one can obtain by repeated application of R, C or any combination of thereof, namely f, Rf, CF and RCf. Depending on f, some of these may be identical. The aforementioned group divides elementary cellular automata into 88 equivalence classes shown in Table 1.1. Rules in that table are represented by their Wolfram number. The rule with the smallest Wolfram number, to be called *the minimal rule*, will be used as the representative of each class. We can see that the number of rules in each class varies between 1 and 4, and this is because sometimes $Rf = f$ or $Cf = f$.

Suppose that F and G are two ECA with corresponding local rules f and g such that $g = Sf$, where S is any element of the group $\{I, R, C, RC\}$. It is straightforward to verify that for any $x \in \{0, 1\}^{\mathbb{Z}}$,

$$SF(x) = GS(x),$$

and therefore

$$SF^n(x) = G^n S(x).$$

This implies

$$F^n(x) = S^{-1} G^n S(x),$$

meaning that the trajectory of x under F and the trajectory of $S(x)$ under G are related to each other by the symmetry S. It is, therefore, sufficient to study trajectories of only one ECA from each equivalence class. We will use the minimal representatives for this purpose.

Table 1.1 Table of 88 equivalence classes of elementary cellular automata

{**0**, 255}	{**26**, 82, 167, 181}	{**56**, 98, 185, 227}	{**132**, 222}
{**1**, 127}	{**27**, 39, 53, 83}	{**57**, 99}	{**134**, 148, 158, 214}
{**2**, 16, 191, 247}	{**28**, 70, 157, 199}	{**58**, 114, 163, 177}	{**136**, 192, 238, 252}
{**3**, 17, 63, 119}	{**29**, 71}	{**60**, 102, 153, 195}	{**138**, 174, 208, 244}
{**4**, 223}	{**30**, 86, 135, 149}	{**62**, 118, 131, 145}	{**140**, 196, 206, 220}
{**5**, 95}	{**32**, 251}	{**72**, 237}	{**142**, 212}
{**6**, 20, 159, 215}	{**33**, 123}	{**73**, 109}	{**146**, 182}
{**7**, 21, 31, 87}	{**34**, 48, 187, 243}	{**74**, 88, 173, 229}	{**150**}
{**8**, 64, 239, 253}	{**35**, 49, 59, 115}	{**76**, 205}	{**152**, 188, 194, 230}
{**9**, 65, 111, 125}	{**36**, 219}	{**77**}	{**154**, 166, 180, 210}
{**10**, 80, 175, 245}	{**37**, 91}	{**78**, 92, 141, 197}	{**156**, 198}
{**11**, 47, 81, 117}	{**38**, 52, 155, 211}	{**90**, 165}	{**160**, 250}
{**12**, 68, 207, 221}	{**40**, 96, 235, 249}	{**94**, 133}	{**162**, 176, 186, 242}
{**13**, 69, 79, 93}	{**41**, 97, 107, 121}	{**104**, 233}	{**164**, 218}
{**14**, 84, 143, 213}	{**42**, 112, 171, 241}	{**105**}	{**168**, 224, 234, 248}
{**15**, 85}	{**43**, 113}	{**106**, 120, 169, 225}	{**170**, 240}
{**18**, 183}	{**44**, 100, 203, 217}	{**108**, 201}	{**172**, 202, 216, 228}
{**19**, 55}	{**45**, 75, 89, 101}	{**110**, 124, 137, 193}	{**178**}
{**22**, 151}	{**46**, 116, 139, 209}	{**122**, 161}	{**184**, 226}
{**23**}	{**50**, 179}	{**126**, 129}	{**200**, 236}
{**24**, 66, 189, 231}	{**51**}	{**128**, 254}	{**204**}
{**25**, 61, 67, 103}	{**54**, 147}	{**130**, 144, 190, 246}	{**232**}

Rule with minimal Wolfram number in each class is shown in bold

References

1. Hedlund, G.A.: Endomorphisms and automorphisms of the shift dynamical system. Mathematical systems theory **3**, 320–375 (1968)
2. Ceccherini-Silberstein, T., Coornaert, M.: Cellular Automata and Groups. Springer Monographs in Mathematics. Springer, Berlin (2010)
3. Lind, D., Marcus, B.: Symbolic dynamics and coding. Cambridge UP, Cambridge (1995)
4. Hadeler, K., Müller, J.: Cellular Automata: Analysis and Applications. Springer Monographs in Mathematics. Springer (2017)
5. Ilachinski, A.: Cellular Automata: A Discrete Universe. World Scientific, Singapore (2001)
6. Wolfram, S.: Cellular Automata and Complexity: Collected Papers. Addison-Wesley, Reading, Mass. (1994)
7. Fatès, N., Regnault, D., Schabanel, N., Thierry, É.: Asynchronous behavior of double-quiescent elementary cellular automata. In: J. Correa, A. A. Hevia, M. Kiwi (eds.) LATIN 2006: Theoretical Informatics, *LNCS*, vol. 3887, pp. 455–466 (2006)
8. Jen, E.: Table of preimage formulae for elementary rules. Report LA-UR-88-3359, Los Alamos National Laboratory (1988)
9. Jen, E.: Enumeration of preimages in cellular automata. Complex Systems **3**, 421–456 (1989)
10. Voorhees, B.H.: Computational analysis of one-dimensional cellular automata. World Scientific, Singapore (1996)
11. McIntosh, H.V.: One Dimensional Cellular Automata. Luniver Press (2009)

Chapter 2
Deterministic Initial Value Problem

As mentioned in the preface, in the theory of partial differential equations (PDE), an initial value problem (sometimes also called a Cauchy problem) is frequently considered. In its simplest form it is usually formulated as the problem of finding the function $u(x,t)$ for $t > 0$ subject to

$$\begin{aligned} \frac{\partial u}{\partial t} &= F\left(u, \frac{\partial u}{\partial x}, \frac{\partial^2 u}{\partial x^2}, \ldots\right), \quad \text{for } x \in \mathbb{R}, t > 0, \\ u(x,0) &= G(x) \quad \text{for } x \in \mathbb{R}, \end{aligned} \tag{2.1}$$

where the function $G : \mathbb{R} \to \mathbb{R}$ represents the given initial data and F is the function defining the PDE. In this chapter we will define an analogous problem for cellular automata and we will seek its solution starting from the simplest ECA rules.

The orbit of a point in $\mathcal{A}^{\mathbb{Z}}$ under the action of a cellular automaton F can be viewed as a sequence of points in $\mathcal{A}^{\mathbb{Z}}$. Let $x(n)$ denote the nth point of the orbit, with the convention that $x(0)$ is the initial point. Then

$$x(n+1) = F(x(n)). \tag{2.2}$$

Definition 2.1 Given a cellular automaton F and $v \in \mathcal{A}^{\mathbb{Z}}$, the problem of finding $x(n)$ satisfying Eq. (2.2) subject to $x(0) = v$ is called the *initial value problem for CA*.

It seems at first that the initial value problem (IVP) has an immediate solution,

$$x(n) = F^n(v).$$

However, this form of the solution is not very useful, just like it would not be very useful to say that the solution of the difference equation

H. Fukś, *Solvable Cellular Automata*, Understanding Complex Systems,
https://doi.org/10.1007/978-3-031-38700-5_2

$$z(n+1) = 4z(n)(1 - z(n))$$

is given by

$$z(n) = g^n(z(0)),$$

where $g(x) = 4x(1-x)$. In what follows, therefore, we will attempt to express $F^n(x)$ as and explicit function of n and x, without referring to powers of F. If this is possible, we will say that the rule is *solvable*. We will start from the simplest solvable rules, namely those with a single input.

2.1 Single Input Rules

A single input ECA is a rule for which the local function $f(x_0, x_1, x_2)$ depends only on one of the arguments x_0, x_1, x_2. There are six such rules, 15, 51, 85, 170, 204 and 240, and among them four are minimal,

$$\begin{aligned} f_{15}(x_1, x_2, x_3) &= 1 - x_1, \\ f_{51}(x_1, x_2, x_3) &= 1 - x_2, \\ f_{170}(x_1, x_2, x_3) &= x_3, \\ f_{204}(x_1, x_2, x_3) &= x_2. \end{aligned}$$

Lower indices following f indicate Wolfram numbers. Obviously f_{204} represents identity, f_{170} is the local function of the shift σ, and f_{51} represents the Boolean conjugation. Function f_{15} represent composition of σ^{-1} and conjugation. Expressions for $F^n(x)$ for the above functions are rather easy to find. For the identity and shift these are obvious and trivial,

$$[F_{204}^n(x)]_j = x_j, \tag{2.3}$$
$$[F_{170}^n(x)]_j = x_{j+n}. \tag{2.4}$$

Function F_{51} flips each bit of x every iteration, and after two iterations each bit returns to its initial value. We have, therefore,

$$[F_{51}^n(x)]_j = \begin{cases} x_j & \text{if } n \text{ is even,} \\ 1 - x_j & \text{if } n \text{ is odd.} \end{cases} \tag{2.5}$$

This can be written as a single expression,

$$[F_{51}^n(x)]_j = \frac{(-1)^{n+1} + 1}{2} + (-1)^n x_j. \tag{2.6}$$

Rule 15 is a composition of inverse shift and conjugation, therefore

$$[F_{15}^n(x)]_j = \begin{cases} x_{j-n} & \text{if } n \text{ is even}, \\ 1 - x_{j-n} & \text{if } n \text{ is odd}. \end{cases} \tag{2.7}$$

As before, this can be combined into a single formula,

$$[F_{15}^n(x)]_j = \frac{(-1)^{n+1} + 1}{2} + (-1)^n x_{j-n}. \tag{2.8}$$

Two other single input rules, 240 and 85, are not minimal. Rule 240, or σ^{-1}, is the spatial reflection of rule 170. Similarly, rule 85 is the spatial reflection of rule 15. Solutions of the corresponding initial value problems are, therefore,

$$[F_{240}^n(x)]_j = x_{j-n}, \tag{2.9}$$

$$[F_{85}^n(x)]_j = \frac{(-1)^{n+1} + 1}{2} + (-1)^n x_{j+n}. \tag{2.10}$$

2.2 Polynomial Representation of Local Rules

Before we consider more complex rules than those discussed in the previous section, we need to introduce a very useful method for representing a rule with a given Wolfram number.

Let the *density polynomial* associated with a symbol $a \in \{0, 1\}$ be defined as

$$\Psi_a(p) = \begin{cases} p & \text{if } a = 1, \\ 1 - p & \text{if } a = 0. \end{cases} \tag{2.11}$$

The density polynomial associated with a block $\mathbf{b} = b_1 b_2 \ldots b_n$ is a product of density polynomials of individual symbols,

$$\Psi_{\mathbf{b}}(p_1, p_2, \ldots, p_n) = \prod_{i=1}^{n} \Psi_{b_i}(p_i). \tag{2.12}$$

If A is a set of binary strings of the same length n, we define density polynomial associated with A as

$$\Psi_A(p_1, p_2, \ldots, p_n) = \sum_{\mathbf{a} \in A} \Psi_{\mathbf{a}}(p_1, p_2, \ldots, p_n). \tag{2.13}$$

Example 2.1 The density polynomial associated with $\mathbf{b} = 010$ is

$$\Psi_{\mathbf{b}}(p_1, p_2, p_3) = (1 - p_1) p_2 (1 - p_3).$$

If $A = \{1100, 1111, 1010\}$, then

$$\Psi_A(p_1, p_2, p_3, p_4) = p_1 p_2(1 - p_3)(1 - p_4) + p_1 p_2 p_3 p_4 + p_1(1 - p_2)p_3(1 - p_4).$$

Proposition 2.1 *If f is a local function of ECA, then*

$$f(x_1, x_2, x_3) = \Psi_A(x_1, x_2, x_3),$$

where A is the set of preimages of 1 under the block mapping $\mathbf{f}$,

$$A = \mathbf{f}^{-1}(1).$$

Proof The set A is the set of 3-blocks which are mapped to 1 by the local rule. Note that

$$\Psi_a(x) = \begin{cases} x & \text{if } a = 1, \\ 1 - x & \text{if } a = 0, \end{cases} \tag{2.14}$$

and therefore, for $a, x \in \{0, 1\}$,

$$\Psi_a(x) = \begin{cases} 1 & \text{if } x = a, \\ 0 & \text{if } x \neq a, \end{cases} \tag{2.15}$$

which is easy to check by trying all possibilities for the values of a and x. By Eq. (2.12) we then conclude that $\Psi_{b_1 b_2 b_3}(x_1, x_2, x_3)$ will be equal to 1 if and only if $x_1 = b_1, x_2 = b_2, x_3 = b_3$, otherwise it will be zero. This means that $\Psi_A(x_1, x_2, x_3)$ behaves exactly as the local function f, returning 1 only when $x_1 x_2 x_3$ is one of the elements of $A = \mathbf{f}^{-1}(1)$, otherwise returning 0, as desired. □

Example 2.2 Consider rule 18 defined in Eq. (1.7). For this rule the 3-blocks which return 1 are $\mathbf{f}^{-1}(1) = \{001, 100\}$. We can construct the desired density polynomial directly by replacing in strings of $\mathbf{f}^{-1}$ each 0 by $1 - x_i$ and each 1 by x_i, where i indicates the position of the given symbol in the string. We thus have

$$001 \longrightarrow (1 - x_1)(1 - x_2)x_3,$$

$$100 \longrightarrow x_1(1 - x_2)(1 - x_3).$$

The final density polynomial $\Psi_A(x_1, x_2, x_3)$ is then a sum of all polynomials obtained this way, and we have

$$f(x_1, x_2, x_3) = (1 - x_1)(1 - x_2)x_3 + x_1(1 - x_2)(1 - x_3), \tag{2.16}$$

or, after expanding and reordering,

$$f(x_1, x_2, x_3) = x_1 + x_3 - x_1 x_2 - x_2 x_3 - 2x_1 x_3 + 2x_1 x_2 x_3. \tag{2.17}$$

The reader can verify that functions defined in Eqs. (1.7) and (2.17) return the same values for all $(x_1, x_2, x_3) \in \{0, 1\}^3$.

Example 2.3 In some cases density polynomials simplify to shorter expressions than in the above example even if $\mathbf{f}^{-1}(1)$ contains more elements. A good example is rule 172 which we have encountered in the previous chapter, with $\mathbf{f}^{-1}(1) = \{010, 011, 101, 111\}$. We proceed as before,

$$
\begin{aligned}
010 &\longrightarrow (1 - x_1)x_2(1 - x_3), \\
011 &\longrightarrow (1 - x_1)x_2x_3. \\
101 &\longrightarrow x_1(1 - x_2)x_3, \\
111 &\longrightarrow x_1x_2x_3.
\end{aligned}
$$

This yields

$$f(x_1, x_2, x_3) = (1 - x_1)x_2(1 - x_3) + (1 - x_1)x_2x_3 + x_1(1 - x_2)x_3 + x_1x_2x_3, \tag{2.18}$$

which simplifies to

$$f(x_1, x_2, x_3) = x_2 - x_1x_2 + x_1x_3. \tag{2.19}$$

Again, one can verify that Eqs. (1.6) and (2.19) return the same values for all $(x_1, x_2, x_3) \in \{0, 1\}^3$.

Expressing local rules by density polynomials as in the above examples will often be advantageous in solving initial value problems. For all minimal ECA, polynomial representations of local functions are given in Appendix A.

2.3 Emulation

In orbits of some cellular automata, after a finite number of steps, it is possible to switch to another rule and the orbit remains the same. If this happens we will say that one rule emulates the other [1].

Definition 2.2 Let F and G be two cellular automata rules. If there exists $k > 0$ such that

$$F^{k+1} = GF^k, \tag{2.20}$$

we say that F *emulates* G *in* k *steps* or that F *is kth order emulator of* G.

Note that if F emulates G in k steps, it follows immediately that for any $n > k$,

$$F^n = G^{n-k}F^k.$$

This means that the orbit

$$\{x, F(x), F^2(x), \ldots\}$$

is identical to

$$\{x, F(x), \ldots F^k(x), GF^k(x), G^2F^k(x), G^3F^k(x), \ldots\}$$

for any initial point $x \in \mathcal{A}^{\mathbb{Z}}$. If G is solvable, then it is possible to obtain solution of the initial value problem for F as well, as we will see below.

Since in the previous section we found that single input rules are solvable, we will now check if there are any emulators of these rules among other minimal ECA. We need, therefore, a method to determine if one rule emulates another. First note that if F and G are rules of radius r, then in the condition for emulation given in Eq. (2.20), by the virtue of Proposition 1.1, the left hand side is a cellular automaton with the local rule given by $\mathbf{f}^{k+1}$ restricted to blocks of length $2r(k+1)+1$. Since both F and G are continuous and shift-commuting, the right had side of Eq. (2.20) is a cellular automaton as well. Its local function can be constructed in a similar way as in the proof of Proposition 1.1, and it is straightforward to show that it is given by $\mathbf{g}\mathbf{f}^k$ restricted to blocks of length $2r(k+1)+1$. This yields a practical method for finding emulators.

Proposition 2.2 *Let F and G be two cellular automata with radius r. F emulates G in k steps if and only if their corresponding block mappings* $\mathbf{f}$ *and* $\mathbf{g}$ *satisfy*

$$\mathbf{f}^{k+1}(\mathbf{b}) = \mathbf{g}\mathbf{f}^k(\mathbf{b})$$

for any block $\mathbf{b} \in \{0,1\}^{2r(k+1)+1}$.

For ECA we need to check equality of $\mathbf{f}^{k+1}$ and $\mathbf{g}\mathbf{f}^k$ on all blocks of length $2k+3$, and this can be easily done by a computer program, for example using the HCELL library developed by the author [2]. Table 2.1 shows all ECA emulators of single input rules obtained this way.

Table 2.1 Emulators of single input rules

Rules satisfying $F^{k+1} = GF^k$

F	G	k
0, 4, 12, 68, **76, 200**, 205, 207, 221, 223, 236, 255	204	1
8, 36, 64, **72**, 219, 237, 239, 253	204	2
0, 2, 10, 34, 42, 138, 171, 174, 175, 187, 191, 255	170	1
8, 46, 64, 66, 139, 189, 239, 253	170	2
8, 24	240	2
19, 55	51	2

Minimal rules are shown in bold

2.3.1 Emulators of Identity

First we will consider emulators of identity I, that is, ECA rule 204. If F emulates identity in 1 step, then the orbit of any $x \in \mathcal{A}^{\mathbb{Z}}$ is

$$\{x, F(x), F(x), F(x)\ldots\},$$

because $I^j = I$ for any j. Solution of the initial value problem follows immediately,

$$[F^n(x)]_j = f(x_{j-1}, x_j, x_{j+1}). \tag{2.21}$$

Among 88 minimal ECA, there are 5 first order emulators of identity, and these are rules 0, 4, 12, 76 and 200. Using polynomial representation of their local functions, also obtainable by the HCELL software, we can construct solutions of the initial value problem of all of them, as shown below.

$$\begin{aligned}
[F_0^n(x)]_j &= 0,\\
[F_4^n(x)]_j &= x_j - x_j x_{j+1} - x_{j-1}x_j + x_{j-1}x_j x_{j+1},\\
[F_{12}^n(x)]_j &= x_j - x_{j-1}x_j,\\
[F_{76}^n(x)]_j &= x_j - x_{j-1}x_j x_{j+1},\\
[F_{200}^n(x)]_j &= x_{j-1}x_j + x_j x_{j+1} - x_{j-1}x_j x_{j+1}.
\end{aligned} \tag{2.22}$$

In addition to the above, there also exist three second order emulators of identity, namely rules 8, 36 and 72. For the second order emulators, the orbit of x is

$$\{x, F(x), F^2(x), F^2(x), F^2(x), \ldots\},$$

therefore

$$[F^n(x)]_j = \begin{cases} f(x_{j-1}, x_j, x_{j+1}) & \text{if } n = 1,\\ f(f(x_{j-2}, x_{j-1}, x_j), f(x_{j-1}, x_j, x_{j+1}), f(x_j, x_{j+1}, x_{j+2})) & \text{if } n > 1. \end{cases} \tag{2.23}$$

Applying this to rule 8 yields

$$[F_8^n(x)]_j = \begin{cases} x_j x_{j+1} - x_{j-1}x_j x_{j+1} & \text{if } n = 1,\\ 0 & \text{if } n > 1. \end{cases} \tag{2.24}$$

In the above, $f(f(x_{j-2}, x_{j-1}, x_j), f(x_{j-1}, x_j, x_{j+1}), f(x_j, x_{j+1}, x_{j+2}))$ simplified to 0 because rule 8 is also a first order emulator of rule 0.

For rule 36 the resulting expressions are rather complicated,

$$\begin{aligned}
[F_{36}^1(x)]_j &= -x_{j-1}x_j - x_jx_{j+1} + x_{j-1}x_{j+1} + x_j,\\
[F_{36}^n(x)]_j &= x_j - x_jx_{j-2}x_{j-1}x_{j+1} - x_jx_{j-2}x_{j-1}x_{j+2} - x_jx_{j-2}x_{j+1}x_{j+2}\\
&\quad - x_jx_{j-1}x_{j+1}x_{j+2} + x_{j-2}x_{j-1}x_{j+1}x_{j+2} + x_jx_{j-2}x_{j-1} + x_jx_{j-2}x_{j+1}\\
&\quad + x_jx_{j-2}x_{j+2} + x_jx_{j-1}x_{j+1} + x_jx_{j-1}x_{j+2} + x_jx_{j+1}x_{j+2}\\
&\quad - x_{j-2}x_j - x_{j-1}x_j - x_jx_{j+1} - x_jx_{j+2} \quad \text{for } n > 1.
\end{aligned}$$

The third rule, ECA 72, is a bit simpler,

$$\begin{aligned}
[F_{72}^1(x)]_j &= x_{j-1}x_j + x_jx_{j+1} - 2x_{j-1}x_jx_{j+1},\\
[F_{72}^n(x)]_j &= x_jx_{j-2}x_{j-1}x_{j+1} + x_jx_{j-1}x_{j+1}x_{j+2} - x_{j-2}x_{j-1}x_j - 2x_{j-1}x_jx_{j+1}\\
&\quad - x_jx_{j+1}x_{j+2} + x_{j-1}x_j + x_jx_{j+1} \quad \text{for } n > 1.
\end{aligned}$$

2.3.2 Emulators of Shift and Its Inverse

Let us turn our attention to emulators of σ, or ECA rule 170. If F emulates σ in one step, then the orbit of any $x \in \mathcal{A}^{\mathbb{Z}}$ is

$$\{x, F(x), \sigma F(x), \sigma^2 F(x), \sigma^3 F(x) \ldots\}.$$

Solution of the initial value problem, therefore, is given by

$$\begin{aligned}
[F^1(x)]_j &= f(x_{j-1}, x_j, x_{j+1}),\\
[F^n(x)]_j &= [\sigma^{n-1}(F(x))]_j = [F(x)]_{j+n-1} \quad \text{for } n > 1.
\end{aligned}$$

This yields

$$[F^n(x)]_j = \begin{cases} f(x_{j-1}, x_j, x_{j+1}) & \text{if } n = 1,\\ f(x_{j+n-2}, x_{j+n-1}, x_{j+n}) & \text{if } n > 1. \end{cases} \tag{2.25}$$

Since the $n > 1$ expression also works for $n = 1$, we can write the final solution as a single formula valid for all $n \geq 1$,

$$[F^n(x)]_j = f(x_{j+n-2}, x_{j+n-1}, x_{j+n}). \tag{2.26}$$

The following minimal rules emulate σ in one step: 0, 2, 10, 34, 42 and 138. Rule 0 has been considered before, and the solution of the initial value problem for others can be easily obtained using polynomial representations of their local functions,

$$\begin{aligned}
[F_2^n(x)]_j &= x_{j+n-2}x_{j+n-1}x_{j+n} - x_{j+n-2}x_{j+n} - x_{j+n-1}x_{j+n} + x_{j+n},\\
[F_{10}^n(x)]_j &= x_{j+n} - x_{j+n-2}x_{j+n},\\
[F_{34}^n(x)]_j &= x_{j+n} - x_{j+n-1}x_{j+n},\\
[F_{42}^n(x)]_j &= x_{j+n} - x_{j+n-2}x_{j+n-1}x_{j+n},\\
[F_{138}^n(x)]_j &= x_{j+n-2}x_{j+n-1}x_{j+n} - x_{j+n-2}x_{j+n} + x_{j+n}.
\end{aligned}$$

Since the equivalence class containing σ is not a singleton, we need to check if there are any emulators of the other member of the equivalence class, namely rule 240, or σ^{-1}. Computer check reveals that there are no minimal rules emulating σ^{-1} except rule 0, already covered. We thus turn to 2-step emulators. Among 2nd order emulators of σ there is one rule not considered before, namely rule 46. For 2nd order emulators of σ the orbit is of the form

$$\{x, F(x), F^2(x), \sigma F^2(x), \sigma^2 F^2(x), \sigma^3 F^2(x) \ldots\},$$

thus the solution of the initial value problem is of the form

$$\begin{aligned}
[F^1(x)]_j &= f(x_{j-1}, x_j, x_{j+1}),\\
[F^2(x)]_j &= f(f(x_{j-2}, x_{j-1}, x_j), f(x_{j-1}, x_j, x_{j+1}), f(x_j, x_{j+1}, x_{j+2})),\\
[F^n(x)]_j &= [\sigma^{n-2}(F^2(x))]_j = [F^2(x)]_{j+n-2} \quad \text{for } n > 2.
\end{aligned}$$

Since the third line of the above also works for $n = 2$, the final solution is given by

$$\begin{aligned}
[F^1(x)]_j &= f(x_{j-1}, x_j, x_{j+1}), \qquad (2.27)\\
[F^n(x)]_j &= f(f(x_{j+n-4}, x_{j+n-3}, x_{j+n-2}), f(x_{j+n-3}, x_{j+n-2}, x_{j+n-1}),\\
&\qquad f(x_{j+n-2}, x_{j+n-1}, x_{j+n})) \quad \text{for } n > 1.
\end{aligned}$$

Applying the above to rule 46 one obtains

$$\begin{aligned}
[F_{46}^1(x)]_j &= x_j + x_{j+1} - x_{j-1}x_j - x_jx_{j+1}\\
[F_{46}^n(x)]_j &= x_{j+n} + x_{j+n-1} - x_{j+n}x_{j+n-3}x_{j+n-2}x_{j+n-1}\\
&\quad - x_{j+n-4}x_{j+n-3}x_{j+n-2}x_{j+n-1} + x_{j+n}x_{j+n-3}x_{j+n-2}\\
&\quad + x_{j+n}x_{j+n-2}x_{j+n-1} + x_{j+n-4}x_{j+n-3}x_{j+n-1}\\
&\quad + x_{j+n-3}x_{j+n-2}x_{j+n-1} - x_{j+n}x_{j+n-2} - x_{j+n-1}x_{j+n}\\
&\quad - x_{j+n-3}x_{j+n-1} - x_{j+n-2}x_{j+n-1} \quad \text{for } n > 1.
\end{aligned}$$

Finally, among 2nd order emulators of σ^{-1} there is also one new rule, namely rule 24. The derivation of the solution formula for second order emulators of σ^{-1} is analogous to derivation of Eq. (2.27), thus we will not repeat it here and only write the result,

$$[F^1(x)]_j = f(x_{j-1}, x_j, x_{j+1}), \tag{2.28}$$

$$[F^n(x)]_j = f(f(x_{j-n}, x_{j-n+1}, x_{j-n+2}), f(x_{j-n+1}, x_{j-n+2}, x_{j-n+3}), f(x_{j-n+2}, x_{j-n+3}, x_{j-n+4})) \quad \text{for } n > 1. \tag{2.29}$$

Applying the above to rule 24, we obtain

$$\begin{aligned}
[F_{24}^1(x)]_j &= x_{j-1} - x_{j-1}x_j + x_j x_{j+1} - x_{j-1}x_{j+1} \\
[F_{24}^n(x)]_j &= x_{j-n} - x_{j-n}x_{j-n+1}x_{j-n+2}x_{j-n+3} - x_{j-n}x_{j-n+1}x_{j-n+2}x_{j-n+4} \\
&\quad + x_{j-n}x_{j-n+1}x_{j-n+3}x_{j-n+4} + x_{j-n}x_{j-n+2}x_{j-n+3}x_{j-n+4} \\
&\quad - x_{j-n+1}x_{j-n+2}x_{j-n+3}x_{j-n+4} + x_{j-n}x_{j-n+1}x_{j-n+2} \\
&\quad - x_{j-n}x_{j-n+3}x_{j-n+4} + x_{j-n+1}x_{j-n+2}x_{j-n+3} + x_{j-n+1}x_{j-n+2}x_{j-n+4} \\
&\quad - x_{j-n}x_{j-n+1} - x_{j-n}x_{j-n+2} \quad \text{for } n > 1.
\end{aligned}$$

2.3.3 *Emulators of Other Single Input Rules*

The remaining single input rules (15, 51 and 85) have no first order emulators at all. The only minimal second order emulator is rule 19 which emulates rule 51 in two steps. If F is the global function of rule 19 and C the global function of rule 51, then the orbit of x under F will be

$$\{x, F(x), F^2(x), CF^2(x), C^2F^2(x), C^3F^2(x), \ldots\}.$$

Since $C^2 = I$, this becomes

$$\{x, F(x), F^2(x), CF^2(x), F^2(x), CF^2(x), \ldots\},$$

therefore

$$\begin{aligned}
[F^1(x)]_j &= f(x_{j-1}, x_j, x_{j+1}) \\
[F^n(x)]_j &= f(f(x_{j-2}, x_{j-1}, x_j), f(x_{j-1}, x_j, x_{j+1}), f(x_j, x_{j+1}, x_{j+2})) \\
&\qquad \text{for even } n > 1, \\
[F^n(x)]_j &= 1 - f(f(x_{j-2}, x_{j-1}, x_j), f(x_{j-1}, x_j, x_{j+1}), f(x_j, x_{j+1}, x_{j+2})) \\
&\qquad \text{for odd } n > 2.
\end{aligned}$$

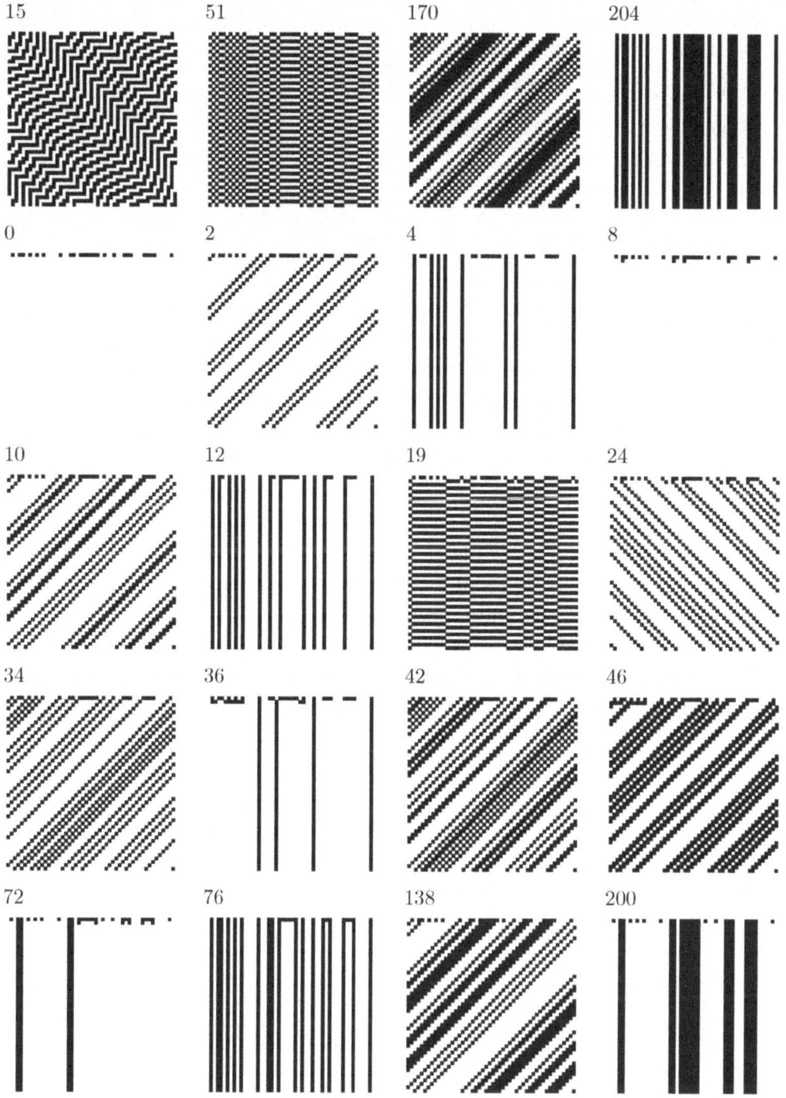

Fig. 2.1 Spatiotemporal patterns for single input rules (top row) and their 16 emulators (rows below)

For rule 19 this becomes

$$
\begin{aligned}
[F_{19}^1(x)]_j &= x_{j-1}x_jx_{j+1} - x_{j-1}x_{j+1} - x_j + 1, \\
[F_{19}^n(x)]_j &= -x_jx_{j-2}x_{j-1}x_{j+1}x_{j+2} + x_jx_{j-2}x_{j-1}x_{j+1} + x_jx_{j-2}x_{j-1}x_{j+2} \\
&\quad + x_jx_{j-2}x_{j+1}x_{j+2} + x_jx_{j-1}x_{j+1}x_{j+2} - x_{j-2}x_{j-1}x_j - x_jx_{j-2}x_{j+1} \\
&\quad - x_jx_{j-2}x_{j+2} - 2x_{j-1}x_jx_{j+1} - x_jx_{j-1}x_{j+2} - x_jx_{j+1}x_{j+2} \\
&\quad + x_{j-2}x_j + x_jx_{j-1} + x_jx_{j+1} + x_jx_{j+2} + x_{j-1}x_{j+1} \\
&\qquad\qquad \text{for even } n > 1, \\
[F_{19}^n(x)]_j &= x_jx_{j-2}x_{j-1}x_{j+1}x_{j+2} - x_jx_{j-2}x_{j-1}x_{j+1} - x_jx_{j-2}x_{j-1}x_{j+2} \\
&\quad - x_jx_{j-2}x_{j+1}x_{j+2} - x_jx_{j-1}x_{j+1}x_{j+2} + x_jx_{j-2}x_{j-1} + x_jx_{j-2}x_{j+1} \\
&\quad + x_jx_{j-2}x_{j+2} + 2x_{j-1}x_jx_{j+1} + x_jx_{j-1}x_{j+2} + x_jx_{j+1}x_{j+2} \\
&\quad - x_jx_{j-2} - x_{j-1}x_j - x_jx_{j+1} - x_jx_{j+2} - x_{j-1}x_{j+1} + 1 \\
&\qquad\qquad \text{for odd } n > 2.
\end{aligned}
$$

This concludes our discussion of single input rules and their emulators. Exploiting the concept of emulation, we were able to solve the initial value problem for 16 emulators of single input rules. Samples of their spatiotemporal diagrams are shown in Fig. 2.1. We can see that these patterns are rather simple, thus the fact that they are all solvable is not very surprising. Now we turn our attention to more complex rules.

References

1. Rogers, T., Want, C.: Emulation and subshifts of finite type in cellular automata. Physica D **70**, 396–414 (1994)
2. Fukś, H.: HCELL library for cellular automata simulations (2005). http://hcell.sourceforge.net

Chapter 3
Multiplicative and Additive Rules

In Proposition 1.1, we demonstrated that F^n is a cellular automaton with local rule given by the block mapping $\mathbf{f}^n$ restricted to blocks of length $2rn + 1$. It will be convenient to denote this local function by $f^n : \mathcal{A}^{2rn+1} \to \mathcal{A}$, so that

$$f^n(x_1, x_2, \ldots, x_{2rn+1}) = \mathbf{f}^n(x_1 x_2 \ldots x_{2rn+1}).$$

For binary rules, if the value of the local function $f(x_1, x_2, \ldots, x_{2r+1})$ can be expressed as a product or sum of some of its arguments $x_1, x_2, \ldots, x_{2r+1}$, then the value of f^n will also be a product or sum of some of its arguments. This can be exploited to produce the solution of the initial value problem for such rules.

3.1 Multiplicative Rules

Among minimal ECA there are three rules for which the polynomial representation of their local function is a product of the arguments, namely rules 128, 136 and 160,

$$\begin{aligned} f_{128}(x_1, x_2, x_3) &= x_1 x_2 x_3, \\ f_{136}(x_1, x_2, x_3) &= x_2 x_3, \\ f_{160}(x_1, x_2, x_3) &= x_1 x_3. \end{aligned}$$

Let us consider rule 160 first.

Proposition 3.1 *Let f be the local function of ECA 160 and $n \in \mathbb{N}$. Then we have*

$$f^n(x_1 x_2, \ldots x_{2n+1}) = \prod_{i=0}^{n} x_{2i+1}. \tag{3.1}$$

H. Fukś, *Solvable Cellular Automata*, Understanding Complex Systems,
https://doi.org/10.1007/978-3-031-38700-5_3

Proof We will give proof by induction. For $n = 1$ Eq. (3.1) is obviously true by the definition of the local function. Suppose now that Eq. (3.1) holds for some n, and let us compute f^{n+1}. We have

$$\begin{aligned} f^{n+1}&(x_1, x_2, \ldots, x_{2n+3}) \\ &= f\left(f^n(x_1, \ldots, x_{2n+1}), f^n(x_2, \ldots, x_{2n+2}), f^n(x_3, \ldots, x_{2n+3})\right) \\ &= f\left(\prod_{i=0}^{n} x_{2i+1}, \prod_{i=0}^{n} x_{2i+2}, \prod_{i=0}^{n} x_{2i+3}\right) = \prod_{i=0}^{n} x_{2i+1} \prod_{i=0}^{n} x_{2i+3} \\ &= \prod_{i=0}^{n} x_{2i+1} \prod_{i=1}^{n+1} x_{2i+1} = x_1 \left(\prod_{i=1}^{n} x_{2i+1} \prod_{i=1}^{n} x_{2i+1}\right) x_{2n+3} = \prod_{i=0}^{n+1} x_{2i+1}, \end{aligned}$$

where in the last equality we used the fact that $x_{2i+1}^2 = x_{2i+1}$. Formula (3.1) is thus valid for $n + 1$, concluding the proof by induction. □

Having the local function of F^n, we can write the explicit solution of the initial value problem,

$$[F^n(x)]_j = f^n(x_{j-n}, x_{j-n+1}, \ldots, x_{j+n}).$$

Using the result of Proposition 3.1, this yields

$$[F_{160}^n(x)]_j = \prod_{i=0}^{n} x_{2i+j-n}. \tag{3.2}$$

For rules 128 and 136, proofs of analogous results are very similar, thus we will not reproduce them here, giving only final formulae:

$$\begin{aligned} [F_{128}^n(x)]_j &= \prod_{i=-n}^{n} x_{i+j}, \\ [F_{136}^n(x)]_j &= \prod_{i=0}^{n} x_{i+j}. \end{aligned}$$

3.1.1 Rules 32 and 140

Although local functions of rules 32 and 140 are not products of their argument, they can be treated in a similar fashion as the product rules discussed in the previous section. Rule 32 has the local function given by

$$f_{32}(x_1, x_2, x_3) = x_1 x_3 - x_1 x_2 x_3.$$

This can be written as a product involving negation of x_2,

$$f_{32}(x_1, x_2, x_3) = x_1\bar{x}_2x_3,$$

where we used the notation $\overline{x} = 1 - x$. Now let us note that

$$\begin{aligned} f_{32}^2(x_1, x_2, x_3, x_4, x_5) &= x_1\bar{x}_2x_3 \cdot (1 - x_2\bar{x}_3x_4) \cdot x_3\bar{x}_4x_5 = x_1\bar{x}_2x_3 \cdot x_3\bar{x}_4x_5 \\ &= x_1\bar{x}_2x_3\bar{x}_4x_5, \end{aligned} \tag{3.3}$$

where we used the fact that $x_2\bar{x}_2 = 0$. It is easy to see that this result extends to higher powers of f,

$$f_{32}^n(x_1, x_2, \ldots, x_{2n+1}) = x_1\bar{x}_2x_3\bar{x}_4 \ldots \bar{x}_{2n}x_{2n+1} = x_1 \prod_{i=1}^{n} \bar{x}_{2i}x_{2i+1},$$

yielding the solution formula

$$[F_{32}^n(x)]_j = x_{j-n} \prod_{i=1}^{n} \bar{x}_{2i-n-1+j}x_{2i-n+j}. \tag{3.4}$$

Rule 140 has the local function

$$f_{140}(x_1, x_2, x_3) = x_2 - x_1x_2 + x_1x_2x_3 = \bar{x}_1x_2 + x_1x_2x_3.$$

It is again fairly straightforward to "guess" the expression for f^n. We start with f^2, as before:

$$\begin{aligned} f_{140}^2(x_1, x_2, x_3, x_4, x_5) &= f_{140}(f_{140}(x_1, x_2, x_3), f_{140}(x_2, x_3, x_4), f_{140}(x_3, x_4, x_5)) \\ &= (1 - f_{140}(x_1, x_2, x_3))f_{140}(x_2, x_3, x_4) \\ &\quad + f_{140}(x_1, x_2, x_3)f_{140}(x_2, x_3, x_4)f_{140}(x_3, x_4, x_5). \end{aligned} \tag{3.5}$$

Omitting tedious but straightforward algebra, this yields

$$f_{140}^2(x_1, x_2, x_3, x_4, x_5) = \bar{x}_2x_3 + x_2x_3x_4x_5.$$

One can then prove by induction the following general formula,

$$f_{140}^n(x_1, x_2, \ldots, x_{2n+1}) = \bar{x}_nx_{n+1} + x_nx_{n+1} \ldots x_{2n+1}.$$

The solution of the initial value problem is, therefore,

$$[F_{140}^n(x)]_j = \bar{x}_{j-1}x_j + \prod_{i=n-1}^{2n} x_{i-n+j}. \tag{3.6}$$

3.2 Additive Rules

In some cases the polynomial representation of the local function is not the best choice for solving the initial value problem. This is the case for rules 60, 90 and 150, for which the local functions in the polynomial representation have the form

$$\begin{aligned}
f_{60}(x_1, x_2, x_3) &= x_2 + x_1 - 2x_1x_2,\\
f_{90}(x_1, x_2, x_3) &= x_3 + x_1 - 2x_1x_3,\\
f_{150}(x_1, x_2, x_3) &= x_3 + x_2 + x_1 - 2x_2x_3 - 2x_1x_3 - 2x_1x_2 + 4x_1x_2x_3.
\end{aligned}$$

These rule are sometimes called *linear* or *additive* because their local functions can also be represented as addition modulo 2, as follows,

$$\begin{aligned}
f_{60}(x_1, x_2, x_3) &= x_1 + x_2 \mod 2,\\
f_{90}(x_1, x_2, x_3) &= x_1 + x_3 \mod 2,\\
f_{150}(x_1, x_2, x_3) &= x_1 + x_2 + x_3 \mod 2.
\end{aligned}$$

Consider rule 60 first. Writing the first three iterates of F_{60} we obtain

$$\begin{aligned}
[F_{60}^1]_j &= x_{-1+j} + x_j,\\
[F_{60}^2]_j &= x_{-2+j} + 2x_{-1+j} + x_j,\\
[F_{60}^3]_j &= x_{-3+j} + 3x_{-2+j} + 3x_{-1+j} + x_j.
\end{aligned}$$

It is easy to notice that the coefficients of x_i are consecutive binomial coefficients, suggesting the following general formula for $[F_{60}^n(x)]_j$,

$$[F_{60}^n(x)]_j = \sum_{i=0}^{n} \binom{n}{i} x_{i-n+j} \mod 2. \tag{3.7}$$

To prove this, let us note that F_{60} can be represented as a sum (modulo 2) of identity I and inverse shift σ^{-1},

$$F_{60} = I + \sigma^{-1}.$$

Since both I and σ^{-1} are linear and commute, we can compute power of F as follows,

$$F_{60}^n = (I + \sigma^{-1})^n = \sum_{i=0}^{n} \binom{n}{i} I^i \sigma^{-(n-i)} = \sum_{i=0}^{n} \binom{n}{i} \sigma^{-n+i}, \tag{3.8}$$

where all sums are modulo 2. This implies

$$[F_{60}^n(x)]_j = \sum_{i=0}^{n} \binom{n}{i} [\sigma^{-n+i}(x)]_j \mod 2 = \sum_{i=0}^{n} \binom{n}{i} x_{i-n+j} \mod 2,$$

verifying our "guess" in Eq. (3.7).

For rule 90 we have $F_{90} = \sigma + \sigma^{-1}$, thus the expressions for $[F^n(x)]_j$ follow the same pattern and the solution is only slightly different,

$$[F_{90}^n(x)]_j = \sum_{i=0}^{n} \binom{n}{i} x_{2i-n+j} \mod 2. \tag{3.9}$$

It is worth noting that since the sums in the solution formulae for rules 60 and 90 are taken modulo 2, the terms with even binomial coefficients $\binom{n}{i}$ do not matter and can be omitted. Moreover, when the binomial coefficient is odd, it can be replaced by 1 without affecting the value of the sum. The first three iterates of F_{60} could thus be written as

$$\begin{aligned}
[F_{60}^1]_j &= x_{-1+j} + x_j \mod 2,\\
[F_{60}^2]_j &= x_{-2+j} + x_j \mod 2,\\
[F_{60}^3]_j &= x_{-3+j} + x_{-2+j} + x_{-1+j} + x_j \mod 2.
\end{aligned}$$

In general, for rule 60 we have

$$[F_{60}^n(x)]_j = \sum_{\substack{0 \le i \le n \\ \binom{n}{i} \text{ odd}}} x_{i-n+j} \mod 2, \tag{3.10}$$

and for rule 90

$$[F_{90}^n(x)]_j = \sum_{\substack{0 \le i \le n \\ \binom{n}{i} \text{ odd}}} x_{2i-n+j} \mod 2. \tag{3.11}$$

Rule 150 is slightly more complicated. Let us list the first four expressions for $[F_{150}^n(x)]_j$, starting from $n = 0$,

$$
\begin{aligned}
[F^0_{150}]_j &= x_j,\\
[F^1_{150}]_j &= x_{-1+j} + x_j + x_{1+j} \mod 2,\\
[F^2_{150}]_j &= x_{-2+j} + 2x_{-1+j} + 3x_j + 2x_{1+j} + x_{2+j} \mod 2,\\
[F^3_{150}]_j &= x_{-3+j} + 3x_{-2+j} + 6x_{-1+j} + 7x_j + 6x_{1+j} + 3x_{2+j} + x_{3+j} \mod 2.
\end{aligned}
$$

If we write only the coefficients of the terms of these expressions we obtain

```
      1
    1 1 1
  1 2 3 2 1
1 3 6 7 6 3 1
```

One can verify that any number in this triangular table is the sum of three numbers directly above, for example, $6 = 1 + 2 + 3$. It resembles Pascal's triangle and its entries are called *trinomial coefficients*. The trinomial coefficient $\binom{n}{k}_2$, with $n \geq 0$ and $-n \leq k \leq n$, is given by the coefficient of z^{n+k} in the expansion of $(1 + z + z^2)^n$,

$$
\left(1 + z + z^2\right)^n = \sum_{k=-n}^{n} \binom{n}{k}_2 z^{n+k}. \tag{3.12}
$$

Trinomial coefficients satisfy the recurrence formula

$$
\binom{n+1}{k}_2 = \binom{n}{k-1}_2 + \binom{n}{k}_2 + \binom{n}{k+1}_2,
$$

which is exactly the reason why they appear in the solution formula for rule 150,

$$
[F^n_{150}(x)]_j = \sum_{i=0}^{2n} \binom{n}{i-n}_2 x_{i+j-n} \mod 2. \tag{3.13}
$$

In order to formally derive this expression, let us note that

$$
F_{150} = \sigma^{-1} + I + \sigma = \sigma^{-1}(I + \sigma + \sigma^2).
$$

Using Eq. (3.12) with $z = \sigma$ we obtain

$$
F^n_{150} = \sigma^{-n}(I + \sigma + \sigma^2)^n = \sigma^{-n} \sum_{k=-n}^{n} \binom{n}{k}_2 \sigma^{n+k} = \sum_{k=-n}^{n} \binom{n}{k}_2 \sigma^k, \tag{3.14}
$$

and therefore

$$[F_{150}^n(x)]_j = \sum_{k=-n}^{n} \binom{n}{k}_2 [\sigma^k(x)]_j \mod 2 = \sum_{k=-n}^{n} \binom{n}{k}_2 x_{j+k} \mod 2. \quad (3.15)$$

Changing the index k in the last sum to $i = k + n$ we obtain the desired Eq. (3.13).

Similarly as in the case of rules 60 and 90, the terms in the sum in Eq. (3.13) corresponding to even coefficients do not matter, and odd coefficients can be replaced by 1, therefore the alternative expression for the solution of rule 150 could be

$$[F_{150}^n(x)]_j = \sum_{\substack{0 \le i \le 2n \\ \binom{n}{i-n}_2 \text{ odd}}} x_{i+j-n} \mod 2. \quad (3.16)$$

We will add that the the trinomial coefficient can be expressed in a closed form [1] as

$$\binom{n}{k}_2 = \sum_{j=0}^{n} \frac{n!}{(j)!(j+k)!(n-2j-k)!}. \quad (3.17)$$

3.3 Rules 132 and 77

For rule 77, the polynomial representation of the local function is

$$f_{77}(x_1, x_2, x_3) = 1 - x_3 - x_1 + x_2x_3 + x_1x_3 + x_1x_2 - 2x_1x_2x_3,$$

which can be also written as

$$f_{77}(x_1, x_2, x_3) = x_2 - x_1x_2x_3 + \overline{x}_1\overline{x}_2\overline{x}_3. \quad (3.18)$$

This is neither multiplicative nor additive rule, yet the high degree of symmetry in the formula for the local function suggest that there is a chance of discovering a pattern in f^n, just like we did for multiplicative and additive rules. Indeed, the first three iterates of f are

$$\begin{aligned}
f_{77}^1(x_1, x_2, x_3) &= x_2 - x_1x_2x_3 + \overline{x}_1\overline{x}_2\overline{x}_3, \\
f_{77}^2(x_1, x_2, x_3, x_4, x_5) &= x_3 + x_1x_2x_3x_4x_5 - x_2x_3x_4 + \overline{x}_2\overline{x}_3\overline{x}_4 - \overline{x}_1\overline{x}_2\overline{x}_3\overline{x}_4\overline{x}_5, \\
f_{77}^3(x_1, x_2, x_3, x_4, x_5, x_6, x_7) &= x_4 - x_1x_2x_3x_4x_5x_6x_7 + x_2x_3x_4x_5x_6 - x_3x_4x_5 \\
&\quad + \overline{x}_3\overline{x}_4\overline{x}_5 - \overline{x}_2\overline{x}_3\overline{x}_4\overline{x}_5\overline{x}_6 + \overline{x}_1\overline{x}_2\overline{x}_3\overline{x}_4\overline{x}_5\overline{x}_6\overline{x}_7.
\end{aligned}$$

This suggests that the general expression might be

$$f_{77}^n(x_1, x_2, \ldots, x_{2n+1}) = x_{n+1} + \sum_{r=1}^{n} (-1)^r \prod_{i=-r}^{r} x_{n+1+i} + \sum_{r=1}^{n} (-1)^{r+1} \prod_{i=-r}^{r} \overline{x}_{n+1+i},$$

leading to the solution formula

$$[F_{77}^n(x)]_j = x_j + \sum_{r=1}^{n} (-1)^r \prod_{i=-r}^{r} x_{j+i} + \sum_{r=1}^{n} (-1)^{r+1} \prod_{i=-r}^{r} \overline{x}_{j+i}. \tag{3.19}$$

Proof of correctness of this formula can be obtained by direct verification that

$$[F_{77}^{n+1}(x)]_j = [F_{77}^n(z)]_j, \tag{3.20}$$

where $z_i = f_{77}(x_{i-1}, x_i, x_{i+1})$. We start from the right hand side,

$$[F_{77}^n(z)]_j = f_{77}(x_{j-1}, x_j, x_{j+1}) + \sum_{r=1}^{n} (-1)^r \prod_{i=-r}^{r} f_{77}(x_{i+j-1}, x_{i+j}, x_{i+j+1})$$
$$+ \sum_{r=1}^{n} (-1)^{r+1} \prod_{i=-r}^{r} (1 - f_{77}(x_{i+j-1}, x_{i+j}, x_{i+j+1})).$$

One can show (for example, by induction) that

$$\prod_{i=-r}^{r} f_{77}(x_{i+j-1}, x_{i+j}, x_{i+j+1}) = \prod_{i=-r-1}^{r+1} \overline{x}_{i+j}, \tag{3.21}$$

and similarly

$$\prod_{i=-r}^{r} (1 - f_{77}(x_{i+j-1}, x_{i+j}, x_{i+j+1})) = \prod_{i=-r-1}^{r+1} x_{i+j}. \tag{3.22}$$

We obtain, therefore

$$[F_{77}^n(z)]_j = x_j - x_{j-1}x_jx_{j+1} + \overline{x}_{j-1}\overline{x}_j\overline{x}_{j+1} + \sum_{r=1}^{n} (-1)^r \prod_{i=-r-1}^{r+1} \overline{x}_{i+j}$$
$$+ \sum_{r=1}^{n} (-1)^{r+1} \prod_{i=-r-1}^{r+1} x_{i+j}.$$

Changing the index r to $s = r + 1$ this becomes

$$[F_{77}^n(z)]_j = x_j - x_{j-1}x_jx_{j+1} + \overline{x}_{j-1}\overline{x}_j\overline{x}_{j+1} + \sum_{s=2}^{n}(-1)^{s-1}\prod_{i=-s}^{s}\overline{x}_{i+j}$$
$$+ \sum_{s=2}^{n}(-1)^{s}\prod_{i=-s}^{s}x_{i+j}.$$

Using $(-1)^{s-1} = (-1)^{s+1}$, we can incorporate the third term into the second sum and the second term into the third sum obtaining

$$[F_{77}^n(z)]_j = x_j + \sum_{s=1}^{n}(-1)^{s+1}\prod_{i=-s}^{s}\overline{x}_{i+j} + \sum_{s=1}^{n}(-1)^{s}\prod_{i=-s}^{s}x_{i+j} = [F_{77}^{n+1}(x)]_j,$$

as desired.

Another rule where the solution can be obtained by discovering the pattern in expressions for f^n is rule 132 with the local function

$$f_{132}(x_1, x_2, x_3) = x_2 - x_2x_3 - x_1x_2 + 2x_1x_2x_3, \tag{3.23}$$

which, exploiting indempotency of Boolean variables, can be written as

$$f_{132}(x_1, x_2, x_3) = x_2 - (x_1 - x_3)^2 x_2. \tag{3.24}$$

Inspecting the first three iterates we find the they can be put into a similar form,

$$\begin{aligned}
f_{132}^1(x_1, x_2, x_3) &= x_2 - (x_1 - x_3)^2 x_2,\\
f_{132}^2(x_1, x_2, x_3, x_4, x_5) &= x_3 - (x_2 - x_4)^2 x_3 - (x_1 - x_5)^2 x_2x_3x_4,\\
f_{132}^3(x_1, x_2, x_3, x_4, x_5, x_6, x_7) &= x_4 - (x_3 - x_5)^2 x_4 - (x_2 - x_6)^2 x_3x_4x_5\\
&\quad - (x_1 - x_7)^2 x_2x_3x_4x_5x_6.
\end{aligned}$$

This suggest that

$$f_{132}^n(x_1, x_2, \ldots, x_{2n+1}) = x_{n+1} - \sum_{r=1}^{n}\left((x_{n-r+1} - x_{n+1+r})^2 \prod_{i=-r+1}^{r-1} x_{i+n+1}\right).$$

Indeed, as in the case of rule 77, the solution formula

$$[F_{132}^n(x)]_i = x_j - \sum_{r=1}^{n}\left((x_{j-r} - x_{j+r})^2 \prod_{i=-r+1}^{r-1} x_{i+j}\right) \tag{3.25}$$

can be verified by direct substitution of the above to the equation $[F_{132}^{n+1}(x)]_j = [F_{132}^n(z)]_j$ with $z_i = f_{132}(x_{i-1}, x_i, x_{i+1})$.

There are several more rules for which the patterns in density polynomials for f^n can be easily discovered and the general formula for the solution "guessed" and then proved by induction, just like we did in the case of rule 77. For example, rule 162 has the local function

$$f_{162}(x_1, x_2, x_3) = x_3 - x_2x_3 + x_1x_2x_3.$$

Powers of f_{162} have a similar structure, namely the sum of products with alternating signs. The general solution formula is not hard to find,

$$[F_{162}^n(x)]_j = x_{j+n} + \sum_{p=-n+1}^{n} \left((-1)^{p+n} \prod_{i=j-p}^{j+n} x_i \right).$$

Rules 23, 27, 44, 50, 130 and 232 also exhibit "discoverable" patterns in their density polynomials which can be exploited to construct solutions. In some cases, like rules 27, 50 and 130, slightly different patterns are followed for odd and even iterations. Inspecting the list of density polynomials with even n separately from those with odd n helps to reveal these patterns. For rule 44, there are different patterns for n divisible by 3 and for the remaining n's. These patterns are actually not very obvious and were rather difficult to find. In fact, they were found with the help of some other tools like finite state machines, to be discussed in Chap. 6. This also applies to rule 232, which has a rather complicated solution formula.

We show the relevant solution formulae for all the rules mentioned above in Appendix B.

Reference

1. Andrews, G.E.: Euler's "Exemplum Memorabile Inductionis Fallacis" and q-trinomial coefficients. Journal of the American Mathematical Society **3**(3), 653–669 (1990)

Chapter 4
More Complex Rules

In the previous chapter we were able to find patterns in polynomial representations of consecutive powers of f by simply inspecting the first few powers, f, f^2, f^3 etc. If there is no obvious regularity in such polynomials, another strategy to try for constructing the solution formulae is as follows. First, we construct n-step preimages of 1 (i.e., sets $\mathbf{f}^{-n}(1)$) for various n. We can then try to find patterns in these sets which would allow us to give a combinatorial description of them, as a set of binary strings satisfying certain conditions. Once this is done, we can construct a Boolean function which is an indicator function of $\mathbf{f}^{-n}(1)$. Such function will then be used to construct an explicit expression for $[F^n(x)]_j$.

4.1 Rule 172

We will apply this method first to rule 172. Its local function can be defined as

$$f_{172}(x_1, x_2, x_3) = \begin{cases} x_2 & \text{if } x_1 = 0, \\ x_3 & \text{if } x_1 = 1, \end{cases} \tag{4.1}$$

or, using polynomial representation, as

$$f_{172}(x_1, x_2, x_3) = x_2 + x_1 x_3 - x_1 x_2.$$

It will be helpful to start with the inspection of the spatiotemporal pattern generated by rule 172, as shown in Fig. 4.1. It reveals three facts, each of them easily provable in a rigorous way:

(i) A cluster of two or more zeros keeps its right boundary in the same place for ever.

H. Fukś, *Solvable Cellular Automata*, Understanding Complex Systems,
https://doi.org/10.1007/978-3-031-38700-5_4

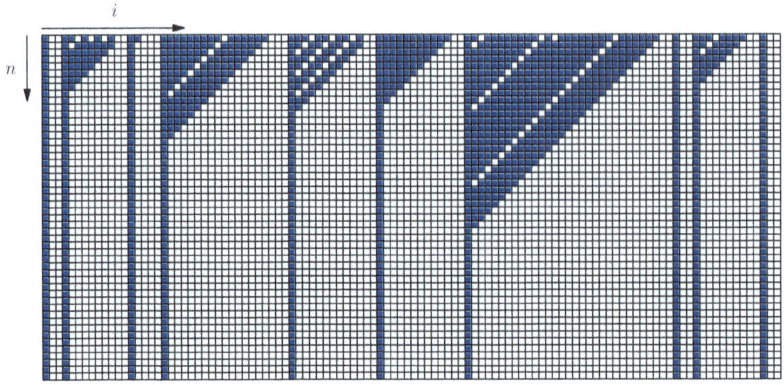

Fig. 4.1 Example of a spatiotemporal pattern produced by rule 172

(ii) A cluster of two or more zeros extends its left boundary to the left one unit per time step as long as the left boundary is preceded by two or more ones. If the left boundary of the cluster of zeros is 01, the cluster does not grow.
(iii) Isolated zero moves to the left one step at a time as long as it has at least two ones on the left. If an isolated zero is preceded by 01, it disappears in the next time step.

Suppose now that we have a string $\mathbf{b}$ of length $2n+1$ and we want to find out the necessary and sufficient conditions for $\mathbf{f}^n(\mathbf{b}) = 1$. From (i) it is clear that the word 001 will remain in the same position forever, which means that if

$$\mathbf{b} = \underbrace{\star\star\ldots\star}_{n-2} 001 \underbrace{\star\star\ldots\star}_{n}, \tag{4.2}$$

then $\mathbf{f}^n(\mathbf{b}) = 1$. What are the other possibilities for b which would result in $\mathbf{f}^n(\mathbf{b}) = 1$?

From (ii) we deduce that if there is no cluster of two or more zeros somewhere in the last $n+3$ bits of b, then there is no possibility of the growth of cluster of zeros producing $\mathbf{f}^n(b) = 0$. The only way to get zero after n iterations of $\mathbf{f}$ in such a case would be having zero at the end of b preceded by 11. This means that in order to avoid this scenario, the last 3 bits of b must be 010, 011, 101 or 111, or, in other words, the last three bits must be $01\star$ or $1\star 1$. Putting this together, the block b must have the form

$$b = \underbrace{\star\star\ldots\star}_{n-2} a_1 a_2 \ldots a_{n+1} c_1 c_2, \tag{4.3}$$

where $a_1 a_2 \ldots a_n$ is a binary string which does not contain any pair of adjacent zeros, and

$$c_1 c_2 = \begin{cases} 1\star, & \text{if } a_{n+1} = 0, \\ \star 1, & \text{otherwise.} \end{cases} \tag{4.4}$$

4.1.1 Solving Rule 172

We are now ready to construct the solution of rule 172. Without the loss of generality, to simplify the algebra, we will find the expression for $[F^n(x)]_0$ first. We have already discovered that in order to get $[F^n(x)]_0 = 1$, we need $x_{-n}x_{-n+1}, \ldots x_n$ to be in one of two forms, the first of them being

$$x_{-n}x_{-n+1}, \ldots x_n = \underbrace{\star\star\ldots\star}_{n-2} 001 \underbrace{\star\star\ldots\star}_{n}. \tag{4.5}$$

The above means that $x_{-2}x_{-1}x_0 = 001$, and it will be true if and only if

$$\overline{x}_{-2}\overline{x}_{-1}x_0 = 1. \tag{4.6}$$

The second possibility is given by

$$x_{-n}x_{-n+1}, \ldots x_n = \underbrace{\star\star\ldots\star}_{n-2} a_1a_2\ldots a_{n+1}c_1c_2, \tag{4.7}$$

where $a_ia_{i+1} \neq 00$ for $i = 1, 2, \ldots, n$ and c_1, c_2 satisfy condition of Eq. (4.4). This means that $a_1a_2\ldots a_{n+1} = x_{-2}x_{-1}\ldots x_{n-2}$, and therefore $x_ix_{i+1} \neq 00$ for $i = -2, -1, \ldots, n-3$, as well as

$$x_{n-1}x_n = \begin{cases} 1\star, & \text{if } x_{n-2} = 0, \\ \star 1, & \text{otherwise.} \end{cases} \tag{4.8}$$

The second possibility will be realized if and only if

$$\left(\prod_{i=-2}^{n-3} (1 - \bar{x}_i\bar{x}_{i+1}) \right) (\bar{x}_{n-2}x_{n-1} + x_{n-2}x_n) = 1, \tag{4.9}$$

where we used the notation $\bar{x}_i = 1 - x_i$. Combining Eqs. (4.6) and (4.9) we obtain

$$[F^n(x)]_0 = \bar{x}_{-2}\bar{x}_{-1}x_0 + \left(\prod_{i=-2}^{n-3} (1 - \bar{x}_i\bar{x}_{i+1}) \right) (\bar{x}_{n-2}x_{n-1} + x_{n-2}x_n). \tag{4.10}$$

This is the desired solution expressing the value of the central site after n iterations of rule F starting from an initial configuration x. Of course, we can now obtain analogous expression for any other site $[F^n(x)]_j$, by simply translating the above formula from $j = 0$ to an arbitrary position j.

Proposition 4.1 *Let F be the global function of elementary CA rule 172 and $x \in \{0,1\}^{\mathbb{Z}}$. Then, after $n \in \mathbb{N}$ iterations of F, for any $j \in \mathbb{Z}$,*

$$[F^n(x)]_j = \bar{x}_{j-2}\bar{x}_{j-1}x_j + \left(\prod_{i=j-2}^{j+n-3}(1-\bar{x}_i\bar{x}_{i+1})\right)(\bar{x}_{j+n-2}x_{j+n-1} + x_{j+n-2}x_{j+n}). \tag{4.11}$$

Proof Since we formulated the above proposition from somewhat informal observation of the features of spatiotemporal patterns of rule 172, we will now present a formal proof of the correctness of Eq. (4.11). We need to show that $[F^n(x)]_j$ defined by Eq. (4.11) satisfies

$$[F^{n+1}(x)]_j = [F^n(y)]_j, \tag{4.12}$$

where $y_j = f_{172}(x_{j-1}x_j, x_{j+1}) = x_j + x_{j-1}x_{j+1} - x_{j-1}x_j$. Without a loss of generality, we will do it for $j = 0$. We start with the left had side of Eq. (4.12),

$$[F^{n+1}(x)]_0 = \bar{x}_{-2}\bar{x}_{-1}x_0 + \left(\prod_{i=-2}^{n-2}(1-\bar{x}_i\bar{x}_{i+1})\right)(\bar{x}_{n-1}x_n + x_{n-1}x_{n+1}).$$

For reasons which will soon become clear, we will decrease the upper bound of the index i to $n-4$ and label the terms with letters A, B, C, as follows,

$$[F^{n+1}(x)]_0 = \underbrace{\bar{x}_{-2}\bar{x}_{-1}x_0}_{A} + \underbrace{\left(\prod_{i=-2}^{n-4}(1-\bar{x}_i\bar{x}_{i+1})\right)}_{B}$$
$$\underbrace{(1-\bar{x}_{n-3}\bar{x}_{n-2})(1-\bar{x}_{n-2}\bar{x}_{n-1})(\bar{x}_{n-1}x_n + x_{n-1}x_{n+1})}_{C}.$$

The right hand side of Eq. (4.12) is given by

$$[F^n(y)]_0 = \bar{y}_{-2}\bar{y}_{-1}y_0 + \left(\prod_{i=-2}^{n-3}(1-\bar{y}_i\bar{y}_{i+1})\right)(\bar{y}_{n-2}y_{n-1} + y_{n-2}y_n),$$

and we can write it is in a similar form as the left hand side,

$$[F^n(y)]_0 = \underbrace{\bar{y}_{-2}\bar{y}_{-1}y_0}_{A'}$$
$$+ \underbrace{\left(\prod_{i=-2}^{n-4}(1-\bar{y}_i\bar{y}_{i+1})\right)}_{B'}\underbrace{(1-\bar{y}_{n-3}\bar{y}_{n-2})(\bar{y}_{n-2}y_{n-1} + y_{n-2}y_n)}_{C'}.$$

Remembering that $y_j = x_j + x_{j-1}x_{j+1} - x_{j-1}x_j$, we have

$$\begin{aligned} A' &= \bar{y}_{-2}\bar{y}_{-1}y_0 \\ &= (1 - x_{-3}x_{-1} - x_{-2} + x_{-3}x_{-2})\,(1 - x_{-2}x_0 - x_{-1} + x_{-2}x_{-1}) \\ &\quad (x_0 - x_{-1}x_0 + x_{-1}x_1)\,. \end{aligned}$$

It is straightforward to expand and simplify the above expression using the usual property of Boolean variables, $x_i^k = x_i$. The result is

$$A' = \bar{x}_{-2}\bar{x}_{-1}x_0 = A.$$

In a similar fashion one can show that $C = C'$, although it is rather tedious and therefore best done using symbolic algebra software. We now need to handle B' and B. Let us first notice that

$$\begin{aligned} 1 - \bar{y}_i\bar{y}_{i+1} &= 1 - (x_i x_{i-1} - x_{i-1}x_{i+1} - x_i + 1)\,(x_i x_{i+1} - x_i x_{i+2} - x_{i+1} + 1) \\ &= x_i + x_{i+1} - x_i x_{i-1} x_{i+1} x_{i+2} + x_i x_{i-1} x_{i+1} \\ &\quad + x_i x_{i-1} x_{i+2} - x_i x_{i-1} - x_i x_{i+1}. \end{aligned}$$

It is easy to verify that this can be written as

$$1 - \bar{y}_i\bar{y}_{i+1} = 1 - \bar{x}_i\bar{x}_{i+1} - x_{i-1}x_i\bar{x}_{i+1}\bar{x}_{i+2},$$

therefore

$$B' = \prod_{i=-2}^{n-4} (1 - \bar{x}_i\bar{x}_{i+1} - x_{i-1}x_i\bar{x}_{i+1}\bar{x}_{i+2})\,.$$

The term $x_{i-1}x_i\bar{x}_{i+1}\bar{x}_{i+2}$ is non-zero only when $x_{i-1}x_i x_{i+1} x_{i+2} = 1100$. If this happens for $i < n-4$, then we will have $x_{i+1}x_{i+2} = 00$, making B' equal to 0. If $x_{i-1}x_i x_{i+1} x_{i+2} = 1100$ for $i = n-4$, then $x_{n-3}x_{n-2} = 00$, making $C = C' = 0$. This means that the term $x_{i-1}x_i\bar{x}_{i+1}\bar{x}_{i+2}$ has no effect on the overall value of $[F^n(y)]_0$, therefore it can be omitted, so that

$$B'C' = BC.$$

This proves that

$$A' + B'C' = A + BC,$$

verifying Eq. (4.12), as desired. □

4.2 Rule 168

Rule 168 is defined by $f_{168}(1, 1, 1) = f_{168}(1, 0, 1) = f_{168}(0, 1, 1) = 1$, and $f_{168}(x_1, x_2, x_3) = 0$ for all other values of x_1, x_2, x_3, or, using polynomial representation, by

$$f_{168}(x_1, x_2, x_3) = x_2x_3 + x_1x_3 - x_1x_2x_3. \tag{4.13}$$

Upon close examination of preimages of 1, it is possible to discover a pattern in these preimages, as described in the following proposition [1].

Proposition 4.2 *The set of n-step preimages of 1 under the rule 168 is the union of sets A_n and B_n, defined as follows.*

- *The set A_n is the set of binary strings of length $2n + 1$ ending with 1 such that, counting from the right, the first pair of zeros begins at kth position from the right, and the number of isolated zeros in the substring to the right of this pair of zeros is m, satisfying $m < k - n - 1$.*
- *The set B_n is the set of binary strings of length $2n + 1$ ending with 1 which do not contain 00.*

Before we give the outline of the proof, let us note that elements of the set A_n described in the proposition have the structure

$$\underbrace{\star \dots \star}_{2n-k} 00 a_1 a_2 \dots a_{k-1}, \tag{4.14}$$

where the string $a_1a_2 \dots a_{k-1}$ starts from $a_1 = 1$, ends with $a_{k-1} = 1$, has no 00 pair and contains m isolated zeros, where $m < k - n - 1$. The elements of set B_n have the structure

$$b_1b_2 \dots b_{2n}1, \tag{4.15}$$

where the string $b_1b_2 \dots b_{2n}$ contains only isolated zeros.

Let us now explain how the spatiotemporal dynamics of rule 168 leads to the above result. We start from observing that $\mathbf{f}_{168}^{-1}(1) = \{011, 101, 111\}$. This means that if a given block ends with 1, then its preimage must also end with 1, and, by induction, its n-step preimage must end with 1 as well. Ending with 1, therefore, is a necessary condition for being a preimage of 1.

Let us note that one can treat a block $\mathbf{b}$ as a collection of blocks of zeros of various lengths separated by blocks of ones of various lengths. Suppose that a given block contains one isolated zero and to the left of it a pair of adjacent zeros, like in Fig. 4.2. When the rule 168 is applied iteratively, the block 00 will increase its length by moving its left boundary to the left, while its right boundary will remain in place. The isolated zero, on the other hand, simply moves to the left, as illustrated in Fig. 4.2. When the boundary of the growing cluster of zeros meets with the isolated zero, the isolated zero disappears, and the boundary of the cluster of zeros moves one position to the right. Two such events as shown in Fig. 4.2, marked by circles.

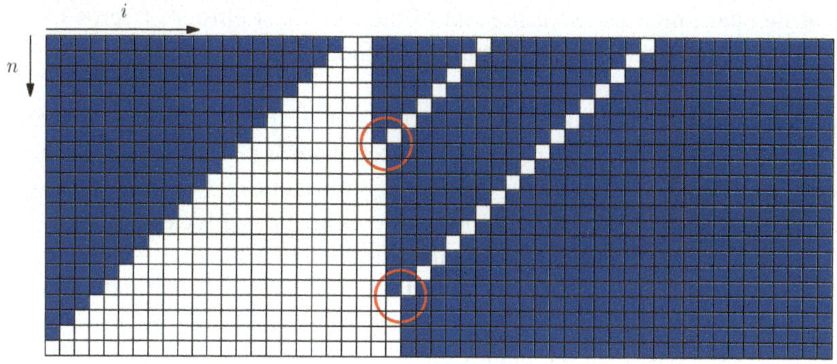

Fig. 4.2 Absorption of isolated zeros by the cluster of zeros in rule 168

```
101101101101011101101    101111001111011101111    101111001111011101011
 1101101101011101101      1111000111011101111      1111000111011101011
  01101101011101101        11000011011101111        11000011011101011
   101101011101101          000001011101111          000001011101011
    1101011101101            0000011101111            0000011101011
     01011101101              00001101111              00001101011
      011101101                000101111                000101011
       1101101                  0001111                  0001011
        01101                    00111                    00011
         101                      011                      001
          1                        1                        0
```

Fig. 4.3 Examples of blocks of length 21 for which 10 iterations of $\mathbf{f}_{168}$ produce 1 (left and center) and 0 (right)

We can now attempt to find conditions which a block must satisfy in order to be an n-step preimage of 1. If a block of length $2n + 1$ is an n-step preimage of 1, then either it contains a block of two or more zeros or not. If it does not, and ends with 1, then it necessarily is a preimage of 1. This is because when the rule is iterated, all isolated zeros move to the left, and after n iterations we obtain 1, as shown in Fig. 4.3 (left). Blocks of this type belong to the set B_n.

If, on the other hand, there is at least one cluster of adjacent zeros in a block of length $2n + 1$, then everything depends on the number of isolated zeros to the right of the rightmost cluster of zeros. Clearly, if there are not too many isolated zeros, and the rightmost cluster of zeros is not too far to the right, then the collisions of isolated zeros with the boundary of the cluster of zeros will not be able to move the boundary sufficiently far to change the final outcome, which will remain 1. This situation is illustrated in Fig. 4.3 (center). Blocks of this type are elements of A_n.

Let us also note that the balance of clusters of zeros and individual zeros must be somewhat preserved, and if there are too many isolated zeros, they may change the final outcome to 0, as in Fig. 4.3 (right). To express the conditions of this balance formally, suppose that we have a string $\mathbf{b} \in \mathcal{A}^{2n+1}$ and the first pair of zeros begins at kth position from the right. If there are no isolated zeros in the substring to the right of this pair, then we want the end of the rightmost cluster of zeros to be not further than just to the right of the center of **b**. Since the center of **b** is at the $n+1$th position from the right, we want $k > n+1$.

If the are m isolated zeros in the substring to the right of this pair of zeros, we must push the boundary of the rightmost cluster of zeros m units to the left, because these isolated zeros, after colliding with the rightmost cluster of zeros, will move the boundary to the right. The condition should, therefore, be in this case $k > n+1+m$, or, equivalently, $m < k-n-1$, as required for elements of A_n. □

Patterns for the structure of preimage strings given in Eqs. (4.14) and (4.15) allow us to construct the solution formula, in exactly the same fashion as it was done for rule 172. We will only give the final result,

$$[F_{168}^n(x)]_j = \sum_{k=n+2}^{2n} \left(H_k \overline{x}_{n-k+j} \overline{x}_{n-k+1+j} x_{n-k+2+j} x_{n+j} \prod_{i=2n-k+4}^{2n} (1-\overline{x}_{i-n-1+j}\overline{x}_{i-n+j}) \right)$$
$$+ x_{n+j} \prod_{i=1}^{2n-1} (1-\overline{x}_{i-n-1+j}\overline{x}_{i-n+j}),$$

where

$$H_k = \begin{cases} 1 & \text{if } \displaystyle\sum_{i=2n-k+4}^{2n} x_{i-n-1+j} \geq n-1, \\ 0 & \text{otherwise.} \end{cases} \tag{4.16}$$

The sum is the density polynomial for A_n and the product following it is the density polynomial for B_n. The factor H_k is necessary to allow only those strings form A_n for which the number of zeros m in $a_1 a_1 \ldots a_{k-1}$ satisfies $m < k-n-1$. Since $a_{k-1} = 1$, this condition means

$$\sum_{i=2n-k+4}^{2n} (1 - x_{i-n-1+j}) < k-n-1, \tag{4.17}$$

and because $\sum_{i=2n-k+4}^{2n} 1 = k-3$, it becomes

$$-\sum_{i=2n-k+4}^{2n} (x_{i-n-1+j}) < -n+2. \tag{4.18}$$

This is equivalent to $H_k = 1$ with H_k defined in Eq. (4.16) once we realize that for integers the relation $> n-2$ is the same as $\geq n-1$.

4.3 Rule 164

Rule 164 has the local function

$$f_{164}(x_1, x_2, x_3) = x_2 - x_2x_3 + x_1x_3 - x_1x_2 + x_1x_2x_3.$$

It differs from rule 165 only on (0, 0, 0) input,

$$f_{164}(x_1, x_2, x_3) = f_{165}(x_1, x_2, x_3) - \overline{x}_1\overline{x}_2\overline{x}_3.$$

This is illustrated in Fig. 4.4 where spatiotemporal patterns of both rules 164 and 165 are shown. We can see that in the areas where 000 blocks do not occur both patterns are identical. This property is very useful because rule 165 is linear,

$$f_{165} = (x_1, x_2, x_3) = 1 + x_1 + x_3 \mod 2,$$

and as such it is a solvable rule. It is actually a conjugated version of rule 90, for which we found solution in Eq. (3.9),

$$[F_{90}^n(x)]_j = \sum_{i=0}^{n} \binom{n}{i} x_{2i-n+j} \mod 2.$$

Solution of rule 165, therefore, is

$$[F_{165}^n(x)]_j = 1 - [F_{90}^n(\overline{x})]_j = 1 - \sum_{i=0}^{n} \binom{n}{i} (1 - x_{2i-n+j}) \mod 2.$$

Using $\sum_{i=0}^{n} \binom{n}{i} = 2^n$ and properties of mod 2 we obtain

$$[F_{165}^n(x)]_j = 1 + \sum_{i=0}^{n} \binom{n}{i} x_{2i-n+j} \mod 2. \tag{4.19}$$

One can easily notice that $f_{164}(x_1, 1, x_3) = f_{165}(x_1, 1, x_3)$, and $f_{164}(1, x_2, 1) = f_{165}(1, x_2, 1) = 1$. For this reason, for any string $\mathbf{b} \in \{0, 1\}^{2n+1}$ which has 1's on all even positions or on all odd positions we have

$$\mathbf{f}_{164}^n(\mathbf{b}) = \mathbf{f}_{165}^n(\mathbf{b}).$$

Define the set A_n to be the set of all binary strings which have 1's on all odd positions, of the form

$$\underbrace{1 \star 1 \star \ldots \star 1}_{2n+1}.$$

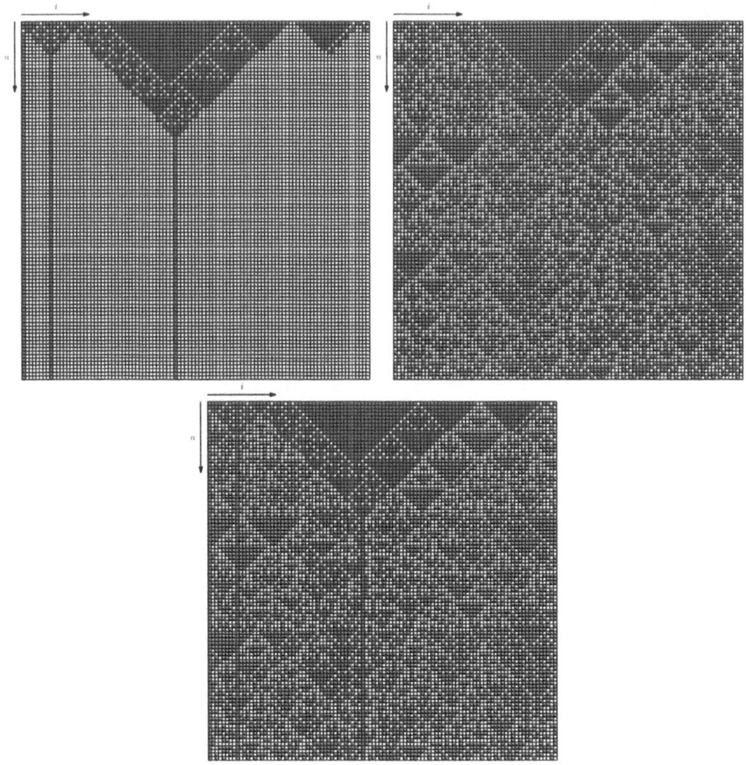

Fig. 4.4 Comparison or patterns generated by rule 164 (left) and 165 (right). Bottom picture shows the pattern of rule 164 (green) imposed on top of the pattern of rule 165 (blue)

All elements of A_n belong to $\mathbf{f}_{164}^{-n}(1)$, the set of n-step preimages of 1 under the rule 164.

Let us also consider the set B_n consisting of binary strings of length $2n + 1$ and of the form

$$b_1 1 b_2 1 b_3 \ldots 1 b_{n+1},$$

satisfying

$$\sum_{p=0}^{n} \binom{n}{i} b_{p+1} \mod 2 = 0. \tag{4.20}$$

By the virtue of Eq. (4.19) elements of B_n also belong to $\mathbf{f}_{164}^{-n}(1)$. We thus have two sets A_n and B_n consisting of only preimages of 1, but these are not all preimages of 1. In [2] it has been demonstrated that strings of the form

$$\underbrace{** \ldots *}_{n+i-1} 0 c_1 1 c_2 1 \ldots 1 c_i 1 c_{i+1} 0 \underbrace{** \ldots *}_{n+i-1} \tag{4.21}$$

where $i \in \{1, 2, \ldots, n-1\}$ are also the preimages of 1 as long as at least one of $c_1 \ldots c_{i+1}$ is zero and

$$\sum_{p=0}^{n} \binom{i}{p} c_{p+1} \mod 2 = 0. \tag{4.22}$$

We will call the set containing these strings C_n. One can further demonstrate [2] that there are no other preimages of 1, therefore $\mathbf{f}_{164}^{-n}(1) = A_n \cup B_n \cup C_n$. Sets A_n and C_n as well as B_n and C_n are disjoint, and the only common element of A_n and B_n is the string consisting of all ones, $11\ldots1$.

We will now proceed to find a condition which $x_{j-n}x_{j-n+1}\ldots x_{j+n}$ must satisfy in order to obtain $[F_{168}^n(x)]_j = 1$. We can do this for $j = 0$ without any loss of generality due to shift invariance of cellular automata. Let us first construct the indicator functions of elements of A_n, B_n, and C_n, i.e., the functions of $x_{-n}, x_{-n+1}, \ldots, x_n$ which take value 1 if the string $x_{-n}, x_{-n+1}, \ldots, x_n$ belongs to the respective set, and 0 otherwise.

If $x_{-n}x_{-n+1}\ldots x_n \in A_n$, then every other element x_i starting from x_{-n} must be equal to 1. This will happen when

$$\prod_{p=0}^{n} x_{-n+2p} = 1,$$

thus the left hand side of the above is the desired indicator function.

If $x_{-n}x_{-n+1}\ldots x_n \in B_n$, then by a similar argument we need

$$\prod_{p=0}^{n-1} x_{-n+2p+1} = 1,$$

as well as

$$\sum_{p=0}^{n} \binom{n}{p} x_{j-n+2p} \mod 2 = 0. \tag{4.23}$$

These two conditions will be satisfied together when

$$\left(1 + \sum_{p=0}^{n} \binom{n}{p} x_{j-n+2p} \mod 2\right) \prod_{p=0}^{n-1} x_{j-n+2p+1} = 1. \tag{4.24}$$

Again, the left hand side of the above is the desired indicator function for elements of B_n.

We now need to construct the indicator function for C_n, that is, the condition for

$$x_{-n}x_{-n+1}\ldots x_n = \underbrace{**\ldots*}_{n+i-1}0c_1 1c_2 1\ldots 1c_i 1c_{i+1}0\underbrace{**\ldots*}_{n+i-1} \tag{4.25}$$

subject to Eq. (4.22) and the requirement of least one $c_1 \dots c_{i+1}$ being zero. For a fixed value of i we need

$$\left(1+\sum_{p=0}^{i}\binom{i}{p}x_{-i+2p} \bmod 2\right)\left(\prod_{p=0}^{i-1}x_{-i+2p+1}\right)=1, \tag{4.26}$$

which is similar to Eq. (4.24) and takes care of the part $c_1 1 c_2 1 \dots 1 c_i 1 c_{i+1}$. We must also make sure that $c_1 1 c_2 1 \dots 1 c_i 1 c_{i+1}$ is preceded and followed by zero, thus the left hand side should be multiplied by $\overline{x}_{-1-i}\overline{x}_{i+1}$. Moreover, we need to make sure that at least one $c_1 \dots c_{i+1}$ is zero, and the factor $\left(1-\prod_{p=0}^{i}x_{-i+2p}\right)$ will take care of this. The final condition for belonging to C_n for a given i will be

$$\left(1-\prod_{p=0}^{i}x_{-i+2p}\right)\overline{x}_{-1-i}\overline{x}_{+i+1}$$
$$\times\left(\prod_{p=0}^{i-1}x_{-i+2p+1}\right)\left(1+\sum_{p=0}^{i}\binom{i}{p}x_{-i+2p} \bmod 2\right)=1.$$

The sum of the left hand side of the above over $i \in \{1, 2 \dots n-1\}$ will be the desired indicator function for C_n. Summing the indicator functions for all three sets A_n, B_n and C_n and adding j to each index we obtain the solution of the initial value problem for rule 164,

$$[F_{164}^{n}]_j=\prod_{p=0}^{n}x_{j-n+2p}-\prod_{p=-n}^{n}x_{j+p}+\left(1+\sum_{p=0}^{n}\binom{n}{p}x_{j-n+2p} \bmod 2\right)$$
$$\prod_{p=0}^{n-1}x_{j-n+2p+1}+\sum_{i=1}^{n-1}\left[\left(1-\prod_{p=0}^{i}x_{j-i+2p}\right)\overline{x}_{j-1-i}\overline{x}_{j+i+1}\right.$$
$$\left.\times\left(\prod_{p=0}^{i-1}x_{j-i+2p+1}\right)\left(1+\sum_{p=0}^{i}\binom{i}{p}x_{j-i+2p} \bmod 2\right)\right] \tag{4.27}$$

Note that we subtracted $\prod_{p=-n}^{n}x_{j+p}$ from the final expression, this is to avoid including $11\dots1$ twice—it belongs to both A_n and B_n, as already remarked.

4.4 Pattern Decomposition

Consider rules 156 and 140, with spatiotemporal patterns shown in the top and in the middle of Fig. 4.5. We can see that the characteristic black triangles with attached vertical strips, noticeable in the pattern of rule 140, seem to be also present in the

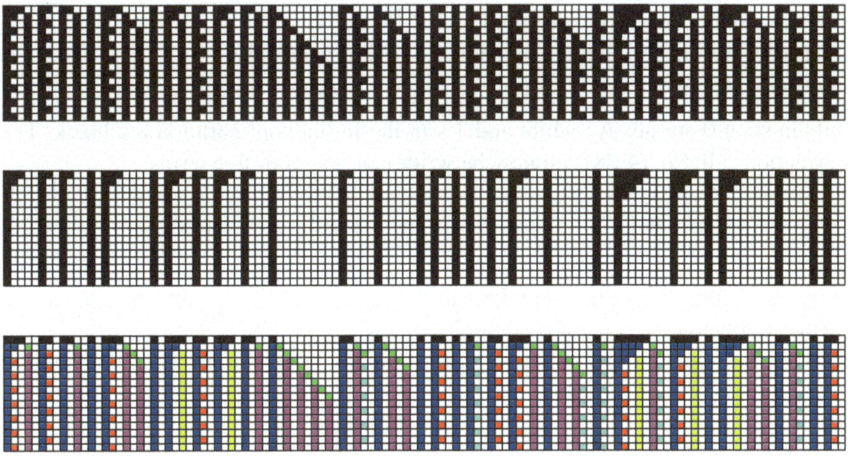

Fig. 4.5 Spatiotemporal pattern of rule 156 (top) decomposed into six elements shown in different colors (bottom). Middle figure is the pattern of rule 140

pattern of rule 156. A closer examination of Fig. 4.5 actually reveals that every black square appearing in the pattern of rule 140 also appears in the pattern of rule 156. Since we have already obtained solution for rule 140, this might be helpful in constructing of the solution for rule 156. The local functions of the aforementioned two rules are

$$f_{156}(x_1, x_2, x_3) = x_2 + x_1 - x_1x_3 - 2x_1x_2 + 2x_1x_2x_3,$$
$$f_{140}(x_1, x_2, x_3) = x_2 - x_1x_2 + x_1x_2x_3.$$

It is easy to check that

$$f_{156}(x_1, x_2, x_3) = f_{140}(x_1, x_2, x_3) + x_1\bar{x}_2\bar{x}_3,$$

meaning that these two rules differ only on block $x_1x_2x_3 = 100$, otherwise they are equal. This suggests that a more general relationship may hold,

$$[F^n_{156}(x)]_j = [F^n_{140}(x)]_j + \text{“corrections”}. \tag{4.28}$$

These "corrections" appear in the spatiotemporal pattern of rule 156 as various additional elements not present in the pattern or rule 140, namely vertical lines, diagonal lines and "blinkers" (vertical lines of alternating 0's and 1's). In Fig. 4.5 (bottom) we marked them with different colors, as follows:

- solid triangles and attached vertical strips are blue;
- diagonal lines (to be denoted by D) are green;
- blinkers under diagonal lines (denoted $B^{(0)}$) are cyan;

- blinkers under solid triangles (denoted $B^{(1)}$) are red;
- vertical lines under diagonals (denoted $S^{(0)}$) are magenta;
- vertical lines under solid triangles (denoted $S^{(1)}$) are yellow.

Cells in state 0 are always white and 1's in the initial configuration are black. The "corrections" in Eq. (4.28) can now be written as a sum of five terms,

$$[F_{156}^n(x)]_j = [F_{140}^n(x)]_j + D_{n,j} + B_{n,j}^{(0)} + B_{n,j}^{(0)} + S_{n,j}^{(0)} + S_{n,j}^{(1)}, \tag{4.29}$$

where the expressions corresponding to $D_{n,j}$, $B_{n,j}^{(0)}$, $B_{n,j}^{(1)}$, $S_{n,j}^{(0)}$ and $S_{n,j}^{(1)}$ are to be constructed below.

The expression corresponding to F_{140} is already known, it has been delivered in Sect. 3.1,

$$[F_{140}^n(x)]_j = \overline{x}_{j-1}x_j + \prod_{i=n-1}^{2n} x_{i-n+j}. \tag{4.30}$$

Looking at Fig. 4.5, we can easily conclude that the cell j after n iterations will belong to the diagonal (green) line if $x_{j-n} = 1$ and if it lies below the cluster of continuous zeros, so that $x_{-i+n+j} = 0$ for all $i \in \{n-1, n, \dots, 2n-1\}$. This corresponds to the expression

$$D_{n,j} = x_{j-n} \prod_{i=n-1}^{2n-1} \overline{x}_{-i+n+j}. \tag{4.31}$$

We will now deal with blinkers. The first type are those below the diagonal line, $B^{(0)}$. Let us define the "selector" of even numbers as

$$\text{ev}(n) = \frac{1}{2}(-1)^n + \frac{1}{2},$$

so that $\text{ev}(n) = 1$ if n is even, otherwise $\text{ev}(n) = 0$. Blinkers $B^{(0)}$ occur below clusters of zeros in the initial configuration, and we need to deal separately with clusters of zeros of even and odd length. The relevant expression is

$$\begin{aligned} B_{n,j}^{(0)} &= \text{ev}\,(n) \sum_{r=2}^{n} \left(\text{ev}\,(r+1)\, x_{j-r}x_{j-r+1}x_{2+j} \prod_{m=j-r+2}^{1+j} \overline{x}_m \right) \\ &+ \text{ev}\,(n+1) \sum_{r=2}^{n} \left(\text{ev}\,(r)\, \overline{x}_{j-r}x_{j-r+1}x_{2+j} \prod_{m=j-r+2}^{1+j} \overline{x}_m \right). \end{aligned} \tag{4.32}$$

The product $\prod_{m=j-r+2}^{1+j} \overline{x}_m$ is needed to ensure that we have a cluster of zeros above, terminated by 1 (hence factor x_{2+j}). The factors $x_{j-r}x_{j-r+1}$ and $\overline{x}_{j-r}x_{j-r+1}$ distinguish two different types of blinkers, those taking value 1 on even iterations and those taking value 1 on odd iterations (hence $\text{ev}(n)$ and $\text{ev}(n+1)$ factors in front of the sums). The other blinkers, $B_{n,j}^{(1)}$, yield a similar expression,

$$B_{n,j}^{(1)} = \text{ev}\,(n+1) \sum_{m=0}^{n-1} \left(\text{ev}\,(m+1)\, \overline{x}_{j-2} \overline{x}_{j+m+1} \overline{x}_{j+m+2} \prod_{k=-1}^{m} x_{j+k} \right)$$
$$+ \text{ev}\,(n) \sum_{m=0}^{n-1} \left(\text{ev}\,(m)\, \overline{x}_{j-2} \overline{x}_{j+m+1} x_{j+m+2} \prod_{k=-1}^{m} x_{j+k} \right). \tag{4.33}$$

The final two items we need to consider are the two types of vertical strips. The first one corresponds to the expression

$$S_{n,j}^{(0)} = \sum_{k=1}^{n} \left(\text{ev}\,(k)\, x_{j-k} x_{j-k+1} \prod_{m=j-k+2}^{1+j} \overline{x}_m \right)$$
$$+ \sum_{k=2}^{n} \left(\text{ev}\,(k+1)\, \overline{x}_{j-k} x_{j-k+1} \prod_{m=j-k+2}^{1+j} \overline{x}_m \right). \tag{4.34}$$

Here we again deal with structures occurring below clusters of 0's, hence the products $\prod_{m=j-k+2}^{1+j} \overline{x}_m$. There are two sums because there are two different expressions depending on the parity, and ev(k) and ev$(k+1)$ take care of this. The other type of strips, $S_{n,j}^{(1)}$, is only slightly more complicated,

$$S_{n,j}^{(1)} = \sum_{k=2}^{n} \left((-1)^k \left(x_{j+k} + \text{ev}\,(k+1) \left(x_{j-1+k} - 1 \right) \right) \prod_{i=j-2}^{j-2+k} x_i \right)$$
$$- \sum_{k=2}^{n} \left(x_j x_{j-2} x_{j-1} (-1)^k \prod_{m=1}^{k} x_{j+m} \right). \tag{4.35}$$

Equation (4.29) together with expressions defined in Eqs. (4.31)–(4.35) provide complete solution of the initial value problem for rule 156. A formal proof of the correctness of this formula can be obtained by verifying that

$$[F_{156}^{n}(x)]_j = [F_{156}^{n-1}(y)]_j,$$

where $y_i = f_{156}(x_{i-1}, x_i, x_{i+1})$. This is tedious but straightforward, and it is best done with the help of computer algebra system.

4.4.1 Rule 78

In the case of rule 156, we decomposed the pattern into elements which were added together. In some cases, not only adding but also subtracting pattern elements is required. Rule 78 is an example of such a case. Its pattern, show in the top of Fig. 4.6, can be constructed from the pattern of rule 206, shown in blue in the second picture of Fig. 4.6. Starting from the pattern of rule 206, we first remove some strips lying under clusters of zeros (third pattern of Fig. 4.6, red) together with some strips lying under clusters of ones (fourth pattern of Fig. 4.6, red), and then add some correction (last pattern of Fig. 4.6, blue). We will not go into all details of this construction here, but we will outline the main points. Rule 206, which serves as our starting point, is a conjugated version of rule 140, for which the solution is known. The formula for $[F_{206}^n(x)]_j$ can be obtained form the formula for rule $[F_{140}^n(x)]_j$ by taking expression for $1 - [F_{140}^n(x)]_j$ and replacing every occurrence of x_i by $\overline{x}_i$, hence

$$[F_{206}^n(x)]_j = 1 - x_{j-1}\overline{x}_j - \prod_{i=n-1}^{2n} \overline{x}_{i-n+j}. \tag{4.36}$$

The strips removed in the first step are strips under clusters of zeros preceded by 10, 110 or 111, resulting in the expression

$$\begin{aligned} S_{n,j}^{(A)} &= \sum_{k=2}^{n-1} \left(\mathrm{ev}(k) x_{j+k-1}\overline{x}_{j+k} \prod_{j-2}^{j-2+k} \overline{x}_i \right) \\ &+ \sum_{k=1}^{n-1} \left(\mathrm{ev}(k+1) x_{j+k-1} x_{j+k} \overline{x}_{j+k+1} \prod_{j-2}^{j-2+k} \overline{x}_i \right) \\ &+ \sum_{k=2}^{n-1} \left(\mathrm{ev}(k) x_{j+k-1} x_{j+k} x_{j+k+1} \prod_{j-2}^{j-2+k} \overline{x}_i \right). \end{aligned} \tag{4.37}$$

The corresponding pattern is shown in Fig. 4.6 as third from the top. We always remove every other strip only, hence the $\mathrm{ev}(k)$ or $\mathrm{ev}(k+1)$ factors. Removal of strips under 111 clusters is the next step, and the corresponding expression is straightforward,

$$S_{n,j}^{(B)} = x_{j-1} x_j x_{j+1}. \tag{4.38}$$

This corresponds to the fourth picture from the top in Fig. 4.6.

Detailed analysis of the pattern of rule 78, which we will not include here, reveals that when we remove strips given by expressions $S_{n,j}^{(A)}$ and $S_{n,j}^{(B)}$, too much is removed, and we need to add a corrective term. The form of this term is rather complicated, and we will only give the final expression for it,

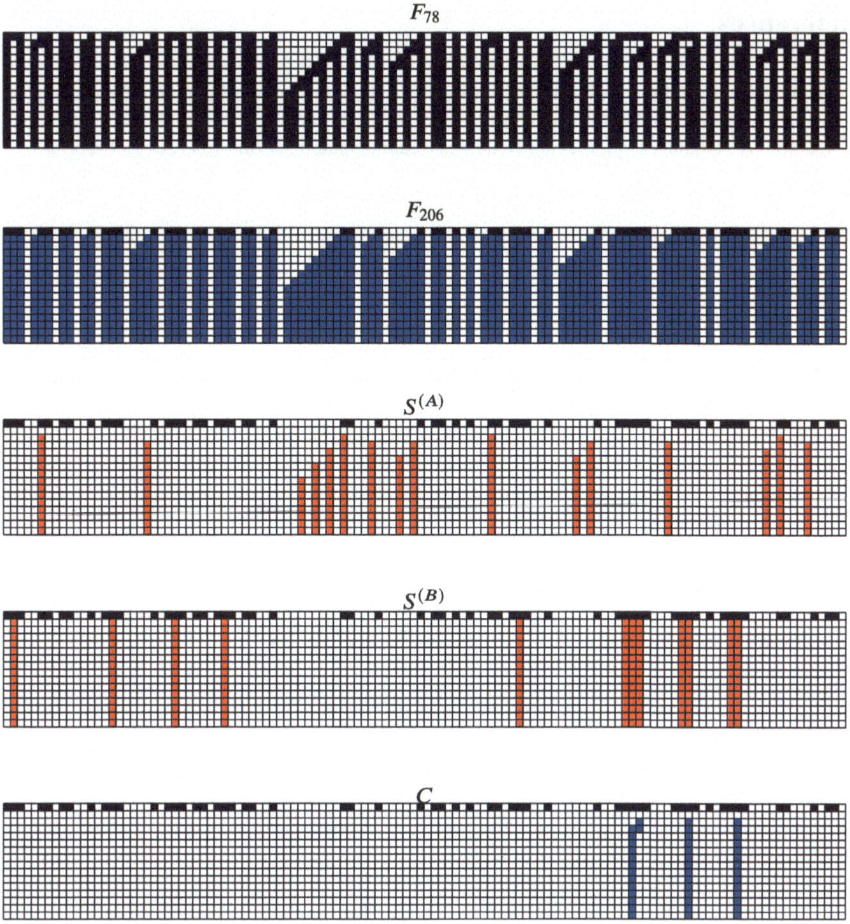

Fig. 4.6 Spatiotemporal pattern of rule 78 (top) decomposed into four sub-patterns (below). Blue patterns are added, red ones are subtracted, so that $F_{78} = F_{206} - S^{(A)} - S^{(B)} + C$. Initial configuration is black

$$C_{n,j} = x_{j-3}x_{j-2}x_{j-1}x_j x_{j+1} \sum_{r=2}^{2\lfloor n/2 \rfloor - 2} \left((-1)^r \prod_{k=2}^{r} x_{j+k} \right) - \prod_{i=-2}^{n} x_{j+i}$$
$$+ \overline{x}_{j-3}x_{j-2}x_{j-1}x_j x_{j+1} + I_3(n) x_{j-3}x_{j-2}x_{j-1}x_j x_{j+1}, \tag{4.39}$$

where $I_3(n) = 1$ for $n \leq 3$, otherwise $I_3(n) = 0$. The final formula for the solution of the initial value problem for rule 78 is, therefore, given by

$$[F_{78}^n(x)]_j = [F_{206}^n(x)]_j - S_{n,j}^{(A)} - S_{n,j}^{(B)} + C_{n,j}.$$

References

1. Fukś, H., Soto, J.M.G.: Exponential convergence to equilibrium in cellular automata asymptotically emulating identity. Complex Systems **23**, 1–26 (2014)
2. Skelton, A.: Gould's sequence and elementary cellular automata rule 164 (2017). Preprint

Chapter 5
Exploiting Rule Identities

In addition to emulation identities discussed in Chap. 2, there exist various other identities satisfied by elementary rules which can be helpful in solving the initial value problem. We will explore some of these identities below.

5.1 Identities of $F^{n+1} = G^k F^{n-k}$ Type

We have already considered emulations which are identities of the form $F^{n+1} = GF^n$. A more general version of such identity could be $F^{n+1} = G^k F^{n-k}$, where $0 < k < n$. Table 5.1 shows all identities of this type for minimal ECA for $n \leq 3$, excluding emulations already presented in Table 2.1.

When $G = F_{204}$, we find three rules in Table 5.1 which we have not yet solved. These are rules 1, 5, 29 and 108, the first three satisfying the identity $F^3 = F$. This identity implies that the orbit of x is $\{x, F(x), F^2(x), F(x), F^2(x), \ldots\}$, therefore

$$[F^n(x)]_j = \begin{cases} [F(x)]_j & n \text{ odd}, \\ \left[F^2(x)\right]_j & n \text{ even}. \end{cases} \tag{5.1}$$

For rule 1 we have

$$f_1(x_1, x_2, x_3) = \overline{x}_1\overline{x}_2\overline{x}_3,$$

thus

$$f_1^2(x_1, x_2, x_3, x_4, x_5) = (1 - \overline{x}_1\overline{x}_2\overline{x}_3)(1 - \overline{x}_2\overline{x}_3\overline{x}_4)(1 - \overline{x}_3\overline{x}_4\overline{x}_5).$$

This yields the solution

$$[F_1^n(x)]_j = \begin{cases} \overline{x}_{j-1}\overline{x}_j\overline{x}_{j+1} & n \text{ odd}, \\ (1 - \overline{x}_{j-2}\overline{x}_{j-1}\overline{x}_j)(1 - \overline{x}_{j-1}\overline{x}_j\overline{x}_{j+1})(1 - \overline{x}_j\overline{x}_{j+1}\overline{x}_{j+2}) & n \text{ even}. \end{cases} \tag{5.2}$$

H. Fukś, *Solvable Cellular Automata*, Understanding Complex Systems,
https://doi.org/10.1007/978-3-031-38700-5_5

Table 5.1 Minimal rules satisfying $F^{n+1} = G^k F^{n-k}$ where G is a single input rule, $n \le 3$ and $0 < k < n$

Rules satisfying $F^{n+1} = G^k F^{n-k}$		
$G = F_{15}$	$G = F_{51}$	$G = F_{85}$
$F_8^4 = F_{15}^2 F_8^2$	$F_1^3 = F_{51}^2 F_1$	$F_2^3 = F_{85}^2 F_2$
$F_{24}^4 = F_{15}^2 F_{24}^2$	$F_4^3 = F_{51}^2 F_4$	$F_8^4 = F_{85}^2 F_8^2$
$G = F_{170}$	$F_5^3 = F_{51}^2 F_5$	$F_{10}^3 = F_{85}^2 F_{10}$
$F_{38}^4 = F_{170}^2 F_{38}^2$	$F_8^4 = F_{51}^2 F_8^2$	$F_{34}^3 = F_{85}^2 F_{34}$
$G = F_{204}$	$F_{12}^3 = F_{51}^2 F_{12}$	$F_{38}^4 = F_{85}^2 F_{38}^2$
$F_1^3 = F_{204}^2 F_1$	$F_{29}^3 = F_{51}^2 F_{29}$	$F_{42}^3 = F_{85}^2 F_{42}$
$F_5^3 = F_{204}^2 F_5$	$F_{36}^4 = F_{51}^2 F_{36}^2$	$F_{46}^4 = F_{85}^2 F_{46}^2$
$F_{19}^4 = F_{204}^2 F_{19}^2$	$F_{72}^4 = F_{51}^2 F_{72}^2$	$F_{138}^3 = F_{85}^2 F_{138}$
$F_{29}^3 = F_{204}^2 F_{29}$	$F_{76}^3 = F_{51}^2 F_{76}$	$F_{170}^2 = F_{85}^2$
$F_{51}^2 = F_{204}^2$	$F_{108}^4 = F_{51}^2 F_{108}^2$	
$F_{108}^4 = F_{204}^2 F_{108}^2$	$F_{200}^3 = F_{51}^2 F_{200}$	
$G = F_{240}$	$F_{204}^2 = F_{51}^2$	
$F_{15}^2 = F_{240}^2$		

Emulators of single input rules from Table 2.1 and trivial cases are not included

For rule 5 the expression for f^2 is quite short,

$$f_5(x_1, x_2, x_3) = 1 - x_1 - x_3 + x_1 x_3,$$
$$f_5^2(x_1, x_2, x_3, x_4, x_5) = x_3 + x_1 x_5 - x_1 x_3 x_5,$$

yielding rather simple formula for the solution,

$$[F_5^n(x)]_j = \begin{cases} 1 - x_{j-1} - x_{j+1} + x_{j-1}x_{j+1} & n \text{ odd}, \\ x_j + x_{j-2}x_{j+2} - x_{j-2}x_j x_{j+2} & n \text{ even}. \end{cases} \tag{5.3}$$

For rule 29 we have

$$f_{29}(x_1, x_2, x_3) = 1 - x_1 x_2 + x_2 x_3 - x_3,$$
$$f_{29}^2(x_1, x_2, x_3, x_4, x_5) = x_3 + (x_4 - 1)\, x_2\, (x_1 x_3 - x_3 x_5 - x_1 + x_3)\,,$$

and the solution becomes

$$[F_{29}^n(x)]_j = \begin{cases} 1 - x_{j-1}x_j + x_j x_{j+1} - x_{j+1} & n \text{ odd}, \\ x_j + \left(x_{j+1} - 1\right) x_{j-1} \left(x_{j-2}x_j - x_j x_{j+2} - x_{j-2} + x_j\right) & n \text{ even}. \end{cases} \tag{5.4}$$

One more rule from Table 5.1 which has not yet been solved is the rule 108 for which

$$F_{108}^4 = F_{108}^2. \tag{5.5}$$

Orbit of x under this rule is given by

$$\{x, F_{108}(x), F_{108}^2(x), F_{108}^3(x), F_{108}^2(x), F_{108}^3(x), \ldots\},$$

hence

$$[F_{108}^n(x)]_j = \begin{cases} [F_{108}(x)]_j & n = 1, \\ \left[F_{108}^2(x)\right]_j & n \text{ even}, \\ \left[F_{108}^3(x)\right]_j & \text{odd } n > 1. \end{cases} \tag{5.6}$$

The explicit solution formula is rather long with $[F_{108}^3(x)]_j$ having over 30 terms, thus we do not reproduce it here. It is listed in Appendix B.

5.2 The Case of $G = \sigma$ or $G = \sigma^{-1}$

Two more rules not previously considered can be found in Table 5.1, namely rule 3 with $G = \sigma^{-1}$ and rule 38 with $G = \sigma$. The first one satisfies the identity

$$F_3^2 = F_{240}F_{236}. \tag{5.7}$$

Rule 240 is the inverse shift σ^{-1}, thus it commutes with all rules. For this reason we can write, for $k > 0$,

$$[F_3^{2k}(x)]_j = [\sigma^{-k}(F_{236}^k(x))]_j,$$
$$[F_3^{2k+1}(x)]_j = [\sigma^{-k}(F_{236}^k(F_3(x)))]_j.$$

Since $[\sigma^{-k}(x)]_j = x_{j-k}$, we obtain

$$[F_3^{2k}(x)]_j = [F_{236}^k(x)]_{j-k},$$
$$[F_3^{2k+1}(x)]_j = [F_{236}^k(F_3(x))]_{j-k}.$$

Rule 236 is a conjugated version of rule 200, thus according to Eqs. (1.14) and (2.22)

$$[F_{236}^k(x)]_j = 1 - [F_{200}^k(x)]_j = 1 - \overline{x}_{j-1}\overline{x}_j - \overline{x}_j\overline{x}_{j+1} + \overline{x}_{j-1}\overline{x}_j\overline{x}_{j+1}.$$

Note that the above expression does not depend on k. This is because rule 236, just as its minimal counterpart rule 200, is indempotent.[1] We further obtain

---

[1] Indempotent rules are those satisfying $F^2 = F$.

$$[F_3^{2k}(x)]_j = 1 - \overline{x}_{j-k-1}\overline{x}_{j-k} - \overline{x}_{j-k}\overline{x}_{j-k+1} + \overline{x}_{j-k-1}\overline{x}_{j-k}\overline{x}_{j-k+1},$$
$$[F_3^{2k+1}(x)]_j = 1 - \overline{y}_{j-k-1}\overline{y}_{j-k} - \overline{y}_{j-k}\overline{y}_{j-k+1} + \overline{y}_{j-k-1}\overline{y}_{j-k}\overline{y}_{j-k+1},$$

where $y = F_3(x)$, so that

$$y_i = f_3(x_{i-1}, x_i, x_{i+1}) = 1 - x_{i-1} - x_i + x_{i-1}x_i. \tag{5.8}$$

After substitution of y_i's and simplification, the final expression for the odd powers of F_3 becomes remarkably short,

$$[F_3^{2k+1}(x)]_j = x_{j-k-1}x_{j-k} - x_{j-k} - x_{j-k-1} + 1. \tag{5.9}$$

Solution of the initial value problem for rule 3 is, therefore,

$$[F_3^n(x)]_j = \begin{cases} 1 - \overline{x}_{j-k-1}\overline{x}_{j-k} - \overline{x}_{j-k}\overline{x}_{j-k+1} + \overline{x}_{j-k-1}\overline{x}_{j-k}\overline{x}_{j-k+1} & n = 2k, \\ x_{j-k-1}x_{j-k} - x_{j-k} - x_{j-k-1} + 1 & n = 2k+1. \end{cases} \tag{5.10}$$

A similar identity holds for rule 38, except that powers are higher,

$$F_{38}^4 = F_{170}^2 F_{38}^2. \tag{5.11}$$

The orbit of $x \in \mathcal{A}^{\mathbb{Z}}$ will be

$$\{x, F_{38}(x), F_{38}^2(x), F_{38}^3(x), \sigma^2 F_{38}^2(x), \sigma^2 F_{38}^3(x), \sigma^4 F_{38}^2(x), \sigma^4 F_{38}^3(x) \ldots\}, \tag{5.12}$$

meaning that

$$[F_{38}^n(x)]_j = \begin{cases} [F_{38}(x)]_j & n = 1, \\ \left[\sigma^{n-2}(F_{38}^2(x))\right]_j & n \text{ even}, \\ \left[\sigma^{n-3}(F_{38}^3(x))\right]_j & \text{otherwise}. \end{cases} \tag{5.13}$$

Since $[\sigma^n(x)]_j = x_{j+n}$, this becomes

$$[F_{38}^n(x)]_j = \begin{cases} [F_{38}(x)]_j & n = 1, \\ \left[F_{38}^2(x)\right]_{j+n-2} & n \text{ even}, \\ \left[F_{38}^3(x)\right]_{j+n-3} & \text{otherwise}. \end{cases} \tag{5.14}$$

Computing the first three powers of F_{38} we obtain

$$[F_{38}^n(x)]_j = \begin{cases} x_j x_{j-1} x_{j+1} - x_j x_{j-1} - 2x_j x_{j+1} + x_j + x_{j+1} & n = 1, \\ A_n & n \text{ even}, \\ B_n & \text{otherwise}, \end{cases} \tag{5.15}$$

where

$$A_n = -x_{j+n}x_{j+n-4}x_{j+n-3}x_{j+n-2}x_{j+n-1} + x_{j+n}x_{j+n-4}x_{j+n-3}x_{j+n-1}$$
$$+ x_{j+n}x_{j+n-3}x_{j+n-2} - x_{j+n}x_{j+n-3}x_{j+n-1} - x_{j+n}x_{j+n-2} + x_{j+n},$$

$$B_n = x_{j+n}x_{j+n-4}x_{j+n-3}x_{j+n-2}x_{j+n-1} - x_{j+n}x_{j+n-4}x_{j+n-3}x_{j+n-1}$$
$$- 2x_{j+n}x_{j+n-3}x_{j+n-2}x_{j+n-1} - x_{j+n-4}x_{j+n-3}x_{j+n-2}x_{j+n-1} + x_{j+n}x_{j+n-3}x_{j+n-2}$$
$$+ x_{j+n}x_{j+n-3}x_{j+n-1} + 2x_{j+n}x_{j+n-2}x_{j+n-1} + x_{j+n-4}x_{j+n-3}x_{j+n-1}$$
$$+ x_{j+n-3}x_{j+n-2}x_{j+n-1} - x_{j+n}x_{j+n-2} - 2x_{j+n}x_{j+n-1} - x_{j+n-3}x_{j+n-1}$$
$$- x_{j+n-2}x_{j+n-1} + x_{j+n} + x_{j+n-1}.$$

5.3 Identities of $F = GH$ Type

In addition to previously discussed identities, one can also consider other useful identities, for example, identities involving three rules. The simplest of these is $F = GH$, where all three rules F, G and H are different. Identities of the type $F = GH$ can be very useful when G and H commute, that is, when $GH = HG$. Minimal rules F which can be decomposed into a product of two commuting rules G and H are shown in Table 5.2. In this case $F^n = G^n H^n$, and if solutions of both G and H are known, the solution for F can be obtained immediately,

$$[F^n(x)]_j = [G^n(H^n(x))]_j. \tag{5.16}$$

This is especially useful if G is a simple rule, for example, a single input rule. The rules for which we have not yet obtained solutions and present in Table 5.2

Table 5.2 Minimal rules F which can be decomposed as $F = GH$ where G and H commute

$F_1 = F_1 F_{236}$	$F_{34} = F_{12} F_{42}$	$F_{136} = F_{170} F_{192}$
$F_2 = F_{172} F_2$	$F_{34} = F_{140} F_{34}$	$F_{142} = F_{113} F_{51}$
$F_2 = F_2 F_{236}$	$F_{43} = F_{212} F_{51}$	$F_{150} = F_{105} F_{51}$
$F_3 = F_{17} F_{240}$	$F_{51} = F_{15} F_{170}$	$F_{170} = F_{51} F_{85}$
$F_3 = F_3 F_{236}$	$F_{51} = F_{240} F_{85}$	$F_{178} = F_{51} F_{77}$
$F_{12} = F_{12} F_{140}$	$F_{60} = F_{102} F_{240}$	$F_{200} = F_{119} F_{63}$
$F_{12} = F_{240} F_{34}$	$F_{77} = F_{178} F_{51}$	$F_{204} = F_{15} F_{85}$
$F_{15} = F_{240} F_{51}$	$F_{90} = F_{102} F_{60}$	$F_{204} = F_{170} F_{240}$
$F_{23} = F_{232} F_{51}$	$F_{105} = F_{150} F_{51}$	$F_{204} = F_{51} F_{51}$
$F_{34} = F_{12} F_{170}$	$F_{128} = F_{136} F_{192}$	$F_{232} = F_{23} F_{51}$
$F_{34} = F_{12} F_{34}$		

Trivial cases and cases already presented in the table of emulators are excluded

are rules 178, 105 and 23. They all involve $G = F_{51}$, for which the local rule is $f_{51}(x_1, x_2, x_3) = 1 - x_2$, and, therefore, for these rules $F = F_{51}H$ implies that

$$\left[F^n(x)\right]_j = \begin{cases} [H^n(x)]_j & n \text{ even}, \\ 1 - [H^n(x)]_j & n \text{ odd}. \end{cases} \tag{5.17}$$

The following identities hold for the aforementioned rules,

$$\begin{aligned} F_{178} &= F_{51} F_{77}, \\ F_{105} &= F_{51} F_{150}, \\ F_{23} &= F_{51} F_{232}. \end{aligned}$$

All three rules on the right hand side, namely 77, 150 and 232, commute with rule 51, and solutions of all of them have been already obtained earlier, thus we can use Eq. (5.17). For rule 178, using Eq. (3.19), this yields

$$[F_{178}^n(x)]_j = \begin{cases} x_j + \sum_{r=1}^{n} (-1)^r \prod_{i=-r}^{r} x_{j+i} + \sum_{r=1}^{n} (-1)^{r+1} \prod_{i=-r}^{r} \overline{x}_{j+i} & n \text{ even}, \\ 1 - x_j - \sum_{r=1}^{n} (-1)^r \prod_{i=-r}^{r} x_{j+i} - \sum_{r=1}^{n} (-1)^{r+1} \prod_{i=-r}^{r} \overline{x}_{j+i} & n \text{ odd}. \end{cases}$$

For rule 105, using Eq. (3.13) we obtain

$$[F_{105}^n(x)]_j = \begin{cases} \sum_{i=0}^{2n} \binom{n}{i-n}_2 x_{i+j-n} \mod 2 & n \text{ even}, \\ 1 - \sum_{i=0}^{2n} \binom{n}{i-n}_2 x_{i+j-n} \mod 2 & n \text{ odd}. \end{cases} \tag{5.18}$$

Finally, for rule 23, we obtain for even n,

$$\begin{aligned} [F_{23}^n(x)]_j &= x_{-n+j} \prod_{i=1}^{n} \overline{x}_{2i-n-1+j} x_{2i-n+j} \\ &+ \sum_{k=0}^{n-1} \left(x_{k+j} \left(x_{-k-1+j} \overline{x}_{k+1+j} + x_{k+1+j} \right) \prod_{i=0}^{k-1} x_{-k+2i+j} \overline{x}_{-k+2i+1+j} \right), \end{aligned}$$

and for odd n,

$$[F_{23}^n(x)]_j = 1 - x_{-n+j} \prod_{i=1}^{n} \overline{x}_{2i-n-1+j} x_{2i-n+j}$$

$$- \sum_{k=0}^{n-1} \left(x_{k+j} \left(x_{-k-1+j} \overline{x}_{k+1+j} + x_{k+1+j} \right) \prod_{i=0}^{k-1} x_{-k+2i+j} \overline{x}_{-k+2i+1+j} \right).$$

Chapter 6
Rules with Additive Invariants

In many cases we will not be able to find the solution of the initial value problem by any of the methods used in previous chapters. One more possible strategy to try in such cases is to inspect the set of strings satisfying $\mathbf{f}^n(\mathbf{a}) = 1$, attempt to "guess" the condition C which the string $\mathbf{a}$ must satisfy in order to be a preimage of 1, and then prove this conditions in a rigorous way. Once we do that, the solution of the initial value problem will be given by

$$[F^n(x)]_j = \sum_{\mathbf{a} \text{ satisfying } C} \Psi_{\mathbf{a}}(x_{j-n}, x_{j-n+1}, \ldots, x_{j+n}), \tag{6.1}$$

where $\Psi_{\mathbf{a}}$ is the density polynomial of $\mathbf{a}$. We will pursue this method for rule 184. Since this rule belongs to the class of so-called *number-conserving rules*, we will start from some preliminary remarks about such rules.

6.1 Number-Conserving Rules

The general theory of additive conserved quantities in cellular automata has been established by Hattori and Takesue [1]. Here we will present only a special case of the simplest type of additive invariants. Suppose that $\mathcal{A}$ is an alphabet with N states and let $F : \mathcal{A}^{\mathbb{Z}} \to \mathcal{A}^{\mathbb{Z}}$ be a cellular automaton with local function $f(x_1, x_2, \ldots, x_{m-1}, x_m)$. If the radius of the rule is r, then $m = 2r + 1$. Let $x \in \mathcal{A}^{\mathbb{Z}}$ be called *periodic* with period L if, for all $i \in \mathbb{Z}$, we have $x_i = x_{i+L}$.

The original version of the chapter has been revised. A correction to this chapter can be found at https://doi.org/10.1007/978-3-031-38700-5_13

H. Fukś, *Solvable Cellular Automata*, Understanding Complex Systems,
https://doi.org/10.1007/978-3-031-38700-5_6

Definition 6.1 Cellular automaton rule with local function f is *number-conserving* if, for all periodic configurations of period $L \geq n$, it satisfies

$$\begin{aligned} & f(x_1, x_2, \dots, x_{m-1}, x_m) + f(x_2, x_3, \dots, x_m, x_{m+1}) + \cdots \\ & \quad + f(x_L, x_1 \dots, x_{m-2}, x_{m-1}) = x_1 + x_2 + \cdots + x_L. \end{aligned} \tag{6.2}$$

The following result [2] characterizes number-conserving rules, and it is a special case of a general theorem established in [1].

Theorem 6.1 *Cellular automaton rule with local function f with m inputs and N states is number-conserving if, and only if, for all $(x_1, x_2, \dots, x_m) \in \mathcal{A}^m$, it satisfies*

$$\begin{aligned} f(x_1, x_2, \dots, x_m) = x_1 + \sum_{k=1}^{m-1} \Bigg(& f(\underbrace{0, 0, \dots, 0}_{k}, x_2, x_3, \dots, x_{m-k+1}) \\ & - f(\underbrace{0, 0, \dots, 0}_{k}, x_1, x_2, \dots, x_{m-k}) \Bigg). \end{aligned} \tag{6.3}$$

Proof We start the proof with the observation that if f is a number-conserving rule, then $f(0, 0, \dots, 0) = 0$. To see it, just write Eq. (6.2) for a periodic configuration of length $L \geq n$ whose all elements are equal to zero.

To prove that Eq. (6.3) is necessary, consider a periodic configuration with period $L \geq 2m - 1$ which is the concatenation of a sequence $(x_1, x_2, \dots, x_m)$ and a sequence of $L - m$ zeros, and write Eq. (6.2) for this configuration. This yields

$$\begin{aligned} & f(0, 0, \dots, 0, x_1) + f(0, 0, \dots, 0, x_1, x_2) + \cdots \\ & \quad + f(x_1, x_2, \dots, x_m) + f(x_2, x_3, \dots, x_m, 0) + \cdots \\ & \quad + f(x_m, 0, \dots, 0) = x_1 + x_2 + \cdots + x_m, \end{aligned} \tag{6.4}$$

where all the terms of the form $f(0, 0, \dots, 0)$ are omitted because they vanish. Replacing x_1 by 0 in Eq. (6.4) we obtain

$$\begin{aligned} & f(0, 0, \dots, 0, x_2) + \cdots + f(0, x_2, \dots, x_m) \\ & \quad + f(x_2, x_3, \dots, x_m, 0) + \cdots + f(x_m, 0, \dots, 0) \\ & = x_2 + \cdots + x_m. \end{aligned} \tag{6.5}$$

Subtracting Eq. (6.5) from Eq. (6.4) yields Eq. (6.3).

To show that the condition (6.3) is sufficient, let us write it for site i of a periodic configuration of period L,

$$f(x_i, x_{i+1}, \dots, x_{i+m-1}) = x_i + \sum_{k=1}^{m-1} \Big(f(\underbrace{0, 0, \dots, 0}_{k}, x_{i+1}, x_{i+2}, \dots, x_{i+m-k}) - f(\underbrace{0, 0, \dots, 0}_{k}, x_i, x_{i+1}, \dots, x_{i+m-k-1})\Big). \tag{6.6}$$

When we sum both sides of the above over i from $i = 0$ to $i = L - 1$, all the right-hand side terms except those with x_i will cancel, and we will obtain Eq. (6.2). □

The following corollary will be useful in subsequent considerations.

Corollary 6.1 *If f is a local function of a cellular automaton with m inputs and N states and it is number-conserving, then*

$$\sum_{(x_1,x_2,\dots,x_m)\in\mathcal{A}^m} f(x_1, x_2, \dots, x_m) = \frac{1}{2}(N-1)\,N^m. \tag{6.7}$$

Proof When we the sum Eq. (6.3) over $(x_1, x_2, \dots, x_n) \in \mathcal{A}^m$, all the right-hand side terms except the first cancel, and the sum over the remaining terms is equal to $(0 + 1 + 2 + \cdots + (N-1))N^{m-1} = \frac{1}{2}(N-1)\,N^m$. □

For binary rules $N = 2$, thus

$$\sum_{(x_1,x_2,\dots,x_m)\in\{0,1\}^m} f(x_1, x_2, \dots, x_m) = 2^{m-1}. \tag{6.8}$$

For elementary CA $m = 3$, thus $2^{m-1} = 4$, meaning that among all possible inputs (x_1, x_2, x_3), four of them must return $f(x_1, x_2, x_3) = 1$, and the other four $f(x_1, x_2, x_3) = 0$. We will call such rules *balanced* because 50% of inputs return 0 and 50% return 1.

Note that if the elementary rule f is number conserving, then f^n is also number conserving and therefore must also be balanced, meaning that

$$\sum_{(x_1,x_2,\dots,x_m)\in\{0,1\}^m} f^n(x_1, x_2, \dots, x_{2n+1}) = 2^{2n}. \tag{6.9}$$

Another way of expressing this is that for a number conserving elementary rule with block mapping $\mathbf{f}$ we have

$$\operatorname{card} \mathbf{f}^{-n}(1) = \operatorname{card} \mathbf{f}^{-n}(0) = 2^{2n}. \tag{6.10}$$

Among elementary CA, there are exactly five rules which are number conserving. The first three are trivial rules, namely the identity (204), the shift (170) its inverse (240). The remaining ones are the pair 184 and 226 belonging to the same equivalence class, rule 226 being both reflected and conjugated-reflected version of rule 184. For this reason, in what follows we will only consider rule 184.

6.2 Rule 184—Preliminary Remarks

Rule 184 has the local function given by

$$f_{184}(x_1, x_2, x_3) = x_1 + x_2 x_3 - x_1 x_2.$$

We can directly verify that it is number conserving. Suppose that $x \in \{0, 1\}^{Z} Z$ is periodic with period L. Then

$$\begin{aligned}\sum_{j=0}^{L-1}[F_{184}(x)]_j &= \sum_{j=0}^{L-1} f_{184}(x_{j-1}, x_j, x_{j+1}) = \sum_{j=0}^{L-1} x_{j-1} + \sum_{j=0}^{L-1} x_j x_{j+1} - \sum_{j=0}^{L-1} x_{j-1} x_j \\ &= \sum_{j=0}^{L-1} x_{j-1} = \sum_{j=0}^{L-1} x_j, \end{aligned} \tag{6.11}$$

meaning F_{184} conserves the number of 1's in a single period, as well as the number of 0's. For this reason, we can interpret a single period of x as a one-dimensional lattice of size L with periodic boundary conditions, where each lattice site is either occupied by a particle ($x_i = 1$) or empty ($x_i = 0$). Application of F_{184} is equivalent to the movement of particles such that those particles which have empty site directly on the right synchronously move one unit to the right, and all other particles stay in the same place. This is sometimes represented as

$$\overset{\curvearrowright}{1}0, \quad \overset{\circlearrowleft}{1}1, \tag{6.12}$$

meaning that 1 followed by 0 moves to the right and 1 followed by 1 stays in the same place. Rule 184 is also called a "traffic rule" because we can interpret 1's as cars instead of particles.

6.3 Finite State Machines

In order to "guess" the condition mentioned at the beginning of this chapter, one can generate sets $\mathbf{f}^{-n}(1)$ for various values of n and see if they follow any obvious pattern or regularity. A convenient method for doing this is to build the so-called minimal finite state machines accepting only words of $\mathbf{f}^{-n}(1)$. Finite state machines often reveal patterns in sets of strings much better than simple inspection by eye, and various software tools exist for this purpose, for example, AT&T FSM Library [3].

The process of building the minimal finite state machine is best explained on a concrete example. Consider the set of one-step preimages of 1 under the rule 184 which contains four strings, $\mathbf{f}_{184}^{-1}(1) = \{011, 100, 101, 111\}$. We can portray these

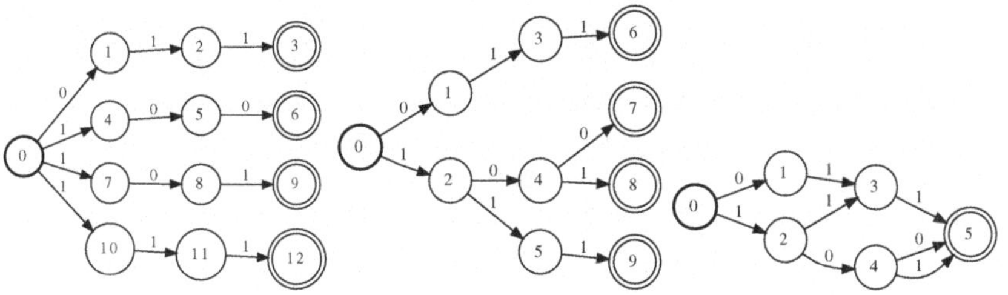

(a) raw (b) determinized (c) minimized

Fig. 6.1 Finite state machines accepting strings belonging to $\mathbf{f}_{184}^{-1}(1)$

strings by a graph shown in Fig. 6.1a. In order to check if a given string **a** belong to preimages of 1, start at ⓪ and follow the arrows with edge labels corresponding to consecutive symbols of **a** until you reach the final state marked by double circle. The final state is called the *accepting state*. If we can reach the accepting state, the string **a** belongs to the set of preimages, otherwise not. For example, there is no way to reach the accepting state for string 000 because we would be stuck at node ① with nowhere to go. We say that this string is *rejected*. If, on the other hand, we can reach the accepting state for a given string, we say that the string is *accepted*.

The graph of Fig. 6.1a represents a model of computation called a *finite state machine* (FSM), an abstract machine that can be in exactly one of a finite number of internal states at any given time. Circled numbers denote these internal states of the FSM. The change from one state to another is called a transition, and each transition is associated with a symbol, marked as label of the relevant arrow in our figure. Obviously there are as many preimage string as paths joining the initial state and one of the final states.

The FSM in Fig. 6.1a is *non-deterministic*, because there are three transitions from state ⓪ labeled with symbol 1, namely ⓪ → ④, ⓪ → ⑦, and ⓪ → ⑩. We thus cannot determine where to go from ⓪ if the fist symbol of the string is 1. There is a way to construct an equivalent machine, accepting the same output strings, but being deterministic, so that no two edges leaving the same state carry the same symbol. This is called determinization of the FSM, and it is known that all acyclic FSM are determinizable. We will not go into the details of the algorithm for determinization, interested readers should consult [4, 5]. It suffices to say that "raw" FSM representing preimage strings, such as the one shown in Fig. 6.1a, are always acyclic, thus determinizable. Deterministic equivalent to the FSM of Fig. 6.1a is shown in Fig. 6.1b. We can see that this FSM has lower number of states than the original, although it still has four final states.

It is also possible to further decrease the number of states and produce the FSM with a single final state. This process is known as minimization of the FSM. Again, details of the minimization algorithms can be found in the aforementioned references, here we will only present the result obtained by software tools included in the AT&T

FSM Library. Figure 6.1c shows the FSM minimized by these tools. We can see that it has only six states, much less than the original "raw" FSM which had 13 states.

Using the same process, one can obtain minimal FSM accepting only strings belonging to $\mathbf{f}^{-n}(1)$ for other values of n. Figure 6.2 shows such FSM's for $n = 2, 3$ and 4. It is clear that they follow a well pronounced pattern. In order to describe this pattern formally, let us take a closer look at the $n = 4$ case. Figure 6.3a shows the same FSM as Fig. 6.2c, except that its nodes are placed on a rectangular grid of size 5×4. We can see four clear features:

(i) all horizontal edges are directed to the right, all vertical edges are directed up;
(ii) vertical edges above the diagonal (dashed) are labeled with 0, otherwise with 1;
(iii) horizontal edges originating from nodes lying below the diagonal (but not on the diagonal) carry label 0, otherwise 1;
(iv) the last column has additional vertical edges with label 0.

In order to formulate a general conjecture about FSM for preimages of 1, it will be convenient to use a different labeling scheme for states of the FSM, labeling each node with its coordinates in the grid, as shown in Fig. 6.3b. This labeling scheme is easy to generalize to FSM representing $\mathbf{f}_{184}^{-n}(1)$ for arbitrary n, resulting in the following proposition.

Proposition 6.1 *Finite state machine representing $\mathbf{f}_{184}^{-n}(1)$ is a directed graph with vertices located at nodes of the rectangular grid. Nodes are labeled with pairs (k, m) corresponding to their coordinates on the grid, where $k \in \{0, 1, \ldots, n+1\}$ and $m \in \{0, 1, \ldots, n\}$. Edges of this graph connect each node to its nearest neighbours with vertical edges directed up and horizontal edges directed to the right. Horizontal edges originating from (k, m) carry label 0 if $k > m$, otherwise they carry label 1. Vertical edges originating from (k, m) carry label 1 if $k > m$, otherwise 0. There are also additional vertical edges originating from $(n + 1, m)$ which carry label 0.*

These edges can be symbolically described as

$$\begin{aligned}
(k, m) &\xrightarrow{0} (k+1, m) \quad \text{if } k > m, \\
(k, m) &\xrightarrow{1} (k+1, m) \quad \text{if } k \leq m, \\
(k, m) &\xrightarrow{1} (k, m+1) \quad \text{if } k > m, \\
(k, m) &\xrightarrow{0} (k, m+1) \quad \text{if } k \leq m, \\
(n+1, m) &\xrightarrow{0} (n+1, m+1),
\end{aligned}$$

for all relevant m, k. We can distinguish three types of edges originating from (k, m), horizontal leading to the right, vertical leading upwards, and "extra" vertical leading upwards (one of the the double edges in the last column). If we denote labels of these edges respectively by $h(k, m)$, $v(k, m)$, and $v_e(m, n)$, we can describe the edges in a concise fashion as

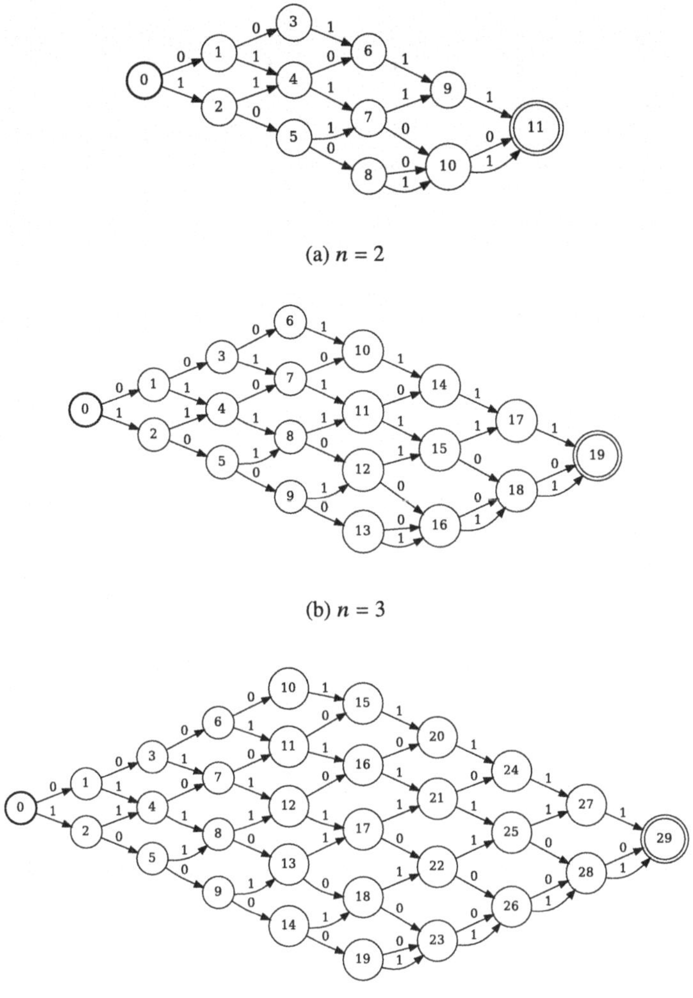

(a) $n = 2$

(b) $n = 3$

(c) $n = 4$

Fig. 6.2 Minimized FSM for preimages $\mathbf{f}_{184}^{-n}(1)$

$$
\begin{aligned}
(k, m) &\xrightarrow{h(k,m)} (k+1, m) \quad \text{for } k \in \{0, \ldots n\}, m \in \{0, \ldots, n\}, \\
(k, m) &\xrightarrow{v(k,m)} (k, m+1) \quad \text{for } k \in \{0, \ldots n+1\}, m \in \{0, \ldots, n-1\}, \\
(k, m) &\xrightarrow{v_e(k,m)} (k, m+1) \quad \text{for } k = n+1, m \in \{0, \ldots, n-1\},
\end{aligned} \tag{6.13}
$$

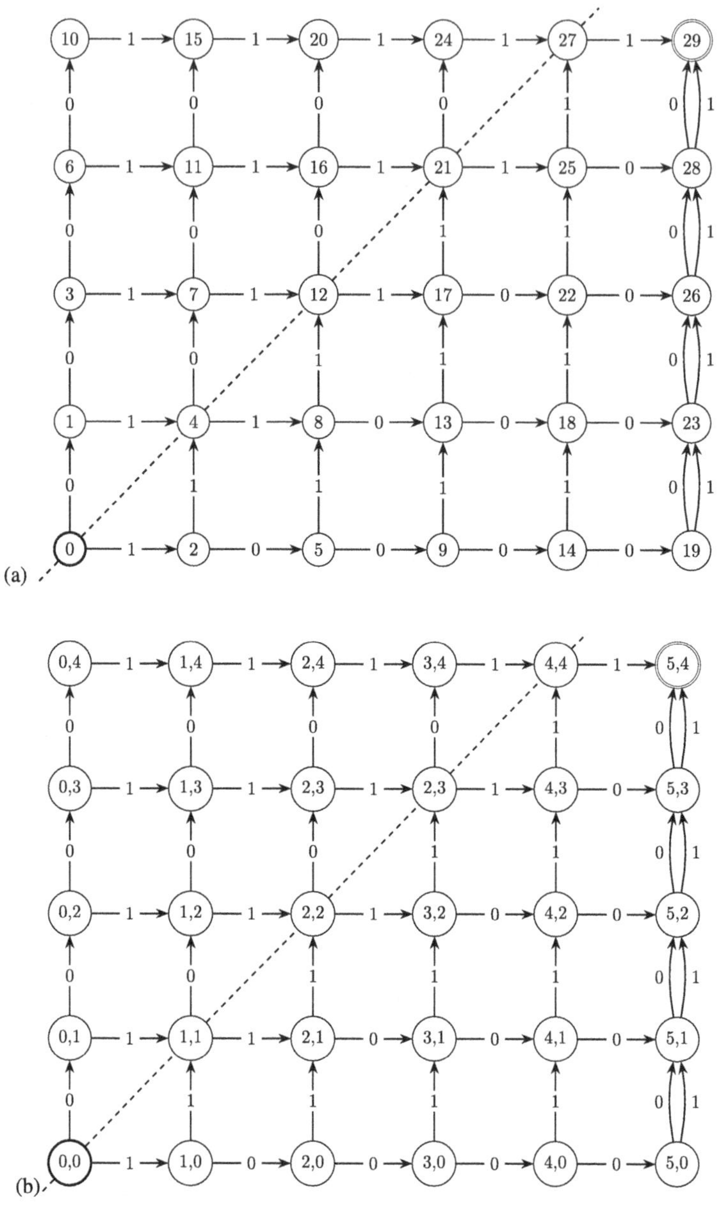

Fig. 6.3 **a** Finite state machine for 4-step preimages of 1 under the rule 184. **b** The same graph with nodes relabeled

where

$$\begin{aligned} h(k,m) &= [k \le m], \\ v(k,m) &= [k > m], \\ v_e(k,m) &= 0. \end{aligned} \tag{6.14}$$

The square bracket in the above is the Iverson bracket,

$$[P] = \begin{cases} 1 & \text{if } P \text{ is true;} \\ 0 & \text{otherwise.} \end{cases}$$

Before we attempt the proof, we need to discuss some further properties of preimages of 1.

6.3.1 Further Properties of Preimages of 1 for Rule 184

Let us rearrange the layout of the FMS shown in Fig. 6.3 in order to reveal some of its regularities which we have not yet discussed. We start by shifting the part below the diagonal slightly. The result of this operation is shown in Fig. 6.4a. The red line corresponds to the path consisting of all 1's. In the next step, we remove from each pair of double edges those carrying 0 label, leaving only those with label 1 (Fig. 6.4b). Such removal eliminates some paths, and, therefore, some preimages, but we will worry about this later.

How to describe all possible paths from ⓪ to ㉙ in the FSM of Fig. 6.4b? We can think of every path as a path along the red line which occasionally (possibly more than once) departs from the red line and then returns to it. In the sample path shown in Fig. 6.5 there are two departures from the red line, represented by strings 01 and 0011. We could say that the block 11 which would be on the red line between nodes ② and ⑧ has been replaced by block 01. Similarly, the block 1111 which would be on the red line between nodes ⑫ and ㉗ has been replaced by 0011. It is easy to see that no matter what is the shape of the departure from the red line, the block representing this departure will always have the property that for every k, its last k symbols have less zeros than ones. We will adopt the following definition to describe such blocks.

Definition 6.2 If a binary string $s_1, s_2 \dots s_m$ for all $k \in \{1, 2, \dots, m\}$ satisfies the condition

$$\sum_{i=1}^{k} s_{m+1-i} \ge k/2,$$

then it is called *tail-dense*.

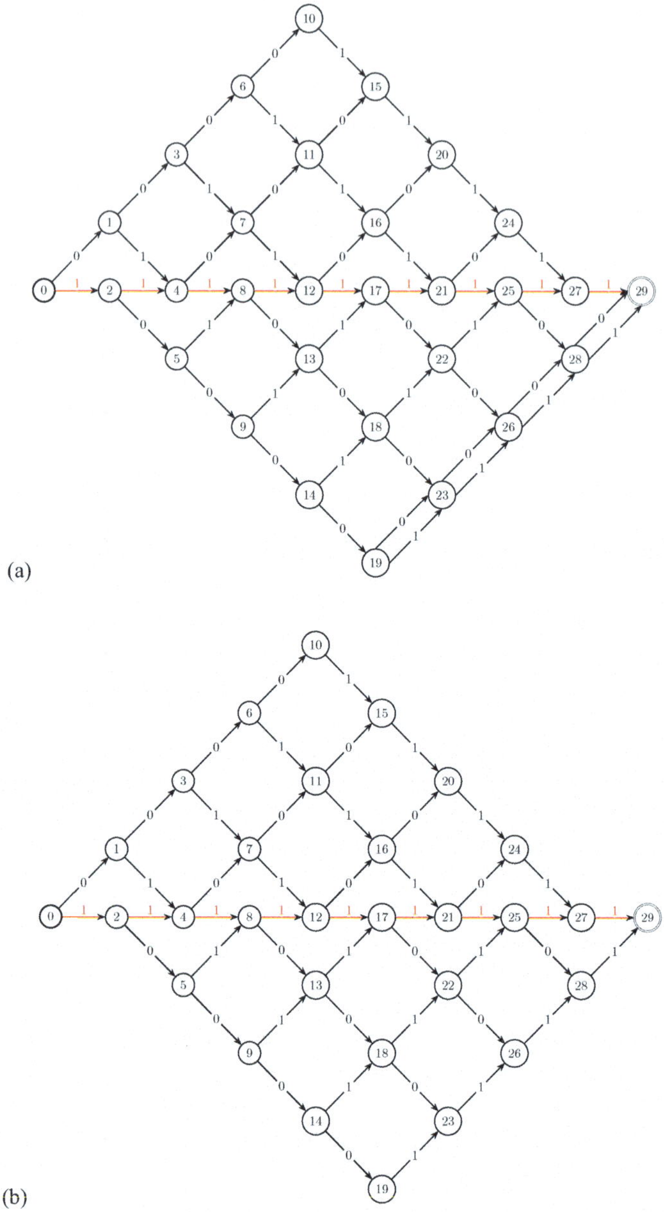

Fig. 6.4 **a** FSM for 4-step preimages of 1 under the rule 184 with "shifted" layout; **b** double edges removed

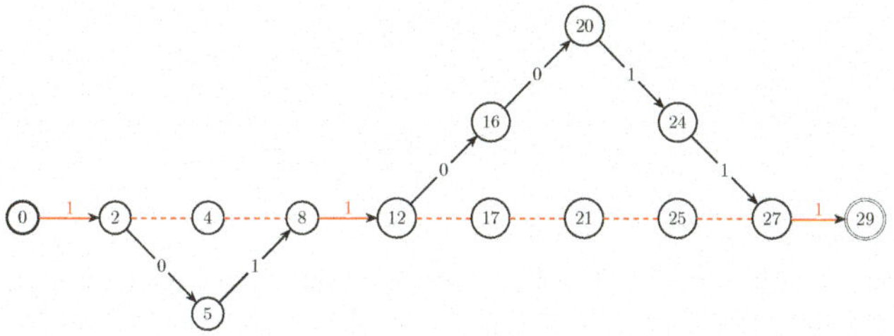

Fig. 6.5 Sample FSM path

Every departure from the red line is thus represented by a tail-dense block. This happens in both cases of the departure below or above the red line. Since each departure replaces a block of ones by a tail-dense block, the path from ⓪ to ㉙ will consist of tail-dense blocks separated by blocks of ones. It is easy to see that such path is itself tail-dense.

All strings accepted by the FSM of Fig. 6.4b are thus tail-dense. Let us now return to the paths which we are missing due to the removal of double edges. We can see from Fig. 6.4a that the paths which were affected by this removal are paths passing through state ㉘. All such paths must end by a sub-path starting somewhere on the red line and ending at ㉙ and lying entirely below the red line. These sub-paths correspond to strings which are balanced (having the same number of zeros and ones). This means that among the strings accepted by the FMS of Fig. 6.4a, if their path passes through the state ㉘, they must end with a substring which is balanced. If that balanced block ends with $11\ldots1$, we can replace as many ones in $11\ldots1$ as we want by zeros, and such replacement(s) will be equivalent to traversing some (or all) of the double edges with 0 label which we removed earlier. Let us introduce another definition to formalize this observation.

Definition 6.3 Let a tail-dense string $s_1s_2\ldots s_m$ end with a balanced substring $s_ks_{k+1}\ldots s_m$, $k > 1$, and $s_p = s_{p+1} = \ldots = s_p = 1$ for some $p > k$. Any string obtained form such $s_1s_2\ldots s_m$ by replacing some (or all) 1's in positions from p to m will be called *diluted tail-dense string*.

For example, consider $\mathbf{a} = 111001011$, corresponding to the path on the graph of Fig. 6.4a given by

$$⓪ \xrightarrow{1} ② \xrightarrow{1} ④ \xrightarrow{1} ⑧ \xrightarrow{0} ⑬ \xrightarrow{0} ⑱ \xrightarrow{1} ㉒ \xrightarrow{0} ㉖ \xrightarrow{1} ㉘ \xrightarrow{1} ㉙.$$

It ends with balanced substring 001011, and the last two symbols of this substring are ones. We can, therefore, obtain diluted strings by replacing any of the ones in the last two position by zeros. Three versions of diluted tail-dense strings can be obtained this way,

1110010**00**,

11100101**0**,

1110010**01**,

where the replaced symbols are bold.

We can thus see that the FSM of Fig. 6.4a would produce either tail-dense strings or diluted tail-dense strings. We have just demonstrated the following result.

Proposition 6.2 *The FSM described in Proposition 6.1 accepts both tail-dense and diluted tail-dense strings.*

6.3.2 Proof of Proposition 6.1

To prove Proposition 6.1, we need to show that both tail-dense and diluted tail-dense strings $\mathbf{a} \in \{0, 1\}^{2n+1}$ have the property $\mathbf{f}_{184}^{n}(\mathbf{a}) = 1$, and that these are all strings having this property.

Tail-dense strings

We start with tail-dense strings and the following useful lemma.

Lemma 6.1 *If the string* $\mathbf{a}$ *is tail-dense, then* $\mathbf{f}_{184}(\mathbf{a})$ *is tail-dense as well.*

To see this, let us assume that $\mathbf{a} = a_1 a_2 \ldots a_{2n+1}$, and define $\mathbf{b} = \mathbf{f}_{184}(\mathbf{a}) = b_2 b_3 \ldots b_{2n}$, so that $b_i = f_{184}(a_{i-1}, a_i, a_{i+1})$. We can write these two strings one under the other as follows,

$$\begin{array}{cccc|cccc} a_1 & a_2 & \ldots & a_{2n+1-k} & a_{2n+2-k} & \ldots & a_{2n} & a_{2n+1} \\ & b_2 & \ldots & b_{2n+1-k} & b_{2n+2-k} & \ldots & b_{2n} & \end{array}$$

In the upper string we have k symbols to the right of the vertical bar, and in the lower string $k-1$ symbols. Rule 184 conserves the number of 1's, and as stated earlier, 1's can be viewed as particles which move to the right if they have 0 on the right. We have $a_{2n+1} = 1$, therefore a_{2n}, even if it is equal to 1, cannot move to $2n+1$ position. For this reason,

$$\sum_{i=2n+2-k}^{2n} b_i = \sum_{i=2n+2-k}^{2n+1} a_i - 1 + c, \tag{6.15}$$

where $c = 1$ if there is a particle which moves across the vertical bar in one iteration, and $c = 0$ otherwise. Since the string $\mathbf{a}$ is tail-dense,

$$\sum_{i=2n+2-k}^{2n+1} a_i \geq k/2,$$

and combining this with Eq. (6.15) we obtain

$$\sum_{i=2n+2-k}^{2n} b_i \geq \frac{k}{2} - 1 + c. \tag{6.16}$$

When $c = 1$,

$$\sum_{i=2n+2-k}^{2n} b_i \geq \frac{k}{2} > \frac{k-1}{2}. \tag{6.17}$$

When $c = 0$, we have either $a_{2n+1+k}a_{2n+2+k} = 00$, 01 or 11. The first two cases imply $a_{2n+1+k} = 0$. Since **a** is tail-dense,

$$\sum_{i-2n+2-k}^{2n+1} a_i = \sum_{i=2n+1-k}^{2n+1} a_i \geq (k+1)/2,$$

and therefore

$$\sum_{i=2n+2-k}^{2n} b_i \geq \frac{k+1}{2} - 1 = \frac{k-1}{2}. \tag{6.18}$$

All what is left to do is to consider the case $a_{2n+1+k}a_{2n+2+k} = 11$. If this is so, we must have $b_{2n+1-k} = 1$. Note that in this case Eq. (6.15) holds if we run the sums from $2n + 1 - k$ still with $c = 0$, because there is no particle flow from the position $2n - k$ to $2n + 1 - k$. We have

$$\sum_{i=2n+1-k}^{2n} b_i = 1 + \sum_{i=2n+2-k}^{2n} b_i = \sum_{i=2n+1-k}^{2n+1} a_i - 1, \tag{6.19}$$

therefore,

$$\sum_{i=2n+2-k}^{2n} b_i = \sum_{i=2n+1-k}^{2n+1} a_i - 2. \tag{6.20}$$

Now

$$\sum_{i=2n+1-k}^{2n+1} a_i = 2 + \sum_{i=2n+3-k}^{2n+1} a_i \geq 2 + \frac{k-1}{2},$$

and the last two lines result in

$$\sum_{i=2n+2-k}^{2n} b_i \geq 2 + \frac{k-1}{2} - 2 = \frac{k-1}{2}, \tag{6.21}$$

which is the same as the inequalities (6.17) and (6.18). For all cases, therefore,

$$\sum_{i=2n+2-k}^{2n} b_i \geq \frac{k-1}{2}, \tag{6.22}$$

and since this holds for any k, string $\mathbf{b}$ is tail-dense, concluding the proof of our lemma. Applying the lemma recursively n times to a tail-dense $\mathbf{a} \in \{0, 1\}^{2n+1}$, we find that $\mathbf{f}_{184}^{n}(\mathbf{a})$ is also tail-dense, and since $\mathbf{f}_{184}^{n}(\mathbf{a})$ is a string of length 1, it necessarily implies $\mathbf{f}_{184}^{n}(\mathbf{a}) = 1$, as required for an n-step preimage of 1.

Diluted tail-dense strings

Let us turn our attention to diluted tail-dense strings. We will first prove the following lemma.

Lemma 6.2 *Let* $\mathbf{a}$ *be tail-dense and balanced, ending with k ones. Then all strings* $\mathbf{f}_{184}(\mathbf{a})$, $\mathbf{f}_{184}^{2}(\mathbf{a})$, …, $\mathbf{f}_{184}^{k-1}(\mathbf{a})$ *are also tail-dense and balanced.*

Proof It is easy to show that tail-dense balanced string must start with 0. The string $\mathbf{f}_{184}(\mathbf{a})$ must have one less 1 than $\mathbf{a}$ because nothing arrived from the left ($\mathbf{a}$ starts with zero), and the last 1 of $\mathbf{a}$ has been deleted by $\mathbf{f}$. Since $|\mathbf{f}_{184}(\mathbf{a})| = |\mathbf{a}| - 2$, this implies that $\mathbf{f}_{184}(\mathbf{a})$ is balanced. It is also tail-dense by Lemma 6.1, and as a consequence it must start with 0. We can thus apply the previous reasoning to it, and we can do it repeatedly as long as we have the guarantee that $\mathbf{f}_{184}^{i}(\mathbf{a})$ ends with 1. This will happen for $i = 1, 2, \ldots k-1$, because the string $\mathbf{a}$ ends with $\underbrace{11, \ldots 1}_{k}$. □

We thus know that if $\mathbf{a}$ is balanced and tail-dense, it must start with zero and if it ends with a cluster of k 1's, we are guaranteed that k consecutive applications of $\mathbf{f}_{184}$ will not change this, i.e., the resulting strings will still start with 0.

Now let us notice that the leading zero in a string $\mathbf{a}$ of length m is separated from the trailing cluster of k 1's by $m-k-1$ symbols. Since $k \leq m/2$, the separation is at least $k-1$ symbols. Information travels in CA with radius 1 with speed of one site per iteration, thus if we make a change to any the last k symbols of $\mathbf{a}$, this will not affect the leading symbol of $\mathbf{f}_{184}^{i}(\mathbf{a})$ for $i < k$. The consequence of this is the following result.

Lemma 6.3 *Let* $\mathbf{b} = 1\mathbf{a}$, *where* $\mathbf{a}$ *is tail-dense and balanced, ending with k ones. If rule 184 is applied k times to a configuration containing* $\mathbf{b}$, *the movement of the leading 1 of* $\mathbf{b}$ *does not change if one replaces any of the ones in the last k positions of* $\mathbf{b}$ *by zeros.*

Proof From the preceding discussion we can see that the first $k-1$ iterations of $\mathbf{f}_{184}$ applied to a balanced and tail-dense string will maintain zero at the beginning no matter what we do to the trailing cluster of ones. This means that 1 preceding such string will move to the right every time step. Since, after $k-1$ iterations, we will also have 0 at the beginning, we can perform one more iteration and the preceding 1 will still move one unit to the right. □

An example illustrating Lemma 6.3 is shown below for $\mathbf{a} = 001011$, with results of consecutive iterations of $\mathbf{f}_{184}$ shown in the first column. The last two 1's have been replaced by 00, 01 and 10 in the next three columns, and the trajectory of the red 1 has not been affected:

```
1001011   1001000   1001001   1001010
 10011     10010     10010     10010
  101       100       100       100
```

Since, according to Lemma 6.3, the movement of the 1 preceding balanced string will not be affected by changes to the end cluster of ones, the movement of any other 1's to the left of the balanced string will not be affected either. One of these 1's will end up in the central position of $\mathbf{a} \in \{0, 1\}^{2n+2}$ after n iterations, producing the desired $\mathbf{f}_{184}^{n}(\mathbf{a}) = 1$. This of course means that diluted tail-dense strings belong to n-step preimages of 1.

Enumeration of paths on the FSM graph

Thanks to Proposition 6.2, we know that any string produced by the FSM of Proposition 6.1 is either tail-dense or diluted tail-dense. In previous sections we also demonstrated that any tail-dense or diluted tail-dense string belongs to preimages of 1. To complete the proof of Proposition 6.1, we need to show that these are *all* preimages of 1. Since, according to Eq. (6.10),

$$\operatorname{card} \mathbf{f}_{184}^{-n}(\mathbf{1}) = 2^{2n},$$

we have to demonstrate that the number of distinct strings which can be accepted by the FMS described in Proposition 6.1 is 2^{2n}.

Let us take a look at the $n = 4$ example again, shown in Fig. 6.3a. We can see that starting from ⓪ we can reach node ㉙ via nodes:

- ⑭ in 2^4 ways;
- ⑱ in $\binom{5}{1} \cdot 2^3$ ways;
- ㉒ in $\binom{6}{2} \cdot 2^2$ ways;
- ㉕ in $\binom{7}{3} \cdot 2^1$ ways;
- ⑰ in $\binom{8}{4} \cdot 2^0$ ways.

The total number of paths is, therefore,

$$\sum_{i=0}^{4} \binom{4+i}{i} 2^{4-i} = 16 + 40 + 60 + 70 + 70 = 256,$$

which equals $2^{2\cdot 4}$, as it should be for $n = 4$. For general n the total number of paths will be

$$\sum_{i=0}^{n} \binom{n+i}{i} 2^{n-i},$$

and this indeed sums to 2^{2n}, concluding the proof of Proposition 6.1.

6.4 Solution of the Initial Value Problem for Rule 184

Having the formal description of the FSM graph given by Proposition 6.1, how do we decide if a given string $a_1 a_2 \dots a_{2n+1}$ is an n-step preimage of 1? Obviously this will be the case if the path corresponding to **a** ends at the accepting state, which is the upper right corner $(n+1, n)$.

We need to have a way to find out what are the coordinates of the node corresponding to a_i. Let us denote these coordinates by (K_i, M_i), so that the edge with label a_i is ending at (K_i, M_i), or symbolically

$$(K_{i-1}, M_{i-1}) \xrightarrow{a_i} (K_i, M_i). \tag{6.23}$$

Recall that in Eq. (6.14) we defined edge labeling functions h, v and v_e. We know that M_i increases if $a_i = v(K_i, M_i)$ or, if we are in the last column, if $a_i = v_e(K_i, M_i)$. K_i increases only when $[a_i = h(K_i, M_i)]$. We can, therefore, write

$$\begin{aligned} K_i &= K_{i-1} + \big[a_i = h(K_{i-1}, M_{i-1})\big] \\ M_i &= M_{i-1} + \big[a_i = v(K_{i-1}, M_{i-1}) \vee K_{i-1} = n+1\big] \end{aligned}$$

Since $M_i + K_i = i$, only one of these variables is needed, let us say M_i, the other one being $K_i = i - M_i$. This yields recurrence relation which can be used to compute M_i,

$$\begin{aligned} M_0 &= 0, \\ M_i &= M_{i-1} + \big[a_i = v(i-1-M_{i-1}, M_{i-1}) \vee i = M_{i-1} + n + 2\big]. \end{aligned} \tag{6.24}$$

Using the definition of v we obtain

$$v(i-1-M_{i-1}, M_{i-1}) = [i-1-M_{i-1} > M_{i-1}] = [i > 2M_{i-1} + 1],$$

therefore the final recurrence equation becomes

$$\begin{aligned} M_0(\mathbf{a}) &= 0, \\ M_i(\mathbf{a}) &= M_{i-1}(\mathbf{a}) + \big[a_i = [i > 2M_{i-1}(\mathbf{a}) + 1] \vee i = M_{i-1}(\mathbf{a}) + n + 2\big]. \end{aligned}$$

We added the argument **a** to M to emphasize that M is a quantity which is computed for a given string **a**. Since the nested bracket notation is a bit difficult to read, we can write this more explicitly,

$$M_i(\mathbf{a}) = \begin{cases} 0 & \text{if } i = 0, \\ M_{i-1}(\mathbf{a}) + 1 & \text{if } a_i = 0 \text{ and } i \le 2M_{i-1}(\mathbf{a}) + 1, \\ M_{i-1}(\mathbf{a}) + 1 & \text{if } a_i = 1 \text{ and } i > 2M_{i-1}(\mathbf{a}) + 1, \\ M_{i-1}(\mathbf{a}) + 1 & \text{if } i = M_{i-1}(\mathbf{a}) + n + 2, \\ M_{i-1}(\mathbf{a}) & \text{otherwise.} \end{cases} \tag{6.25}$$

In order for $\mathbf{a}$ to be a preimage of 1, the corresponding path on the graph must end at the accepting state which has coordinates $(n, n+1)$. This translates to $(K_{2n+1}, M_{2n+1}) = (n, n+1)$, and since only one of K, M variables is independent, it suffices to require $M_{2n+1} = n$.

Proposition 6.3 *Block* $\mathbf{a} \in \{0, 1\}^{2n+1}$ *is an* n*-step preimage of 1 under the rule 184 if and only if* $M_{2n+1}(\mathbf{a}) = n$, *where* $M_i(\mathbf{a})$ *is recursively defined in Eq. (6.25).*

Consequently, the solution of rule 184 is obtained directly from Eq. (6.1),

$$\left[F_{184}^n(x)\right]_j = \sum_{\substack{\mathbf{a}\in\{0,1\}^{2n+1} \\ M_{2n+1}(\mathbf{a})=n}} \Psi_{\mathbf{a}}(x_{j-n}, x_{j-n+1}, \ldots, x_{j+n}), \tag{6.26}$$

where $\Psi_{\mathbf{a}}$ is the density polynomial of $\mathbf{a}$ as defined in Eq. (2.12). Although this solution does not appear to be very useful at the first sight, we will actually see in the subsequent chapters that it can be used to obtain the so-called probabilistic solution of rule 184.

Using the above solution we can also obtain solution of another related rule, namely rule 56 which emulates rule 184,

$$F_{56}^2 = F_{184}F_{56}.$$

This immediately yields

$$\left[F_{56}^n(x)\right]_j = \left[F_{184}^{n-1}(y)\right]_j,$$

where

$$y_j = f_{56}(x_{j-1}, x_j, x_{j+1}) = x_{j+1}x_j + x_{j-1} - x_{j+1}x_{j-1}x_j - x_{j-1}x_j,$$

and where $[F_{184}^{n-1}(y)]_j$ is given by Eq. (6.26).

6.5 Higher Order Invariants

The concept of number-conserving rules introduced in Definition 6.1 can be generalized to include conservation of blocks of length greater than one. Let k be a non-negative integer, and let $\xi = \xi(x_0, x_1, \ldots, x_k)$ be a function of $k+1$ variables which takes values in $\mathbb{R}$. Following [1], we say that ξ is a density function of an additive conserved quantity of cellular automaton F if for every positive integer L and for every periodic configuration $x \in \mathcal{A}^{\mathbb{Z}}$ with period L we have

$$\sum_{i=0}^{L-1} \xi(x_i, x_{i+1}, \ldots, x_{i+k}) = \sum_{i=0}^{L-1} \xi([F(x)]_i, [F(x)]_{i+1}, \ldots, [F(x)]_{i+k}) \tag{6.27}$$

For simplicity, if the above condition is satisfied, we will say that ξ is an additive invariant of F. If F has a local function f with n inputs, it is often more convenient to write (6.27) using the function G defined as

$$\begin{aligned} G(x_0, x_1, \ldots, x_{k+n-1}) = \xi(f(x_0, x_1, \ldots, x_{n-1}), f(x_1, x_2, \ldots, x_n), \ldots, \\ f(x_k, x_{k+1}, \ldots, x_{k+n-1})). \end{aligned} \tag{6.28}$$

With this notation, ξ is an additive invariant of f if

$$\sum_{i=0}^{L-1} G(x_i, x_{i+1}, \ldots, x_{i+k+n-1}) = \sum_{i=0}^{L-1} \xi(x_i, x_{i+1}, \ldots, x_{i+k}) \tag{6.29}$$

for every positive integer L and for all $x_0, x_1, \ldots, x_{L-1} \in \mathcal{A}$.

Note that the number conserving rules introduced earlier are a special case of additive invariant, with $k = 0$ and $\xi(x_0) = x_0$. For this invariant, Eq. (6.27) becomes

$$\sum_{i=0}^{L-1} x_i = \sum_{i=0}^{L-1} [F(x)]_i, \tag{6.30}$$

which means that the CA rule possessing this invariant conserves the number of sites in state 1. Such rules, as we have already remarked, are often referred to as number-conserving rules.

Hattori and Takesue [1] established a very general result which we will write here in a somewhat simplified form, taking into account that this book is concerned with binary rules only.

Theorem 6.2 (Hattori and Takesue '91) *Let $\xi(x_0, x_1, \ldots, x_k)$ be a function of $k + 1$ variables. Then ξ is a density function of an additive conserved quantity of cellular automaton rule with local function with radius r and n inputs, $n = 2r + 1$, if and only if the condition*

$$\begin{aligned} G(x_0, x_1, \ldots, x_{k+n-1}) - \xi(x_r, x_{r+1}, \ldots, x_{k+r}) \\ = J(x_0, x_1, \ldots, x_{k+n-2}) - J(x_1, x_2, \ldots, x_{k+n-1}) \end{aligned} \tag{6.31}$$

holds for all $x_0, x_1, \ldots, x_{k+n-1} \in \{0, 1\}$, where the quantity J, to be referred to as the current, *is defined by*

$$J(x_0, x_1, \ldots, x_{k+n-2}) = -\sum_{i=0}^{k+n-2} G(\overbrace{0, 0, \ldots, 0}^{k+n-1-i}, x_0, x_1, \ldots, x_i)$$
$$+ \sum_{i=r-n+2}^{k-l} \xi(\overbrace{0, 0, \ldots, 0}^{k+1-i}, x_0, x_1, \ldots, x_{i-1}). \tag{6.32}$$

The following convention is used in the definition of J:

$$\xi(\overbrace{0, 0, \ldots, 0}^{k+1-i}, x_0, x_1, \ldots, x_{i-1}) = \xi(x_{i-k-1}, x_{i-k}, \ldots, x_{i-1}) \quad \text{if } i \geq k+1, \tag{6.33}$$

and

$$\xi(\overbrace{0, 0, \ldots, 0}^{k+1-i}, x_0, x_1, \ldots, x_{i-1}) = \xi(0, 0, \ldots, 0) \quad \text{if } i \leq 0. \tag{6.34}$$

Proof of this theorem is very similar to the proof of Theorem 6.1 and will be omitted here.

In order to illustrate the theorem we will first consider the case of number-conserving nearest-neighbour rules, i.e., $k = 0, \xi(x_0) = x_0, r = 1, n = 3$. Condition (6.31) becomes

$$G(x_0, x_1, x_2) - \xi(x_1) = J(x_0, x_1) - J(x_1, x_2), \tag{6.35}$$

where

$$J(x_0, x_1) = -G(0, 0, x_0) - G(0, x_0, x_1) + \xi(0) + \xi(x_0). \tag{6.36}$$

Obviously, $G(x_0, x_1, x_2) = f(x_0, x_1, x_2)$, thus the current becomes

$$J(x_0, x_1) = -f(0, 0, x_0) - f(0, x_0, x_1) - x_0, \tag{6.37}$$

and the conservation condition takes the form

$$f(x_0, x_1, x_2) - x_1 = J(x_0, x_1) - J(x_1, x_2). \tag{6.38}$$

As mentioned earlier, rule 184 and its spatial reflection are the only non-trivial elementary CA rules satisfying (6.38). For rule 184, f is defined by $f(x_0, x_1, x_2) = x_0 - x_0x_1 + x_1x_2$ and the current can be written as $J(x_0, x_1) = x_0(1 - x_1)$. This means that the current is non-zero only if $x_0x_1 = 10$, in agreement with the the interpretation of the dynamics of rule 184 given in Eq. (6.12).

In what follows, we will refer to the number of variables in ξ as the *order of the additive invariant*. Rule 184, therefore, is the only minimal non-trivial rule with the additive invariant of the first order.

6.6 Elementary CA with Second Order Invariants

We will explore second order invariants for elementary cellular automata, following the method of [1].

The arguments x_0, x_1 of the density function ξ take values in the set $\{0, 1\}$, and therefore ξ can be defined in terms of four parameters

$$\xi(0,0) = c_{00}, \quad \xi(0,1) = c_{01}, \quad \xi(1,0) = c_{10}, \quad \xi(1,1) = c_{11},$$

where $c_{00}, c_{01}, c_{10}, c_{11}$ are real parameters. This can be also expressed as

$$\begin{aligned}\xi(x_0, x_1) &= c_{00}(1-x_0)(1-x_1) + c_{01}(1-x_0)x_1 + c_{10}x_0(1-x_1) + c_{11}x_0x_1 \\ &= c_{00} + (c_{10} - c_{00})x_0 + (c_{01} - c_{00})x_1 + (c_{00} - c_{01} - c_{10} + c_{11})x_0x_1.\end{aligned} \tag{6.39}$$

Note that due to periodicity of x, for any function $h(x)$ and any $x_0, x_1, \ldots, x_{L-1} \in \{0, 1\}$ we have

$$\sum_{i=0}^{L-1} \xi(x_i, x_{i+1}) = \sum_{i=0}^{L-1} \left(\xi(x_i, x_{i+1}) + h(x_i) - h(x_{i+1})\right). \tag{6.40}$$

This means that if $\xi(x_0, x_1)$ is a density function of some conserved additive quantity, then $\xi'(x_0, x_1) = \xi(x_0, x_1) + g(x_0) - g(x_1)$ is also a density function of a conserved additive quantity. To remove this ambiguity, we will require that $\xi(0, x_1) = 0$, similarly as done in [1]. This yields $c_{00} + (c_{01} - c_{00})x_1 = 0$, which can be satisfied for all values of x_1 only if $c_{00} = c_{01} = 1$. We are thus left with ξ depending on two parameters only

$$\xi(x_0, x_1) = c_{10}x_0 + (c_{11} - c_{10})x_0x_1. \tag{6.41}$$

Defining $a_1 = -c_{10}$, $a_2 = c_{11} - c_{10}$, we arrive at the final parametrization of ξ

$$\xi(x_0, x_1) = a_1x_0 + a_2x_0x_1, \quad a_1, a_2 \in \mathbb{R}. \tag{6.42}$$

For $k = 1$, $r = 1$, and $n = 3$, Eq. (6.31) becomes

$$G(x_0, x_1, x_2, x_3) - \xi(x_1, x_2) = J(x_0, x_1, x_2) - J(x_1, x_2, x_3), \tag{6.43}$$

where

$$G(x_0, x_1, x_2, x_3) = \xi(f(x_0, x_1, x_2), f(x_1, x_2, x_3)), \tag{6.44}$$

Table 6.1 Density of the invariant ξ and the current J for all minimal elementary CA with second order invariants

Rule number	$\xi(x_0, x_1, x_2)$	$J(x_0, x_1)$
12	$x_0 - x_0x_1$	$-x_0x_1$
14	$x_0 - x_0x_1$	$-x_0x_1$
15	$x_0 - x_0x_1$	$-x_0x_1$
34	$x_0 - x_0x_1$	$-x_1$
35	$x_0 - x_0x_1$	$-x_1$
42	$x_0 - x_0x_1$	$-x_1 + x_1x_2 - x_0x_1x_2$
43	$x_0 - x_0x_1$	$-x_1 + x_1x_2 - x_0x_1x_2$
51	$x_0 - x_0x_1$	$-x_1$
140	$x_0 - x_0x_1$	$-x_0x_1 + x_0x_1x_2$
142	$x_0 - x_0x_1$	$-x_0x_1 + x_0x_1x_2$
200	x_0x_1	0

Rules 204 and 170 are excluded

and

$$J(x_0, x_1, x_2) = -G(0, 0, 0, x_0) - G(0, 0, x_0, x_1) - G(0, x_0, x_1, x_2) + \xi(0, 0) + \xi(0, x_0) + \xi(x_0, x_1). \tag{6.45}$$

Since $\xi(0, 0) = \xi(0, x_0) = 0$, this simplifies to

$$J(x_0, x_1, x_2) = -G(0, 0, 0, x_0) - G(0, 0, x_0, x_1) - G(0, x_0, x_1, x_2) + \xi(x_0, x_1). \tag{6.46}$$

For a given elementary CA rule $f(x_0, x_1, x_2)$, one can write Eq. (6.43) for all 2^4 possible values of the variables $x_0, x_1, x_2, x_3 \in \{0, 1\}$, obtaining a system of linear system of 16 equations with two unknowns a_1, a_2. This system is overdetermined yet homogeneous, therefore the solution, if it exists, is not unique: if (a_1, a_2) is a solution, then (ca_1, ca_2) is also a solution for any $c \in \mathbb{R}$. We will choose c such that the first non-zero number in the pair (a_1, a_2) is set to be equal to 1.

Solving these equations for all 88 minimal CA rules, one finds that for most CA rules solutions do not exist, meaning that they do not possess any second-order additive invariants. Rules for which solutions exist can be divided into two classes. The first class contains rules 204 and 170, and for these rules, any pair (a_1, a_2) is a solution. This is of no particular interest to us as it is obvious that both identity and shift trivially conserve any additive quantity. The second class consists of 10 rules for which a unique solution exists (up to the normalization constant c described earlier). These rules are 12, 14, 15, 34, 35, 42, 43, 51, 140, 142, and 200, as reported in [1]. Table 6.1 shows the density function ξ and the current J, obtained by symbolic algebra software, for all of these rules.

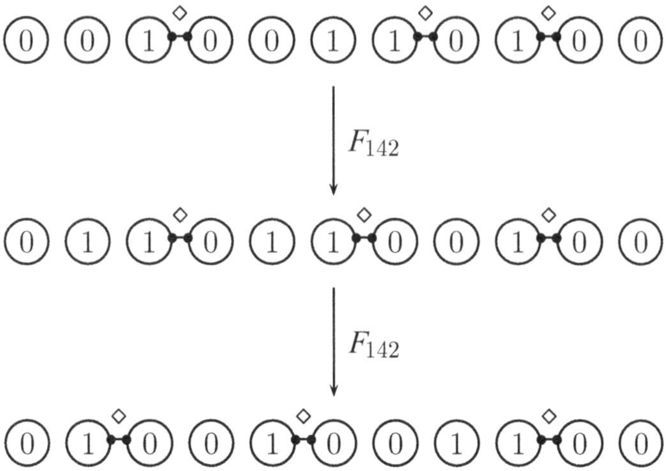

Fig. 6.6 Two iterations of F_{142} applied to a periodic configuration with 00100110100 in a single period. Conserved blocks 10 are marked with diamonds

Example 6.1 Consider as an example rule 142. It has a second-order invariant with density function $\xi(x_0, x_1) = x_0 - x_0x_1 = x_0\overline{x}_1$. Note that $\xi(x_0, x_1) = 1$ if and only if $x_0x_1 = 10$, meaning that rule 142 conserves the number 10 blocks. This is shown in Fig. 6.6, where blocks 10 are marked with diamonds, and we can see that their number remains unchanged when the rule 142 is applied. Blocks 10 can neither appear nor disappear, they can only move. How do they move? This can be figured out from the expression for the current. The current for this invariant is negative,

$$J(x_0, x_1, x_2) = -x_0x_1 + x_0x_1x_2 = -x_0x_1\overline{x}_2,$$

and it will be non-zero if and only if $x_0x_1x_2 = 110$. This means that the block 10 will move only when it is preceded by 1, and it will move to the left (positive current signifies movement to the right). We can represent this graphically in a similar fashion as we did for rule 184 in Eq. (6.12),

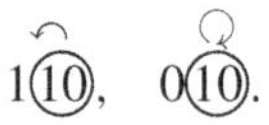

(6.47)

Note that when the block 10 moves to the left, the preceding 1 will disappear. This is to be expected: rule 142 does not conserve the number of 1's, only the number of blocks 10.

Among the rules listed in Table 6.1, for rules 12, 15, 34, 42, 51, 140 and 200 we have already obtained the solution of the initial value problem in previous chapters. This leave rules 14, 35, 43, and 142, and we turn our attention to them now.

6.7 Rules 142 and 43

For rules 142 and 43, the finite state machines representing preimages of 1 have remarkably similar structure as FSM for rule 184. Examples of FSM for $\mathbf{f}_{142}^{-4}(1)$ and $\mathbf{f}_{43}^{-4}(1)$ are shown in Fig. 6.7, and indeed they look similar to Fig. 6.3b, except that edge labels are different. Both can be described as graphs with nodes labeled (k, m) with $k \in \{0, 1, \ldots, n+1\}$ and $m \in \{0, 1, \ldots n\}$, and with edges

$$\begin{aligned}
&(k, m) \xrightarrow{h(k,m)} (k+1, m) \quad \text{for } k \in \{0, \ldots n\}, m \in \{0, \ldots, n\}, \\
&(k, m) \xrightarrow{v(k,m)} (k, m+1) \quad \text{for } k \in \{0, \ldots n+1\}, m \in \{0, \ldots, n-1\}, \\
&(k, m) \xrightarrow{v_e(k,m)} (k, m+1) \quad \text{for } k = n+1, m \in \{0, \ldots, n-1\}.
\end{aligned} \tag{6.48}$$

Edge labels for rule 142 are

$$\begin{aligned}
h(k, m) &= \begin{cases} \frac{1}{2} + \frac{1}{2}(-1)^{m+n} & \text{if } k \leq m, \\ \frac{1}{2} + \frac{1}{2}(-1)^{k+n} & \text{otherwise,} \end{cases} \\
v(k, m) &= 1 - h(k, m), \\
v_e(k, m) &= 0.
\end{aligned}$$

The presence of n in the exponent is due to the fact that for odd n, the bottom edge label sequence is $010101\ldots$ and for the even n (shown in the figure) it is $101010\ldots$. A more concise way of writing the edge labeling function could be

$$h(k, m) = \frac{1}{2} + \frac{1}{2}(-1)^{\max\{k,m\}}, \tag{6.49}$$

$$v(k, m) = \frac{1}{2} - \frac{1}{2}(-1)^{\max\{k,m\}}. \tag{6.50}$$

Equation (6.24) remains valid for rule 142,

$$\begin{aligned}
M_0 &= 0, \\
M_i &= M_{i-1} + \left[a_i = v(i-1-M_{i-1}, M_{i-1}) \vee i = M_{i-1} + n + 2\right].
\end{aligned} \tag{6.51}$$

For rule 43, the edge labels are similar to those of rule 142, except that the role of k and m is reversed,

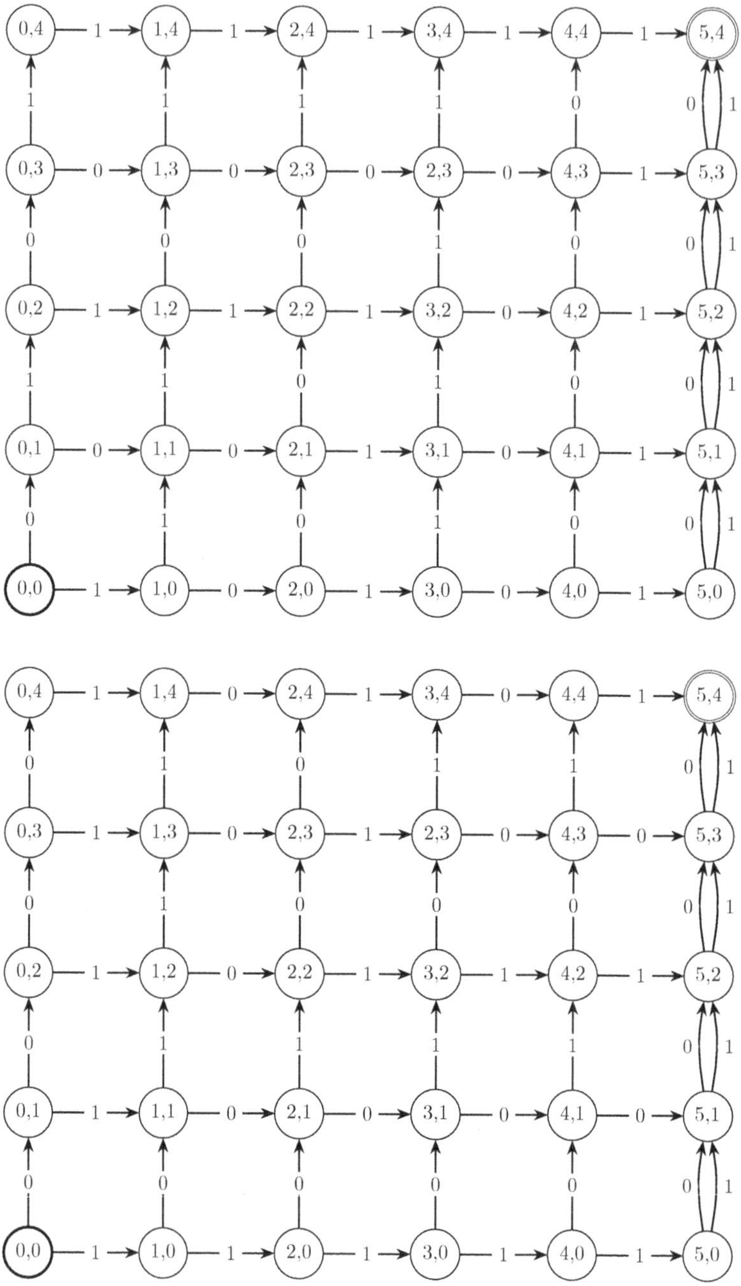

Fig. 6.7 Finite state machines representing 4-step preimages of 1 for rule 142 (top) and 43 (bottom)

$$
\begin{aligned}
h(k,m) &= \begin{cases} \frac{1}{2}+\frac{1}{2}(-1)^{k+n} & \text{if } k \le m,\\ \frac{1}{2}+\frac{1}{2}(-1)^{m+n} & \text{otherwise,}\end{cases}\\
v(k,m) &= 1-h(k,m),\\
v_e(k,m) &= 0.
\end{aligned}
$$

Again, a simpler version would be

$$
h(k,m) = \frac{1}{2}+\frac{1}{2}(-1)^{\min\{k,m\}},
$$

$$
v(k,m) = \frac{1}{2}-\frac{1}{2}(-1)^{\min\{k,m\}}.
$$

Our observations can be summarized as follows.

Proposition 6.4 *Block* $\mathbf{a} \in \{0,1\}^{2n+1}$ *is an* n*-step preimage of 1 under the rule 142 if and only if* $M_{2n+1}(\mathbf{a}) = n$, *where* $M_i(\mathbf{a})$ *is recursively defined in Eq. (6.51) with* $v(k,m) = \frac{1}{2} - \frac{1}{2}(-1)^{\max\{k,m\}}$. *Similarly, block* $\mathbf{a} \in \{0,1\}^{2n+1}$ *is an* n*-step preimage of 1 under the rule 43 if and only if* $M_{2n+1}(\mathbf{a}) = n$, *where* $M_i(\mathbf{a})$ *is recursively defined in Eq. (6.51) with* $v(k,m) = \frac{1}{2} - \frac{1}{2}(-1)^{\min\{k,m\}}$.

Proof of this proposition can be constructed along the lines of the proof of Proposition 6.1. We will not go into details, only remarking that the proof needs to use the fact that blocks 10 move following the rule visualized in Eq. (6.47), similarly as the proof of Proposition 6.1 used (in Lemma 6.1) the motion representation of rule 184 given in Eq. (6.12).

We will now proceed to the construction of the solution of the initial value problem which can be obtained directly from Eq. (6.1). Let us write the complete form of the solution for rule 142,

$$
[F^n_{142}(x)]_j = \sum_{\substack{\mathbf{a}\in\{0,1\}^{2n+1}\\ M_{2n+1}(\mathbf{a})=n}} \Psi_{\mathbf{a}}(x_{j-n}, x_{j-n+1}, \ldots, x_{j+n}), \tag{6.52}
$$

where

$$
\begin{aligned}
M_0 &= 0,\\
M_i &= M_{i-1} + \left[a_i = \frac{1}{2} - \frac{1}{2}(-1)^{\max\{i-1-M_{i-1}, M_{i-1}\}} \vee i = M_{i-1}+n+2 \right].
\end{aligned}
$$

For rule 43 we have

$$
\left[F^n_{43}(x)\right]_j = \sum_{\substack{\mathbf{a}\in\{0,1\}^{2n+1}\\ M_{2n+1}(\mathbf{a})=n}} \Psi_{\mathbf{a}}(x_{j-n}, x_{j-n+1}, \ldots, x_{j+n}), \tag{6.53}
$$

where

$$\begin{aligned}M_0 &= 0,\\ M_i &= M_{i-1} + \left[a_i = \frac{1}{2} - \frac{1}{2}(-1)^{\min\{i-1-M_{i-1}, M_{i-1}\}} \vee i = M_{i-1} + n + 2\right].\end{aligned}$$

In both instances, $\Psi_{\mathbf{a}}$ is the density polynomial of **a** as defined in Eq. (2.12).

The reader may wonder if there is any reason why the solutions of rules 142 and 43 are so similar. The reason is that these rules are very closely related. One can show that the following identity holds,

$$F_{142}^2 = F_{113}^2.$$

Rule 113 appearing in this identity belongs to the same equivalence class as rule 43—in fact, rule 43 is a reflected version of rule 113,

$$F_{113} = RF_{43}R,$$

where $[R(x)]_i = x_i$ is the spatial reflection. This yields

$$F_{142}^2 = RF_{43}RRF_{43}R = RF_{43}^2R,$$

because $R^2 = I$. The following commutative diagram represents the relationship between rules 142 and 43,

$$\begin{array}{ccc} x & \xrightarrow{R} & x' \\ {\scriptstyle F_{142}}\downarrow & & \downarrow{\scriptstyle F_{43}} \\ y & & y' \\ {\scriptstyle F_{142}}\downarrow & & \downarrow{\scriptstyle F_{43}} \\ z & \xrightarrow{R} & z' \end{array}$$

Using the solution of rule 43 we can also obtain solution of yet another related rule, namely ECA 11 which emulates rule 43,

$$F_{11}^2 = F_{43}F_{11}.$$

This immediately yields

$$\left[F_{11}^n(x)\right]_j = \left[F_{43}^{n-1}(y)\right]_j,$$

where

$$y_j = f_{11}(x_{j-1}, x_j, x_{j+1}) = 1 + x_{j-1}x_j - x_{j-1} - x_{j+1}x_{j-1}x_j - x_j + x_{j+1}x_j,$$

and where $[F_{43}^{n-1}(y)]_j$ is given by Eq. (6.53).

6.8 Rule 14

Rule 14 has the local function

$$f_{14}(x_1, x_2, x_3) = x_3 + x_2 - x_2x_3 - x_1x_3 - x_1x_2 + x_1x_2x_3,$$

and, as mentioned, it conserves the number of 10 blocks, similarly as rules 43 and 142. It FSM graph, however, is slightly different then previously discussed graphs, as shown in Fig. 6.8. The main difference is that the top row is not completely present, only the rightmost two vertices are there. For edges which are present, however, the labeling is identical to the labeling of the FSM for rule 142. It can be described as graph with nodes labeled (k, m) with $k \in \{0, 1, \ldots, n+1\}$ and $m \in \{0, 1, \ldots n-1\}$, plus two additional nodes (n, n) and $(n+1, n)$. Its edges are given by

$$\begin{aligned}
(k, m) &\xrightarrow{h(k,m)} (k+1, m) \quad \text{for } (k, m) \in \{0, \ldots n\} \times \{0, \ldots, n-1\} \cup \{(n, n)\}, \\
(k, m) &\xrightarrow{v(k,m)} (k, m+1) \quad \text{for } (k, m) \in \{0, \ldots n+1\} \times \{0, \ldots, n-2\} \\
&\qquad\qquad\qquad\qquad \cup \{(n, n-1), (n+1, n-1)\}, \\
(k, m) &\xrightarrow{v_e(k,m)} (k, m+1) \quad \text{for } k = n+1, m \in \{0, \ldots, n-1\},
\end{aligned} \tag{6.54}$$

where

$$\begin{aligned}
h(k, m) &= \frac{1}{2} + \frac{1}{2}(-1)^{\max\{k,m\}}, \\
v(k, m) &= \frac{1}{2} - \frac{1}{2}(-1)^{\max\{k,m\}}.
\end{aligned}$$

Recall that in the case of rules 184, 43 and 142, the condition for FSM accepting the string $\mathbf{a}$ was $M_{2n+1}(\mathbf{a}) = n$. Here this is not enough, and looking at Fig. 6.8 we can see that the path entering the final state must go through node (4, 3) or (5, 3). Node (4, 3) is two steps before the end, and its second coordinate is $n-1$, which means that we also must require $M_{2n-1}(\mathbf{a}) \leq n-1$.

Proposition 6.5 *Block* $\mathbf{a} \in \{0, 1\}^{2n+1}$ *is an* n*-step preimage of 1 under the rule 14 if and only if* $M_{2n-1}(\mathbf{a}) \leq n-1$ *and* $M_{2n+1}(\mathbf{a}) = n$*, where* $M_i(\mathbf{a})$ *is recursively defined in Eq. (6.51) with* $v(k, m) = \frac{1}{2} - \frac{1}{2}(-1)^{\max\{k,m\}}$.

We will omit the formal proof again, giving only the solution,

$$\left[F_{14}^n(x)\right]_j = \sum_{\substack{\mathbf{a} \in \{0,1\}^{2n+1} \\ M_{2n-1}(\mathbf{a}) \leq n-1 \\ M_{2n+1}(\mathbf{a}) = n}} \Psi_{\mathbf{a}}(x_{j-n}, x_{j-n+1}, \ldots, x_{j+n}), \tag{6.55}$$

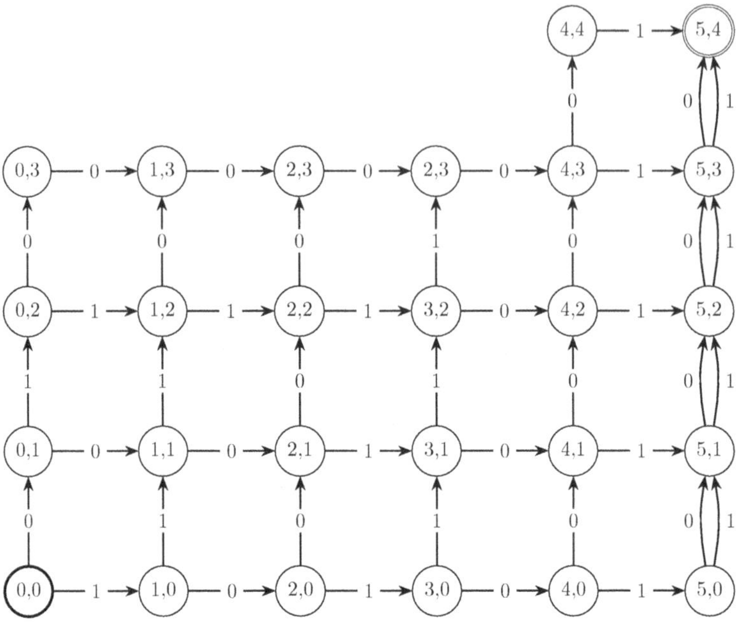

Fig. 6.8 Finite state machine representing 4-step preimages of 1 for rule 14

where

$$
\begin{aligned}
M_0 &= 0, \\
M_i &= M_{i-1} + \left[a_i = \frac{1}{2} - \frac{1}{2}(-1)^{\max\{i-1-M_{i-1}, M_{i-1}\}} \vee i = M_{i-1} + n + 2 \right].
\end{aligned}
$$

As before, the similarity between the solution of rule 14 and solutions of rules 142 and 43 is not accidental. Rule 14 emulates rule 142 as well as rule 113 (reflected version of rule 43),

$$
\begin{aligned}
F_{14}^2 &= F^{142} F^{14}, \\
F_{14}^3 &= F_{113} F_{14}.
\end{aligned}
$$

References

1. Hattori, T., Takesue, S.: Additive conserved quantities in discrete-time lattice dynamical systems. Physica D **49**, 295–322 (1991)
2. Boccara, N., Fukś, H.: Number-conserving cellular automaton rules. Fundamenta Informaticae **52**, 1–13 (2002)

3. Mohri, M., Pereira, F.C.N., Riley, M.D.: AT&T Finite-State Machine Library, version 4.0. http://www3.cs.stonybrook.edu/~algorith/implement/fsm/implement.shtm
4. Mohri, M.: On some applications of finite-state automata theory to natural language processing. Natural Language Engineering **2**(1), 61–80 (1996)
5. Mohri, M.: Finite-state transducers in language and speech processing. Computational Linguistics **23**(2), 269–311 (1997)

Chapter 7
Construction of the Probability Measures

So far we have considered only strictly deterministic cellular automata and deterministic initial conditions. We will now show how to equip the space $\mathcal{A}^{\mathbb{Z}}$ with a probability measure, so we could consider probabilistic initial value problem as well as probabilistic cellular automata. The construction of the probability measure involves several steps, similarly as done in [1]. We will start by defining a class of open sets in $\mathcal{A}^{\mathbb{Z}}$ constituting a semi-algebra. We will then furnish this semi-algebra with a finitely additive measure. Afterwards, we will use the semi-algebra to generate a σ-algebra and extend the measure to this σ-algebra. We will show that if the measure is shift-invariant and the full space has measure 1, then it can be completely characterized by a countable set of block probabilities.

7.1 Cylinder Sets and Their Semi-algebra

In the first chapter we have introduced the Cantor metric on $\mathcal{A}^{\mathbb{Z}}$, defined as $d(\mathbf{x}, \mathbf{y}) = 2^{-k}$, where $k = \min\{|i| : \mathbf{x}_i \neq \mathbf{y}_i\}$. $\mathcal{A}^{\mathbb{Z}}$ with the metric d is a Cantor space, that is, compact, totally disconnected and perfect metric space. Every metric space can be given a metric topology, in which the basic open sets are open balls defined by the metric, so let us consider the concept of open balls in $\mathcal{A}^{\mathbb{Z}}$. Open ball with center $c \in \mathcal{A}^{\mathbb{Z}}$ and radius $r > 0$ is defined as

$$B_r(c) = \{x \in A^{\mathbb{Z}} : d(x, c) < r\}.$$

The condition $d(x, c) < r$ for Cantor metric means that the minimal absolute value of the index i for which x_i and c_i differ is greater than $-\log_2 r$. This means that $x_{-k} = c_{-k}, x_{-k+1} = c_{-k+1}, \ldots, x_{-k} = c_k$, where $k = \lfloor -\log_2 r \rfloor$. Bisequences belonging to the open ball centered at c must, therefore, agree with c on all positions from $-k$ to k, and on other positions they can feature arbitrary symbols from $\mathcal{A}$. We will call

H. Fukś, *Solvable Cellular Automata*, Understanding Complex Systems,
https://doi.org/10.1007/978-3-031-38700-5_7

such open balls *central cylinder sets*. For our subsequent considerations it will be more convenient to define a more general family of open sets.

Definition 7.1 *Cylinder set* generated by the block $\mathbf{b} = b_1 b_2 \ldots, b_n$ anchored at i is defined as

$$[\mathbf{b}]_i = \{\mathbf{x} \in \mathcal{A}^{\mathbb{Z}} : \mathbf{x}_{[i,i+n)} = \mathbf{b}\}. \tag{7.1}$$

If one the of the indices $i, i+1, \ldots, i+n-1$ is equal to zero, or, equivalently, if $-n+1 \leq i \leq 0$, then the cylinder set will be called *elementary*.

For a given elementary cylinder set $[\mathbf{b}]_i$, indices $i, i+1, \ldots, i+n-1$ will be called *fixed*, because the symbols of $x \in [\mathbf{b}]_i$ in these positions must be the same as the corresponding symbols of $\mathbf{b}$. All other indices will be called *free* because there is no restriction on x_i for such indices. Note that the requirement $-n+1 \leq i \leq 0$ means that the origin is always fixed.

Let $[\mathbf{a}]_j$ and $[\mathbf{b}]_i$ be two elementary cylinder sets. We will say that $p \in \mathbb{Z}$ is a *matching* (*mismatching*) index of these cylinder sets if for every $\mathbf{x} \in [\mathbf{a}]_j$, $\mathbf{y} \in [\mathbf{b}]_i$ we have $x_p = y_p$ ($x_p \neq y_p$). An index which is either matching or mismatching will be called *overlapping* index, otherwise it will be called *non-overlapping*.

Example 7.1 Let $\mathcal{A} = \{0, 1\}$, $\mathbf{a} = 10110$ and $\mathbf{b} = 0100$. Cylinder sets $[\mathbf{a}]_{-2}$ and $[\mathbf{b}]_0$ can be represented as

$$\begin{array}{cccccccccc} \cdots & \star & 1 & 0 & 1 & 1 & 0 & \star & \star & \cdots \\ \cdots & \star & \star & \star & 0 & 1 & 0 & 0 & \star & \cdots \end{array}$$

where $\star$ denotes arbitrary symbol in $\mathcal{A}$. The first star from the left corresponds to $i = -3$. For $[\mathbf{a}]_{-2}$ fixed indices are $-2, -1, 0, 1, 2$, and for $[\mathbf{b}]_0$ they are $0, 1, 2, 3$. In both cases 0 is fixed, so both $[\mathbf{a}]_{-2}$ and $[\mathbf{b}]_0$ are elementary cylinder sets. Furthermore, we can see that $i = 1$ and $i = 2$ are matching indices, and $i = 0$ is the only mismatching index. Overlapping indices are $0, 1, 2$, and all other indices are non-overlapping.

A collection of sets S is said to be a semialgebra if it is closed under intersection, and has the property that if A, B are element of S, then $A \setminus B$ is a finite disjoint union of sets in S. The collection of all cylinder sets of $\mathcal{A}^{\mathbb{Z}}$ together with the empty set and the whole space $\mathcal{A}^{\mathbb{Z}}$ will be denoted by $Cyl(\mathcal{A}^{\mathbb{Z}})$. We will use the convention that for $\mathbf{b} = \varnothing$, $[\mathbf{b}]_i = \mathcal{A}^{\mathbb{Z}}$.

Proposition 7.1 *The collection of all elementary cylinder sets together with the empty set and the whole space constitutes a semialgebra over* $\mathcal{A}^{\mathbb{Z}}$.

Proof We need to prove that (i) $Cyl(\mathcal{A}^{\mathbb{Z}})$ is closed under the intersection and (ii) that the set difference of two elementary cylinder sets is a finite union of elementary cylinder sets.

For (i), let $[\mathbf{a}]_j$ and $[\mathbf{b}]_i$ be two elementary cylinder sets, thus they must have some overlapping indices. Suppose that all overlapping indices are matching. In this case,

$[\mathbf{a}]_j \cap [\mathbf{b}]_i$ is just the elementary cylinder set generated by overlapped concatenation of $\mathbf{a}$ and $\mathbf{b}$, thus $[\mathbf{a}]_j \cap [\mathbf{b}]_i \in Cyl(\mathcal{A}^{\mathbb{Z}})$. If, on the other hand, among overlapping indices there is at least one mismatching index, then $[\mathbf{a}]_j \cap [\mathbf{b}]_i = \emptyset \in Cyl(\mathcal{A}^{\mathbb{Z}})$.

For (ii), let us first note that

$$\mathcal{A}^{\mathbb{Z}} \setminus [\mathbf{b}]_i = \bigcup_{j=0}^{n-1} \{\mathbf{x} \in \mathcal{A}^{\mathbb{Z}} : x_{i+j} \neq b_{j+1}\}. \tag{7.2}$$

Each of the sets $\{\mathbf{x} \in \mathcal{A}^{\mathbb{Z}} : x_{i+j} \neq b_{j+1}\}$ can be expressed as a union of elementary cylinder sets, thus $\mathcal{A}^{\mathbb{Z}} \setminus [\mathbf{b}]_i$ is also a union of elementary cylinder sets. The set difference of two cylinder sets can be written

$$[\mathbf{a}]_j \setminus [\mathbf{b}]_i = [\mathbf{a}]_j \cap \left(\mathcal{A}^{\mathbb{Z}} \setminus [\mathbf{b}]_i\right), \tag{7.3}$$

and given what we just said about $\mathcal{A}^{\mathbb{Z}} \setminus [\mathbf{b}]_i$, this is an intersection of a cylinder set with finite union of cylinder sets. Set intersection is distributive over set union, and since $Cyl(\mathcal{A}^{\mathbb{Z}})$ is closed under the intersection, we obtain the desired result (ii). □

One interesting consequence of Proposition 7.1 is the fact that complement of a cylinder set must be a finite union of cylinder sets. Union of open set is open, thus it must be an open set. On the other hand, complement of an open set is closed, so it must be closed as well. Cylinder sets in the Cantor topology, therefore, are both closed and open. We will call them *clopen sets.*

Our next step will be to equip the semi-algebra of elementary cylinder sets with a measure. Let $\mathcal{D}$ be a semialgebra. A map $\mu : \mathcal{D} \to [0, \infty]$ is called a *measure* on $\mathcal{D}$ if $\mu(\varnothing) = 0$ and if for any sequence $\{A_i\}_{i=1}^{\infty}$ of pairwise disjoint sets belonging to $\mathcal{D}$ such that $\bigcup_{i=1}^{\infty} A_i \in \mathcal{D}$,

$$\mu\left(\bigcup_{i=1}^{\infty} A_i\right) = \sum_{i=1}^{\infty} \mu(A_i). \tag{7.4}$$

The above condition is called *countable additivity*. For measures on the semialgebra of cylinder sets, countable additivity is implied by a weaker condition called *finite additivity*. A map μ is *finitely additive* if for any finite sequence $\{A_i\}_{i=1}^{m}$ of pairwise disjoint sets belonging to $Cyl(\mathcal{A}^{\mathbb{Z}})$ such that $\bigcup_{i=1}^{m} A_i \in Cyl(\mathcal{A}^{\mathbb{Z}})$,

$$\mu\left(\bigcup_{i=1}^{m} A_i\right) = \sum_{i=1}^{m} \mu(A_i). \tag{7.5}$$

Proposition 7.2 *Any finitely additive map* $\mu : Cyl(\mathcal{A}^{\mathbb{Z}}) \to [0, \infty]$ *for which* $\mu(\varnothing) = 0$ *is a measure on the semialgebra of elementary cylinder sets.*

Proof Let the map μ satisfy $\mu(\emptyset) = 0$ and be finitely additive. In order to show that μ is a measure on $Cyl(X)$, we need to show that it is countably additive. Let B be a cylinder set and let $\{A_i\}_{i=1}^{\infty}$ be a collection of pairwise disjoint cylinder sets such that $\bigcup_{i=1}^{\infty} A_i = B$. Since B is closed (recall that cylinder sets are clopen), it is also compact. Sets A_i are open, and form a cover of the compact set B. There must exist, therefore, a finite subcover, that is, a finite number of sets A_i covering B. Moreover, since A_i are mutually disjoint, there must exist m such that $A_i = \emptyset$ for all $i > m$, and therefore $B = \bigcup_{i=1}^{m} A_i$. Then by finite additivity of μ and the assumption that $\mu(\emptyset) = 0$ we obtain

$$\mu(B) = \mu\left(\bigcup_{i=1}^{\infty} A_i\right) = \mu\left(\bigcup_{i=1}^{m} A_i\right) = \sum_{i=1}^{m} \mu(A_i) = \sum_{i=1}^{\infty} \mu(A_i), \tag{7.6}$$

which means that μ is countably additive and thus is a measure on $Cyl(X)$, as required. □

7.2 Extension Theorems

Although Proposition 7.2 allows us to construct a measure on the semialgebra of elementary cylinder set, this semialgebra is not sufficient to support the full machinery of probability theory. For this we need a larger class of subsets of $\mathcal{A}^{\mathbb{Z}}$, called a σ-algebra. This is a class of subsets of $\mathcal{A}^{\mathbb{Z}}$ that is closed under the complement and countable unions of its members. Such structure can be defined as an "extension" of the semialgebra $Cyl(X)$. The smallest σ-algebra containing $Cyl(X)$ will be called *σ-algebra generated by $Cyl(X)$*. It is possible to extend a measure on semi-algebra to the σ-algebra generated by it, using the following classic theorem.

Theorem 7.1 (Hahn–Kolmogorov) *Let $\mu : \mathcal{D} \to [0, \infty]$ be a measure on semi-algebra of subsets $\mathcal{D}$. Then μ can be extended to a measure on the σ-algebra generated by $\mathcal{D}$.*

This result has been first obtained by Fréchet [2], and later by Kolmogorov [3] and Hahn [4]. Since the proof bears little relevance to our subsequent considerations, it will be omitted here. Interested readers can find the proof, for example, in Ref. [5]. The proof is based on construction of the so-called outer measure $\mu^\star$ determined by μ, and then applying Carathéodory's extension theorem.

One can also show that the extension is unique if μ satisfies some additional conditions. We will not go into all details of such extensions, only noting that for measures satisfying $\mu(\mathcal{A}^{\mathbb{Z}}) = 1$ (to be called *probabilistic measures*), the extension is always unique [5]. In all further considerations, we will assume that the measure is probabilistic.

The Hahn–Kolmogorov Theorem, together with Proposition 7.2, yields in the following corollary, which summarizes our discussion.

Corollary 7.1 *Any finitely additive map* $\mu : Cyl(\mathcal{A}^{\mathbb{Z}}) \rightarrow [0, 1]$ *satisfying* $\mu(\emptyset)=0$ *and* $\mu(\mathcal{A}^{\mathbb{Z}}) = 1$ *extends uniquely to a measure on the* σ*-algebra generated by elementary cylinder sets of* $\mathcal{A}^{\mathbb{Z}}$.

The last thing we need to do is to characterize finite additivity of maps on $Cyl(X)$ in somewhat simpler terms. Recall that $\mu : Cyl(X) \rightarrow [0, 1]$ is finitely additive if for every $B \in Cyl(X)$ and pairwise disjoint $A_i \in Cyl(X)$, $i = 1, 2, \ldots, m$ such that $B = \bigcup_{i=1}^{m} A_i$, we have $\mu(B) = \sum_{i=1}^{m} \mu(A_i)$. From the definition of the cylinder set, it is clear that if B is a finite union of A_i, then each A_i must be longer than B, and for each pair (B, A_i) all fixed indices of B must be matching. For $B = [\mathbf{b}]_i$ this can happen we attach to $\mathbf{b}$ a postfix word, a prefix word, or both, and take the union over all values of attached word(s), resulting in one of the following three possibilities,

$$[\mathbf{b}]_i = \bigcup_{\mathbf{a}\in\mathcal{A}^k} [\mathbf{ba}]_i, \tag{7.7}$$

$$[\mathbf{b}]_i = \bigcup_{\mathbf{a}\in\mathcal{A}^k} [\mathbf{ab}]_{i-k}, \tag{7.8}$$

$$[\mathbf{b}]_i = \bigcup_{\mathbf{a}\in\mathcal{A}^k,\, \mathbf{c}\in\mathcal{A}^l} [\mathbf{abc}]_{i-k}. \tag{7.9}$$

Note that all cylinder sets on the right hand side of each of the above equations are pairwise disjoint. If we want to test the map for countable additivity, it is thus sufficient to test it on cases described by Eqs. (7.7)–(7.9).

Proposition 7.3 *The map* $\mu : Cyl(\mathcal{A}^{\mathbb{Z}}) \rightarrow [0, 1]$ *is countably additive if and only if for all elementary cylinder sets* $[\mathbf{b}]_i \in Cyl(\mathcal{A}^{\mathbb{Z}})$,

$$\mu([\mathbf{b}]_i) = \sum_{a\in\mathcal{A}} \mu([\mathbf{b}a]_i) = \sum_{a\in\mathcal{A}} \mu([a\mathbf{b}]_{i-1}). \tag{7.10}$$

Proof Suppose that the map is countably additive. If we apply additivity condition to Eqs. (7.7) and (7.8) using $\mathbf{a} = a$, we will obtain the desired result.

Now suppose that the double equality (7.10) holds. Applying it recursively k times we obtain

$$\mu([\mathbf{b}]_i) = \sum_{a_1\in\mathcal{A}} \cdots \sum_{a_k\in\mathcal{A}} \mu([\mathbf{b}a_1a_2\ldots a_k]_i) = \sum_{\mathbf{a}\in\mathcal{A}^k} \mu([\mathbf{ba}]_i), \tag{7.11}$$

which implies additivity of μ for the case covered by Eq. (7.7). One can deal with cases covered by Eqs. (7.8) and (7.9) in a similar fashion. The map μ is thus countably additive on $Cyl(X)$. □

Note that when $\mathbf{b} = \varnothing$, then $[\mathbf{b}]_i = \mathcal{A}^{\mathbb{Z}}$, and Eq. (7.10) reduces to

$$\sum_{a \in \mathcal{A}} \mu([a]_i) = 1, \tag{7.12}$$

where we used the assumption that the measure is probabilistic, $\mu(\mathcal{A}^{\mathbb{Z}}) = 1$.

7.3 Shift-Invariant Measure and Consistency Conditions

We have demonstrated that any map $\mu : Cyl(\mathcal{A}^{\mathbb{Z}}) \to [0, 1]$ satisfying $\mu(\varnothing) = 0$, $\mu(\mathcal{A}^{\mathbb{Z}}) = 1$ and conditions of Eq. (7.10) extends uniquely to a measure on the σ-algebra generated by elementary cylinder sets of X. We will use the same symbol μ for this uniquely extended measure. We will now impose another condition on it, namely translational invariance, by requiring that, for all $\mathbf{b} \in \mathcal{A}^\star$, $\mu([\mathbf{b}]_i)$ is independent of i for all $i \in \mathbb{Z}$. Translational invariance can also be called *shift invariance* for obvious reasons. Note that for arbitrary i, not all cylinder sets $[\mathbf{b}]_i$ will be elementary. This is fine, as by the extension we carried out, all cylinder sets, not only elementary ones, belong to the σ-algebra, and all are measurable with the measure μ. Having independence of i we no longer need the index i, thus to simplify the notation, we define $P : \mathcal{A}^\star \to [0, 1]$ as

$$P(\mathbf{b}) := \mu([\mathbf{b}]_i). \tag{7.13}$$

Values $P(\mathbf{b})$ will be called *block probabilities*. Applying Proposition 7.3 and Hahn–Kolmogorov theorem to the case of shift-invariant μ we obtain the following result.

Theorem 7.2 *Let* $P : \mathcal{A}^\star \to [0, 1]$ *satisfy the conditions*

$$P(\mathbf{b}) = \sum_{a \in \mathcal{A}} P(\mathbf{b}a) = \sum_{a \in \mathcal{G}} P(a\mathbf{b}) \quad \forall \mathbf{b} \in \mathcal{A}^\star, \tag{7.14}$$

$$1 = \sum_{a \in \mathcal{A}} P(a). \tag{7.15}$$

Then P *uniquely determines shift-invariant probability measure on the* σ*-algebra generated by elementary cylinder sets of* $\mathcal{A}^{\mathbb{Z}}$.

The set of shift-invariant probability measures on the σ-algebra generated by elementary cylinder sets of $\mathcal{A}^{\mathbb{Z}}$ will be denoted by $\mathfrak{M}(\mathcal{A}^{\mathbb{Z}})$. Conditions (7.14) and (7.15) are often called *consistency conditions*. It should be stressed, however, they are essentially equivalent to measure additivity conditions. Nevertheless, since the term "consistency conditions" is prevalent in the literature, we will use it in the subsequent considerations.

7.4 Block Probabilities

Since P uniquely determines the probability measure, we can use block probability values to define shift-invariant probability measure. Because of consistency conditions, block probabilities are not independent, thus we do not need *all* of them to define the measure. Which ones do we need then? We will answer this question in what follows.

We will define $\mathbf{P}^{(k)}$ to be the column vector of all probabilities of blocks of length k arranged in lexical order. For $\mathcal{A} = \{0, 1\}$, these are

$$\begin{aligned}
\mathbf{P}^{(1)} &= [P(0), P(1)]^T, \\
\mathbf{P}^{(2)} &= [P(00), P(01), P(10), P(11)]^T, \\
\mathbf{P}^{(3)} &= [P(000), P(001), P(010), P(011), P(100), P(101), P(110), P(111)]^T, \\
&\dots .
\end{aligned}$$

Since Eq. (7.14) are sets of linear equations involving $\mathbf{P}^{(k)}$, we can write them in a matrix form. For example, for $k = 2$ we have

$$\begin{aligned}
P(0) &= P(00) + P(01) = P(00) + P(10), \\
P(1) &= P(10) + P(11) = P(01) + P(11).
\end{aligned}$$

This can be written as

$$\begin{bmatrix} P(0) \\ P(1) \end{bmatrix} = \begin{bmatrix} 1\,1\,0\,0 \\ 0\,0\,1\,1 \end{bmatrix} \begin{bmatrix} P(00) \\ P(01) \\ P(10) \\ P(11) \end{bmatrix},$$

and

$$\begin{bmatrix} P(0) \\ P(1) \end{bmatrix} = \begin{bmatrix} 1\,0\,1\,0 \\ 0\,1\,0\,1 \end{bmatrix} \begin{bmatrix} P(00) \\ P(01) \\ P(10) \\ P(11) \end{bmatrix}.$$

For a general $k > 1$ and an alphabet $\mathcal{A}$ with N symbols we will have

$$\mathbf{P}^{(k-1)} = \mathbf{R}^{(k)}\mathbf{P}^{(k)} = \mathbf{L}^{(k)}\mathbf{P}^{(k)}, \tag{7.16}$$

where $\mathbf{L}^{(k)}$ and $\mathbf{R}^{(k)}$ are binary matrices with N^{k-1} rows and N^k columns, to be described shortly. Let us denote identity matrix $N^{k-1} \times N^{k-1}$ by $\mathbf{I}$, and let $\mathbf{J}_m$ be a matrix with N^{k-1} rows and N columns in which mth row consist of all 1's, and all other entries are 0. Then $\mathbf{L}^{(k)}$ and $\mathbf{R}^{(k)}$ can be written as

$$\mathbf{L}^{(k)} = [\underbrace{\mathbf{I}\ \mathbf{I}\ldots\mathbf{I}}_{N}], \tag{7.17}$$

$$\mathbf{R}^{(k)} = [\mathbf{J}_1\mathbf{J}_2\ldots\mathbf{J}_{N^{k-1}}]. \tag{7.18}$$

For example, for $N = 3$ and $k = 2$ Eq. (7.16) becomes

$$\mathbf{P}^{(1)} = \left[\begin{array}{ccc|ccc|ccc} 1&1&1&0&0&0&0&0&0 \\ 0&0&0&1&1&1&0&0&0 \\ 0&0&0&0&0&0&1&1&1 \end{array}\right]\mathbf{P}^{(2)} = \left[\begin{array}{ccc|ccc|ccc} 1&0&0&1&0&0&1&0&0 \\ 0&1&0&0&1&0&0&1&0 \\ 0&0&1&0&0&1&0&0&1 \end{array}\right]\mathbf{P}^{(2)},$$

where

$$\mathbf{P}^{(2)} = [P(00), P(01), P(02), P(10), P(11), P(12), P(20), P(21), P(22)]^T,$$
$$\mathbf{P}^{(1)} = [P(0), P(1), P(2)]^T,$$

and where the dashed vertical lines illustrate partitioning of matrices $\mathbf{R}^{(3)}$ and $\mathbf{L}^{(3)}$ into blocks of $\mathbf{I}$ and $\mathbf{J}$ type.

We will now point out two important properties of matrices $\mathbf{R}^{(k)}$ and $\mathbf{L}^{(k)}$. First of all, using Eq. (7.16) recursively, we can express every $\mathbf{P}^{(m)}$ for $m \in [1, k)$ by $\mathbf{P}^{(k)}$,

$$\mathbf{P}^{(m)} = \left(\prod_{i=m+1}^{k} \mathbf{L}^{(i)}\right)\mathbf{P}^{(k)}. \tag{7.19}$$

In the above, one could replace all (or only some) $\mathbf{L}$'s by $\mathbf{R}$'s, and the equation would remain valid.

Secondly, note that both $\mathbf{L}^{(1)}$ and $\mathbf{R}^{(1)}$ are single row matrices with all N entries equal to 1. This implies that the product $\mathbf{L}^{(1)}\mathbf{L}^{(2)}$ is a single row matrix with all N^2 entries equal to 1, and, in general, for any $k \geq 1$,

$$\prod_{i=1}^{k} \mathbf{L}^{(i)} = [\underbrace{1\ 1\ldots 1}_{N^k}]. \tag{7.20}$$

Again, one could replace here all (or some) $\mathbf{L}$'s by $\mathbf{R}$'s, and the equation would remain valid. As a consequence of this, normalization condition (7.15) can be written as $\mathbf{L}^{(1)}\mathbf{P}^{(1)} = 1$, or, replacing $\mathbf{P}^{(1)}$ by $\mathbf{L}^{(2)}\mathbf{P}^{(2)}$, as $\mathbf{L}^{(1)}\mathbf{L}^{(2)}\mathbf{P}^{(2)} = 1$, etc. In general, we can write the normalization condition in the form

$$\left(\prod_{i=1}^{k} \mathbf{L}^{(i)}\right)\mathbf{P}^{(k)} = 1, \tag{7.21}$$

which, of course, is equivalent to

$$\sum_{i=1}^{N^k} \mathbf{P}_i^{(k)} = 1. \tag{7.22}$$

Having the consistency conditions written in a matrix form, let us turn our attention to the problem of determining which block probabilities are essential in order to determine the probability measure on $\mathcal{A}^{\mathbb{Z}}$ and which ones can be computed from consistency conditions. The next proposition provides important first step in answering this question.

Proposition 7.4 *Among all block probabilities constituting components of* $\mathbf{P}^{(1)}, \mathbf{P}^{(2)}, \ldots, \mathbf{P}^{(k)}$ *only* $(N-1)N^{k-1}$ *are linearly independent.*

Proof Let us first note that vector $\mathbf{P}^{(i)}$ has N^i components. Collectively, in $\mathbf{P}^{(1)}, \mathbf{P}^{(2)}, \ldots, \mathbf{P}^{(k)}$ we have, therefore, $\sum_{i=1}^{k} N^i = (N^{k+1} - N)/(N-1)$ block probabilities. However, since all $\mathbf{P}^{(i)}$, $i \in [1, k)$, can be expressed in terms of $\mathbf{P}^{(k)}$ with the help of Eq. (7.19), we can treat all of $\mathbf{P}^{(1)}, \mathbf{P}^{(2)}, \ldots, \mathbf{P}^{(k-1)}$ as dependent. This leaves us with $\mathbf{P}^{(k)}$ with N^k components. However, we also have

$$\mathbf{L}^{(k)}\mathbf{P}^{(k)} = \mathbf{R}^{(k)}\mathbf{P}^{(k)}. \tag{7.23}$$

Matrices in the above have N^{k-1} rows, thus we have N^{k-1} equations for N^k variables. Are they all these equations independent? Both L and R have the property that sum of each of their columns is 1. Thus if we add all equations of (7.23), we obtain identity $\sum \mathbf{P}^{(k)} = \sum \mathbf{P}^{(k)}$, meaning that the number of independent equations in Eq. (7.23) is $N^{k-1} - 1$. All of this takes care of consistency conditions (7.14), but we also need to consider normalization condition (7.15) which, as remarked earlier, can be written in equivalent form as equation involving components of $\mathbf{P}^{(k)}$, that is, Eq. (7.22). This additional equation increases our previously obtained number of independent equations back to N^{k-1}. In the end, the number of independent block probabilities, equal to number of variables minus number of independent equations, is $N^k - N^{k-1} = (N-1)N^{k-1}$. □

Once we know how many independent block probabilities are there, we can express the remaining block probabilities in terms of them. We need to choose which block probabilities are deemed to be independent, and we have some freedom in doing so. In what follows we will describe two natural choices, the long block representation and the short block representation.

7.5 The Long Block Representation

A particular choice of independent blocks of probabilities will be subsequently called a *representation*, and the probabilities selected as independent in that representation will be called *fundamental block probabilities*. The first representation we will discuss is constructed as follows. Among various possible choices of independent block probabilities from $\mathbf{P}^{(1)}, \mathbf{P}^{(2)}, \ldots, \mathbf{P}^{(k)}$, one possible choice is to select the longest blocks, that is, blocks of $\mathbf{P}^{(k)}$. There are N^k of them, which is too many, thus we will simply choose the first $(N-1)N^{k-1}$. Note that when probabilities of $\mathbf{P}^{(k)}$ are sorted in lexicographical order, this implies that he fundamental blocks are those which do not start from the last symbol of the alphabet.

Definition 7.2 Let $\mathcal{A} = \{0, 1, \ldots, N-1\}$. In the *long block representation* (LBR) of probabilities $\mathbf{P}^{(1)}, \mathbf{P}^{(2)}, \ldots, \mathbf{P}^{(k)}$ we select as fundamental the probabilities in $\mathbf{P}^{(k)}$ corresponding to blocks which do not start with $N-1$.

Example 7.2 For $\mathcal{A} = \{0, 1, 2\}$ and $k = 3$, among block probabilities $\mathbf{P}^{(1)}, \mathbf{P}^{(2)}, \mathbf{P}^{(3)}$ we have $3^3 - 3^2 = 18$ fundamental ones, probabilities of blocks of length 3 which do not start with 2. These are

$$\begin{aligned}\{&P(000), P(001), P(002), P(010), P(011), P(012), P(020), P(021),\\ &P(022), P(100), P(101), P(102), P(110), P(111), P(112), P(120),\\ &P(121), P(122)\}.\end{aligned}$$

The remaining 21 block are 3-block probabilities which start with 2, all 2-block probabilities and all 1-block probabilities. These are

$$\begin{aligned}\{&P(200), P(201), P(202), P(210), P(211), P(212), P(220), P(221),\\ &P(222), P(00), P(01), P(02), P(10), P(11), P(12), P(20), P(21),\\ &P(22), P(0), P(1), P(2)\},\end{aligned}$$

and they are expressible in terms of the fundamental block probabilities of LBR.

We will now show how to express the non-fundamental blocks in terms of fundamental ones. We need to introduce some additional notation first. As explained in the proof of Proposition 7.4, in the system of equations $\mathbf{R}^{(k)}\mathbf{P}^{(k)} = \mathbf{L}^{(k)}\mathbf{P}^{(k)}$ only $N^{k-1} - 1$ equations are independent. We can, therefore, remove one of them, for example, the last equation, and replace it by normalization condition $\sum \mathbf{P}^{(k)} = 1$. This will result in

$$\mathbf{M}^{(k)}\mathbf{P}^{(k)} = \begin{bmatrix} 0 \\ \vdots \\ 0 \\ 1 \end{bmatrix}, \tag{7.24}$$

where the matrix $\mathbf{M}^{(k)}$ has been obtained from $\mathbf{R}^{(k)} - \mathbf{L}^{(k)}$ by setting every entry in the last row of $\mathbf{R}^{(k)} - \mathbf{L}^{(k)}$ to 1. Let us now partition $\mathbf{M}^{(k)}$ into two submatrices, so that the first $N^k - N^{k-1}$ columns of it are called $\mathbf{A}^{(k)}$, and the remaining N^{k-1} columns are called $\mathbf{B}^{(k)}$, so that

$$\mathbf{M}^{(k)} = [\mathbf{A}^{(k)}\mathbf{B}^{(k)}]. \tag{7.25}$$

Proposition 7.5 *Let $\mathbf{P}^{(k)}$ be partitioned into two subvectors, $\mathbf{P}^{(k)} = (\mathbf{P}^{(k)}_{Top}, \mathbf{P}^{(k)}_{Bot})$, where $\mathbf{P}^{(k)}_{Top}$ contains first $N^k - N^{k-1}$ entries of $\mathbf{P}^{(k)}$, and $\mathbf{P}^{(k)}_{Bot}$ the remaining N^{k-1} entries. Then*

$$\mathbf{P}^{(k)}_{Bot} = \begin{bmatrix} 0 \\ \vdots \\ 0 \\ 1 \end{bmatrix} - \left(\mathbf{B}^{(k)}\right)^{-1} \mathbf{A}^{(k)} \mathbf{P}^{(k)}_{Top}. \tag{7.26}$$

Proof We want to solve

$$[\mathbf{A}^{(k)} \mathbf{B}^{(k)}] \begin{bmatrix} \mathbf{P}^{(k)}_{Top} \\ \mathbf{P}^{(k)}_{Bot} \end{bmatrix} = \begin{bmatrix} 0 \\ \vdots \\ 0 \\ 1 \end{bmatrix} \tag{7.27}$$

for $\mathbf{P}^{(k)}_{Bot}$. Denoting the vector on the right hand side by $\mathbf{c}$ and performing block multiplication we obtain $\mathbf{A}^{(k)}\mathbf{P}^{(k)}_{Top} + \mathbf{B}^{(k)}\mathbf{P}^{(k)}_{Bot} = \mathbf{c}$. The matrix $\mathbf{B}^{(k)}$ is always invertible, and has the property $(\mathbf{B}^{(k)})^{-1}\mathbf{c} = \mathbf{c}$. This leads to $\mathbf{P}^{(k)}_{Bot} = \mathbf{c} - \left(\mathbf{B}^{(k)}\right)^{-1} \mathbf{A}^{(k)}\mathbf{P}^{(k)}_{Top}$, as desired. □

Proposition 7.5 allows us to express non-fundamental block probabilities in $\mathbf{P}^{(k)}$ by the fundamental ones. Probabilities of $\mathbf{P}^{(i)}$ for $i < k$ can be expressed by probabilities of $\mathbf{P}^{(k)}$ by using Eq. (7.19). If such expression includes some k-block probabilities which start with $N - 1$, they can be replaced by fundamental ones by applying Eq. (7.26) again.

Example 7.3 Let us consider $N = 2$ and $k = 3$. Among $\mathbf{P}^{(1)}, \mathbf{P}^{(2)}, \mathbf{P}^{(3)}$ we have only $2^3 - 2^2 = 4$ fundamental blocks, $P(000)$, $P(001)$, $P(010)$ and $P(011)$. The remaining 3-block probabilities can be expressed using Eq. (7.26), as follows. We construct matrices $\mathbf{L}^{(3)}$ and $\mathbf{R}^{(3)}$ first,

$$\mathbf{L}^{(3)} = \begin{bmatrix} 1\,0\,0\,0\,1\,0\,0\,0 \\ 0\,1\,0\,0\,0\,1\,0\,0 \\ 0\,0\,1\,0\,0\,0\,1\,0 \\ 0\,0\,0\,1\,0\,0\,0\,1 \end{bmatrix}, \quad \mathbf{R}^{(3)} = \begin{bmatrix} 1\,1\,0\,0\,0\,0\,0\,0 \\ 0\,0\,1\,1\,0\,0\,0\,0 \\ 0\,0\,0\,0\,1\,1\,0\,0 \\ 0\,0\,0\,0\,0\,0\,1\,1 \end{bmatrix}.$$

Matrix $\mathbf{M}^{(3)}$ is then formed by taking $\mathbf{L}^{(3)} - \mathbf{R}^{(3)}$ and setting all elements of the bottom row to 1,

$$\mathbf{M}^{(3)} = \begin{bmatrix} 0 & -1 & 0 & 0 & 1 & 0 & 0 & 0 \\ 0 & 1 & -1 & -1 & 0 & 1 & 0 & 0 \\ 0 & 0 & 1 & 0 & -1 & -1 & 1 & 0 \\ 1 & 1 & 1 & 1 & 1 & 1 & 1 & 1 \end{bmatrix}$$

Matrices $\mathbf{A}^{(3)}$ and $\mathbf{B}^{(3)}$ are now, respectively, first four and last four columns of $\mathbf{M}^{(3)}$,

$$\mathbf{A}^{(3)} = \begin{bmatrix} 0 & -1 & 0 & 0 \\ 0 & 1 & -1 & -1 \\ 0 & 0 & 1 & 0 \\ 1 & 1 & 1 & 1 \end{bmatrix}, \quad \mathbf{B}^{(3)} = \begin{bmatrix} 1 & 0 & 0 & 0 \\ 0 & 1 & 0 & 0 \\ -1 & -1 & 1 & 0 \\ 1 & 1 & 1 & 1 \end{bmatrix}.$$

Equation (7.26) becomes

$$\begin{bmatrix} P(100) \\ P(101) \\ P(110) \\ P(111) \end{bmatrix} = \begin{bmatrix} 0 \\ 0 \\ 0 \\ 1 \end{bmatrix} - \begin{bmatrix} 0 & -1 & 0 & 0 \\ 0 & 1 & -1 & -1 \\ 0 & 0 & 0 & -1 \\ 1 & 1 & 2 & 3 \end{bmatrix} \begin{bmatrix} P(000) \\ P(001) \\ P(010) \\ P(011) \end{bmatrix}, \tag{7.28}$$

yielding the desired expression for non-fundamental 3-block probabilities in terms of fundamental ones.

For 2-blocks, Eq. (7.19) yields

$$\mathbf{P}^{(2)} = \mathbf{L}^{(3)}\mathbf{P}^{(3)},$$

hence

$$\begin{bmatrix} P(00) \\ P(01) \\ P(10) \\ P(11) \end{bmatrix} = \begin{bmatrix} 1&0&0&0&1&0&0&0 \\ 0&1&0&0&0&1&0&0 \\ 0&0&1&0&0&0&1&0 \\ 0&0&0&1&0&0&0&1 \end{bmatrix} \begin{bmatrix} P(000) \\ P(001) \\ P(010) \\ P(011) \\ P(100) \\ P(101) \\ P(110) \\ P(111) \end{bmatrix}.$$

The last four entries of the 3-block probability vector, $P(100)$, $P(101)$, $P(110)$ and $P(111)$, need to be replaced by appropriate expressions involving only fundamental probabilities, given by Eq. (7.28). For 1 blocks the procedure is similar, with Eq. (7.19) yielding

$$\mathbf{P}^{(1)} = \mathbf{L}^{(2)}\mathbf{L}^{(3)}\mathbf{P}^{(3)}, \tag{7.29}$$

where

$$\mathbf{L}^{(2)} = \begin{bmatrix} 1&0&1&0 \\ 0&1&0&1 \end{bmatrix}.$$

Again, the replacement of non-fundamental blocks probabilities on the right hand side of Eq. (7.29) by appropriate expressions involving only fundamental probabilities is necessary. We will not go into all details of the tedious algebra, presenting only the final results for all non-fundamental components of $\mathbf{P}^{(1)}$, $\mathbf{P}^{(2)}$ and $\mathbf{P}^{(3)}$,

$$\begin{bmatrix} P(100) \\ P(101) \\ P(110) \\ P(111) \end{bmatrix} = \begin{bmatrix} P(001) \\ -P(001) + P(010) + P(011) \\ P(011) \\ 1 - P(000) - P(001) - 2\,P(010) - 3\,P(011) \end{bmatrix},$$

$$\begin{bmatrix} P(00) \\ P(01) \\ P(10) \\ P(11) \end{bmatrix} = \begin{bmatrix} P(000) + P(001) \\ P(010) + P(011) \\ P(010) + P(011) \\ 1 - P(000) - P(001) - 2\,P(010) - 2\,P(011) \end{bmatrix},$$

$$\begin{bmatrix} P(0) \\ P(1) \end{bmatrix} = \begin{bmatrix} P(000) + P(001) + P(010) + P(011) \\ 1 - P(000) - P(001) - P(010) - P(011) \end{bmatrix}. \tag{7.30}$$

7.6 The Short Block Representation

Of course, the long block representation is not the only one possible. We will describe below yet another representation, which is in some sense complementary to the long block one. It declares as independent blocks of shortest possible length, thus it will be called *short block representation*.

Definition 7.3 Let $\mathcal{A} = \{0, 1, \ldots, N-1\}$. In the *short block representation* (SBR) of probabilities $\mathbf{P}^{(1)}, \mathbf{P}^{(2)}, \ldots, \mathbf{P}^{(k)}$ we select as fundamental those entries in $\mathbf{P}^{(1)}, \mathbf{P}^{(2)}, \ldots, \mathbf{P}^{(k)}$ which correspond to blocks neither starting nor ending with $N-1$.

Example 7.4 For $N = 3$ and $k = 3$ we have,

$$\mathbf{P}^{(1)} = [P(0), P(1), P(2)]^T.$$

$$\mathbf{P}^{(2)} = [P(00), P(01), P(02), P(10), P(11), P(12), P(20), P(21), P(22)]^T,$$

$$\begin{aligned} \mathbf{P}^{(3)} = [&P(000), P(001), P(002), P(010), P(011), P(012), P(020), P(021), P(022), \\ &P(100), P(101), P(102), P(110), P(111), P(112), P(120), P(121), P(122), \\ &P(200), P(201), P(202), P(210), P(211), P(212), P(220), P(221), P(222)]^T, \end{aligned}$$

The only entries not starting nor ending with 2 are

$$\begin{aligned} &P(0), P(1), P(00), P(01), P(10), P(11), \\ &\quad P(000), P(001), P(010), P(011), P(020), P(021), \\ &\quad\quad P(100), P(101), P(110), P(111), P(120), P(121), \end{aligned}$$

and we have $3^3 - 3^2 = 18$ of them. These are fundamental probabilities, and all other probabilities of $\mathbf{P}^{(1)}, \mathbf{P}^{(2)}, \mathbf{P}^{(3)}$ can be expressed in terms of them.

A general method for producing such expressions will be illustrated using the above example. We start with block probabilities $\mathbf{P}^{(1)}, \mathbf{P}^{(2)}, \ldots, \mathbf{P}^{(k)}$, and arrange each of the vectors $\mathbf{P}^{(i)}$ in a vertical columns spaced as in Fig. 7.1. From each vector $\mathbf{P}^{(i)}$ we eliminate all entries ending with $N-1$ and starting with $N-1$. In Fig. 7.1, eliminated entries (starting or ending with 2) are underlined and those which are left are boxed. Note that the entries starting with $N-2$ belong to what we called $\mathbf{P}_{Bot}$. The boxed entries are the fundamental ones, and it is straightforward to verify that we have $N^k - N^{k-1}$ of them, as we should.

Now we need to express the underlined entries in terms of the fundamental ones. Figure 7.1 is annotated with arrows and labels which indicates how to do this. In each vector, starting from the left, we can express each underlined entry by a linear combinations of boxed entries from the same column and (possibly) entries from the column on the left hand side, by following the path which starts with $\longmapsto$ arrow and which ends at the underlined entry.

For example, $P(02)$ is underlined, thus non-fundamental. We identify a path which starts at $P(0)$ and ends at $P(02)$

$$P(0) \longmapsto P(00) \to P(01) \to P(02),$$

with arrow labels indicating how to construct the required equation,

$$P(0) - P(00) - P(01) = P(02). \tag{7.31}$$

Note that all arrows are labeled with "−", except the final ones which are labeled with "=". The reason why this works is of course due to consistency conditions. The expression we obtained is simply

$$P(0) = P(00) + P(01) + P(02), \tag{7.32}$$

solved for $P(02)$. We need to repeat the above procedure for all entries underlined with the red color in $\mathbf{P}^{(1)}$, then $\mathbf{P}^{(2)}$, etc. Once we are done, we can express entries marked with blue underline by fundamental block probabilities using Eq. (7.26).

In order to describe the algorithm of Fig. 7.1 in a more formal way, let us define vector of *admissible entries* for short block representation, $\mathbf{P}^{(k)}_{adm}$, as follows. Start with a vector $\mathbf{P}^{(k)}$ in which block probabilities are arranged in lexicographical order, indexed by an index i which runs from 1 to N^k. Vector $\mathbf{P}^{(k)}_{adm}$ consists of all entries which do not start with $N-1$ and which do not end with $N-1$. Vector of fundamental block probabilities in short block representation is now defined as

$$\mathbf{P}^{(k)}_{fund} = \begin{bmatrix} \mathbf{P}^{(1)}_{adm} \\ \mathbf{P}^{(2)}_{adm} \\ \vdots \\ \mathbf{P}^{(k)}_{adm} \end{bmatrix}. \tag{7.33}$$

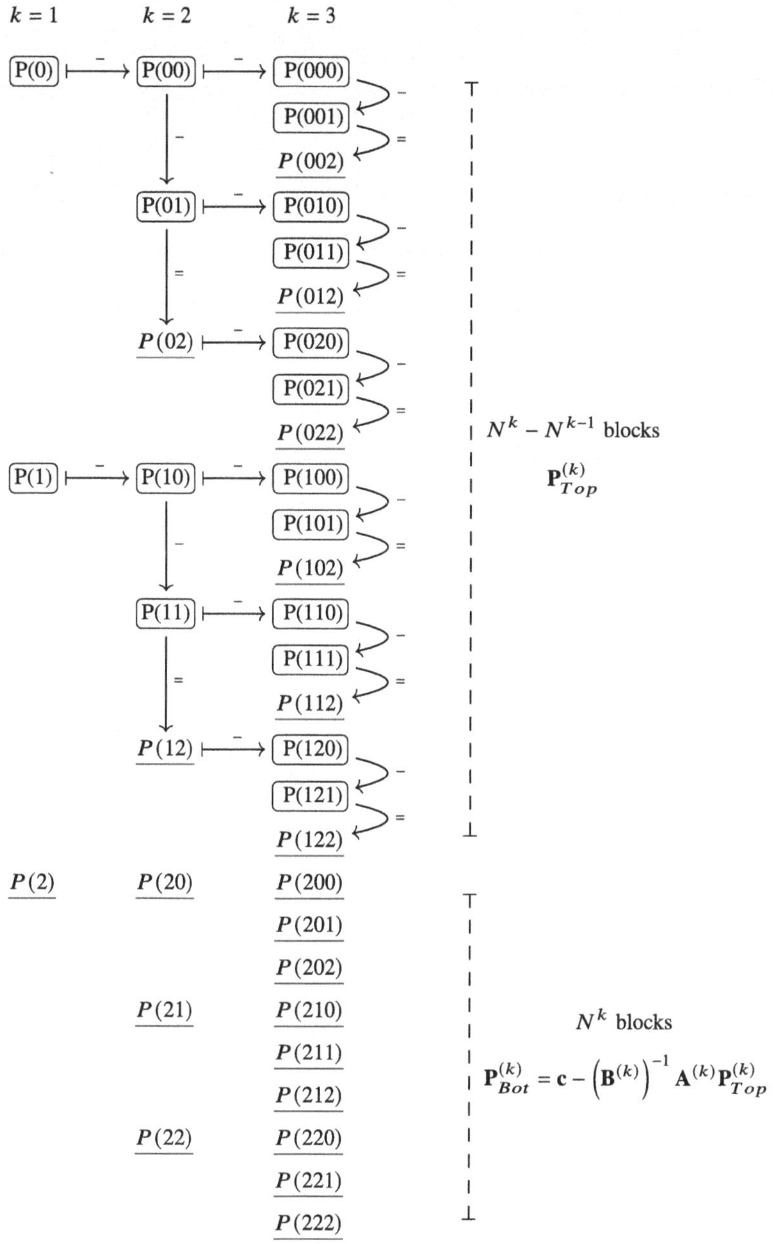

Fig. 7.1 Construction of the short block representation for $N = 3$ and $\mathbf{P}^{(k)}$ for $k = 1, 2, 3$. Fundamental block probabilities are boxed and non-fundamental ones are underlined

Note that the length of $\mathbf{P}^{(k)}_{fund}$ is the same as $\mathbf{P}^{(k)}_{Top}$ in the long block representation. We can, therefore, transform one into the other by a linear transformation. The form of this transformation can be deduced from Fig. 7.1. Consider, for example, $k = 2$, so that $\mathbf{P}^{(2)}_{Top} = [P(00), P(01), P(02), P(10), P(11), P(12)]^T$ and $\mathbf{P}^{(2)}_{adm} = [P(00), P(01), P(10), P(11)]^T$, $\mathbf{P}^{(1)}_{Top} = \mathbf{P}^{(1)}_{adm} = [P(0), P(1)]^T$. From Fig. 7.1, following appropriate paths, we obtain

$$\begin{aligned} P(00) &= P(00), \\ P(01) &= P(01), \\ P(02) &= P(0) - P(00) - P(01), \\ P(10) &= P(00), \\ P(11) &= P(01), \\ P(12) &= P(1) - P(10) - P(11), \end{aligned} \tag{7.34}$$

where, if an element $P(\mathbf{b})$ of $\mathbf{P}^{(2)}_{Top}$ was admissible, we wrote $P(\mathbf{b}) = P(\mathbf{b})$. This can be written in a matrix form,

$$\mathbf{P}^{(2)}_{Top} = \begin{bmatrix} 0\,0 \\ 0\,0 \\ 1\,0 \\ 0\,0 \\ 0\,0 \\ 0\,1 \end{bmatrix} \mathbf{P}^{(1)}_{adm} + \begin{bmatrix} 1 & 0 & 0 & 0 \\ 0 & 1 & 0 & 0 \\ -1 & -1 & 0 & 0 \\ 0 & 0 & 1 & 0 \\ 0 & 0 & 0 & 1 \\ 0 & 0 & -1 & -1 \end{bmatrix} \mathbf{P}^{(2)}_{adm}. \tag{7.35}$$

One can similarly show that for general $k > 1$,

$$\mathbf{P}^{(k)}_{Top} = \mathbf{C}^{(k)} \mathbf{P}^{(k-1)}_{Top} + \mathbf{D}^{(k)} \mathbf{P}^{(k)}_{adm}, \tag{7.36}$$

where

$$\mathbf{C}^{(k)} = \mathrm{diag}(\underbrace{\mathbf{e}_N, \mathbf{e}_N, \ldots, \mathbf{e}_N}_{N^{k-1}-N^{k-2}}), \quad \mathbf{e}_N = \left.\begin{bmatrix} 0 \\ \vdots \\ 0 \\ 1 \end{bmatrix}\right\} N \tag{7.37}$$

$$\mathbf{D}^{(k)} = \mathrm{diag}(\underbrace{\mathbf{D}_N, \mathbf{D}_N, \ldots, \mathbf{D}_N}_{N^{k-1}-N^{k-2}}), \quad \mathbf{D}_N = \underbrace{\begin{bmatrix} \mathbf{I}_{N-1} \\ -1, -1, \ldots, -1 \end{bmatrix}}_{N-1}. \tag{7.38}$$

Note that $\mathbf{C}^{(k)}$ has $N^k - N^{k-1}$ rows and $N^{k-1} - N^{k-2}$ columns, while $\mathbf{D}^{(k)}$ has $N^k - N^{k-1}$ rows and $(N^{k-1} - N^{k-2})(N-1)$ columns. Applying Eq. (7.36) $k-1$ times recursively, one obtains the desired expression of blocks in $\mathbf{P}^{(k)}_{Top}$ by fundamental block probabilities. $\mathbf{P}^{(k)}_{Bot}$ can then be obtained using Eq. (7.19). Let us summarize our findings.

Proposition 7.6 *Among all block probabilities constituting components of* $\mathbf{P}^{(1)}$, $\mathbf{P}^{(2)}, \ldots, \mathbf{P}^{(k)}$, *we can treat probabilities of blocks which do not end with* $N-1$ *and do not start with* $N-1$ *as fundamental probabilities, forming vector* $\mathbf{P}^{(k)}_{fund}$. *We can then express the first* $N^k - N^{k-1}$ *components of* $\mathbf{P}^{(k)}$ *by* $\mathbf{P}^{(k)}_{fund}$ *by means of Eq. (7.36) applied recursively. Remaining components of* $\mathbf{P}^{(k)}$ *can be obtained by using Eq. (7.26), while* $\mathbf{P}^{(1)}, \mathbf{P}^{(2)}, \ldots, \mathbf{P}^{(k-1)}$ *can be obtained by Eq. (7.19).*

Example 7.5 Let us now illustrate this procedure with $N = 2$ and $k = 3$ case. Among components of $\mathbf{P}^{(1)}$, $\mathbf{P}^{(2)}$ and $\mathbf{P}^{(3)}$ we have only four fundamental block probabilities, $\mathbf{P}^{(3)}_{fund} = [P(0), P(00), P(000), P(010)]^T$, and 10 other dependent probabilities. We first partition $\mathbf{P}^{(3)}$ into two subvectors, $\mathbf{P}^{(3)}_{Top} = [P(000), P(001), P(010), P(011)]^T$ and $\mathbf{P}^{(3)}_{Bot} = [P(100), P(101), P(110), P(111)]^T$. We now write Eq. (7.36) for $k = 3$ and $k = 2$,

$$\begin{aligned}\mathbf{P}^{(3)}_{Top} &= \mathbf{C}^{(3)}\mathbf{P}^{(2)}_{Top} + \mathbf{D}^{(3)}\mathbf{P}^{(3)}_{adm},\\ \mathbf{P}^{(2)}_{Top} &= \mathbf{C}^{(2)}\mathbf{P}^{(1)}_{Top} + \mathbf{D}^{(2)}\mathbf{P}^{(2)}_{adm},\end{aligned}$$

therefore

$$\begin{aligned}\mathbf{P}^{(3)}_{Top} &= \mathbf{C}^{(3)}\left(\mathbf{C}^{(2)}\mathbf{P}^{(1)}_{Top} + \mathbf{D}^{(2)}\mathbf{P}^{(2)}_{adm}\right) + \mathbf{D}^{(3)}\mathbf{P}^{(3)}_{adm}\\ &= \mathbf{C}^{(3)}\mathbf{C}^{(2)}\mathbf{P}^{(1)}_{adm} + \mathbf{C}^{(3)}\mathbf{D}^{(2)}\mathbf{P}^{(2)}_{adm} + \mathbf{D}^{(3)}\mathbf{P}^{(3)}_{adm},\end{aligned} \tag{7.39}$$

where we used the fact that $\mathbf{P}^{(1)}_{Top} = \mathbf{P}^{(1)}_{adm}$. The matrices and vectors in the above are

$$\mathbf{C}^{(2)} = \begin{bmatrix} 0\\ 1\end{bmatrix},\ \mathbf{D}^{(2)} = \begin{bmatrix} 1\\ -1\end{bmatrix},\ \mathbf{C}^{(3)} = \begin{bmatrix} 0\,0\\ 1\,0\\ 0\,0\\ 0\,1\end{bmatrix},\ \mathbf{D}^{(3)} = \begin{bmatrix} 1 & 0\\ -1 & 0\\ 0 & 1\\ 0 & -1\end{bmatrix},$$

$$\mathbf{P}^{(3)}_{adm} = [P(000), P(010)]^T,\quad \mathbf{P}^{(2)}_{adm} = [P00],\quad \mathbf{P}^{(1)}_{adm} = P[0].$$

Carrying out the required matrix multiplication and addition in Eq. (7.39), we obtain

$$\mathbf{P}^{(3)}_{Top} = \begin{bmatrix} P(000)\\ P(001)\\ P(010)\\ P(011)\end{bmatrix} = \begin{bmatrix} P(000)\\ P(00) - P(000)\\ P(010)\\ P(0) - P(00) - P(010)\end{bmatrix}.$$

Components of $\mathbf{P}^{(3)}_{Bot}$ can now be obtained from Eq. (7.28),

$$\begin{bmatrix} P(100) \\ P(101) \\ P(110) \\ P(111) \end{bmatrix} = \begin{bmatrix} 0 \\ 0 \\ 0 \\ 1 \end{bmatrix} - \begin{bmatrix} 0 & -1 & 0 & 0 \\ 0 & 1 & -1 & -1 \\ 0 & 0 & 0 & -1 \\ 1 & 1 & 2 & 3 \end{bmatrix} \begin{bmatrix} P(000) \\ P(001) \\ P(010) \\ P(011) \end{bmatrix}, \tag{7.40}$$

and by applying Eq. (7.19) we can obtain $\mathbf{P}^{(2)}$ and of $\mathbf{P}^{(1)}$. This will yield the following 10 dependent blocks probabilities expressed in terms of elements of $\mathbf{P}^{(3)}_{fund}$,

$$\begin{aligned}
\begin{bmatrix} P(001) \\ P(011) \\ P(100) \\ P(101) \\ P(110) \\ P(111) \end{bmatrix} &= \begin{bmatrix} P(00) - P(000) \\ P(0) - P(00) - P(010) \\ P(00) - P(000) \\ P(0) - 2P(00) + P(000) \\ P(0) - P(00) - P(010) \\ 1 - 3P(0) + 2P(00) + P(010) \end{bmatrix}, \\
\begin{bmatrix} P(01) \\ P(10) \\ P(11) \end{bmatrix} &= \begin{bmatrix} P(0) - P(00) \\ P(0) - P(00) \\ 1 - 2P(0) + P(00) \end{bmatrix}, \\
P(1) &= 1 - P(0).
\end{aligned} \tag{7.41}$$

We can see that the resulting expressions are shorter than in the case of long block representation given by Eq. (7.30). Moreover, the short block representation has the advantage that when we increase k, $\mathbf{P}^{(k+1)}_{fund}$ will include all entries of $\mathbf{P}^{(k)}_{fund}$, and we can continue *ad infinitum*, obtaining infinite set of fundamental probabilities. This leads to the following corollary.

Corollary 7.2 *Let $\mathcal{A} = \{0, 1, \ldots, N-1\}$. Any shift-invariant probability measure on $\mathcal{A}^{\mathbb{Z}}$ is completely determined by the set of probabilities $P(\mathbf{a})$, $\mathbf{a} \in \mathcal{A}^\star$, such that the first and the last symbol of $\mathbf{a}$ is different than $N-1$.*

Of course the choice of the $N-1$ symbol in our version of the short block representation is completely arbitrary, we could replace the phrase "different than $N-1$" by "different from 0", or "different from a" where a is any other symbol of $\mathcal{A}$, and the Corollary would remain valid. The choice of $N-1$ is just a convention we will follow in the rest of the book.

One more property of the short block representation should be mentioned. The reader can verify that the short block representation utilizes blocks as short as possible, and it is not possible to declare a larger number of short blocks as fundamental. For example, for $N=2$, probability $P(0)$ is fundamental and $P(1)$ is not. We cannot modify the representation by declaring, for instance, that $P(1)$ is fundamental instead of $P(00)$, because $P(0) + P(1) = 1$.

Furthermore, since there are total $\sum_{i=1}^{k} N^i = (N^{k+1} - N)/(N-1)$ block probabilities in $\mathbf{P}^{(1)}, \mathbf{P}^{(2)}, \ldots, \mathbf{P}^{(k)}$, the fraction of fundamental block probabilities among all probabilities of blocks of length up to k is

$$\text{Fund}(N,k) := \frac{(N-1)(N^k - N^{k-1})}{N^{k+1} - N}. \tag{7.42}$$

For fixed N, $\text{Fund}(N,k)$ decreases as a function of k, and tends to the limit

$$\lim_{k\to\infty} \text{Fund}(N,k) = \frac{(N-1)^2}{N^2}, \tag{7.43}$$

so

$$\text{Fund}(N,k) > \frac{(N-1)^2}{N^2} = \left(1 - \frac{1}{N}\right)^2. \tag{7.44}$$

The above reaches minimum $1/4$ at $N = 2$, thus $\text{Fund}(N,k) > 1/4$ for all $k \geq 1$, $N > 1$. This means that a shift-invariant probability measure cannot be determined by less than 25% of all block probabilities, no matter what the size of the alphabet. The binary alphabet, however, is the most "efficient" of all, requiring the smallest percentage of block probabilities to be fundamental.

References

1. Fukś, H.: Construction of local structure maps for cellular automata. J. of Cellular Automata **7**, 455–488 (2013)
2. Fréchet, M.: Sur l'intégrale d'une fonctionnelle étendue á un ensemble abstrait. Bulletin de la S. M. F. **43**, 248–265 (1915)
3. Kolmogorov, A.: Grundbegriffe der Wahrscheinlichkeitsrechnung. Springer-Verlag, Berlin (1933)
4. Hahn, H.: Über die Mutiplikation total-additiver Mengenfunktionen. Annali della Scuola Normale Superiore di Pisa **2**(4), 429–452 (1933)
5. Parthasarathy, K.R.: Introduction to Probability and Measure. Springer-Verlag, New York (1977)

Chapter 8
Probabilistic Solutions

8.1 Motivation

In the theory of discrete dynamical systems, we investigate properties and behaviour of orbits of points under a map f. Orbits are defined as sequences $\{x_0, x_1, x_2, \ldots\}$ where $x_{n+1} = f(x_n)$ and the initial value problem is the problem of expressing x_n explicitly as a function of n and the initial point x_0. Let us consider three examples of such maps, all of them being quadratic functions.

The first one is $f(x) = x^2 + 2x$. The difference equation defining the orbit is

$$x_{n+1} = x_n^2 + 2x_n,$$

and it is easy to verify by direct substitution that it has the solution

$$x_n = (x_0 + 1)^{2^n} - 1.$$

It is an example of a map for which the initial value problem has explicit solution in terms of elementary functions, because we were able to express x_n as a function of n and the initial value x_0.

Such "solvable" maps are, unfortunately, rather rare. Consider another quadratic function, $f(x) = x - x^2$, so that

$$x_{n+1} = x_n - x_n^2.$$

Solution of the above difference equation is not known, but one can do something else. The map f has a fixed point at 0, $f(0) = 0$, for which $|f'(0)| < 1$. One can show that this fixed point is *stable*, meaning that if x_0 is sufficiently close to 0 then

$$x_n \to 0 \text{ as } n \to \infty$$

(see, for example, [1] for details).

H. Fukś, *Solvable Cellular Automata*, Understanding Complex Systems,
https://doi.org/10.1007/978-3-031-38700-5_8

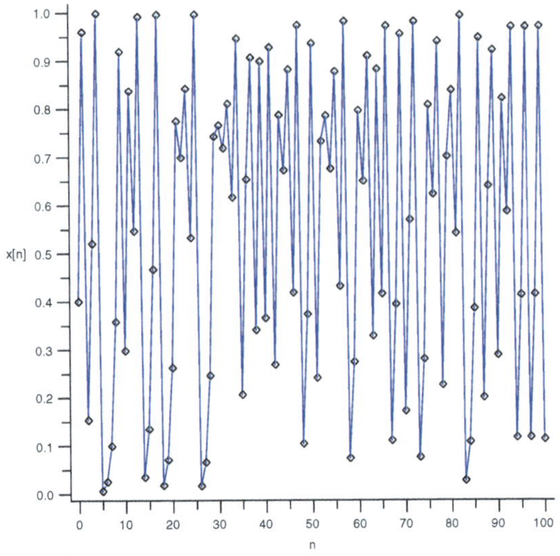

Fig. 8.1 Graph of the orbit of $x_{n+1} = 4x_n - 4x_n^2$

Sometimes, however, even the aforementioned approach fails. Consider $f : [0, 1] \to [0, 1]$ given by $f(x) = 4x - 4x^2$. This map has no stable fixed points, and its orbit is highly irregular, as shown in Fig. 8.1. In this case we can describe the behaviour of the orbit using ergodic theory approach. The idea is to characterize not a single orbit, but a set of many orbits simultaneously. Suppose we start with a set of initial values x_0 distributed evenly in the interval [0, 1]. The distribution of initial points can be visualized by a histogram, and the histogram is initially "flat", as shown in Fig. 8.2a. The bar area in the histogram corresponds to the percentage of points in a given bin. If we apply the map f to all of the initial points, the histogram will no longer be flat, it will look like in Fig. 8.2a. If we keep applying f to all points and observe the histogram after each iteration, we will see that it tends toward a shape resembling the letter "U", as shown in Fig. 8.2c. In fact, the shape of the histogram "stabilizes" rather quickly, changing less and less with each iteration. Its shape after n iterations for very large n will look not much different than as in Fig. 8.2c.

A more formal description of this phenomenon can be constructed as follows. Let $f : [0, 1] \to [0, 1]$, $x_{n+1} = f(x_n)$, and let $\rho_n(x)$ represent the density of points after n-th iterations, so that for small Δx, $\rho_n(x)\Delta x$ is the percentage of points in $[x, x + \Delta x]$. More generally, $\int_a^b \rho_n(x)dx$ represents the percentage of points in the interval $[a, b] \subset [0, 1]$ after n iterations of f. If we start with a finite set of initial points and apply f iteratively, the number of points remains conserved, therefore

$$\text{nr of points } \in [a, b] \text{at step } n + 1 = \text{ nr of points } \in f^{-1}([a, b]) \text{at step } n. \quad (8.1)$$

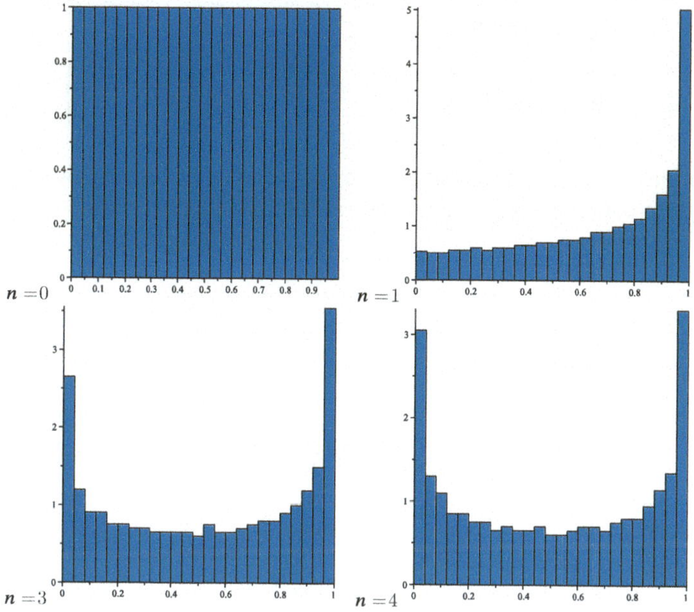

Fig. 8.2 Histograms representing distribution of points after n iterations of the map $f(x) = 4x - 4x^2$

This carries over to the continuous distribution of points with density ρ, thus

$$\int_a^b \rho_{n+1}(y)dy = \int_{f^{-1}([a,b])} \rho_n(y)dy.$$

Taking $[a, b] = [0, x]$, we obtain

$$\int_0^x \rho_{n+1}(y)dy = \int_{f^{-1}([0,x])} \rho_n(y)dy,$$

and after differentiation of both sides with respect to x,

$$\rho_{n+1}(x) = \frac{d}{dx} \int_{f^{-1}([0,x])} \rho_n(y)dy. \tag{8.2}$$

The above equation is a recurrence equation which allows us to construct consecutive ρ_n's if we know ρ_0. It is usually called Ruelle-Frobenius-Perron equation, although it is also known as the equation for pushforward measure, transfer measure, Ruelle-Perron-Frobenius operator, etc.

For $f(x) = 4x - 4x^2$ we have

$$f^{-1}([0,x]) = \left[0, \frac{1}{2} - \frac{1}{2}\sqrt{1-x}\right] \cup \left[\frac{1}{2} + \frac{1}{2}\sqrt{1-x}, 1\right],$$

and, therefore, the Ruelle-Frobenius-Perron equation, after carrying out the integration and differentiation on the right hand side of Eq. (8.2), becomes

$$\rho_{n+1}(x) = \frac{\rho_n\left(\frac{1}{2} + \frac{1}{2}\sqrt{1-x}\right)}{4\sqrt{1-x}} + \frac{\rho_n\left(\frac{1}{2} - \frac{1}{2}\sqrt{1-x}\right)}{4\sqrt{1-x}}.$$

One can show that if $\rho_0 = 1$ then

$$\rho_n(x) \to \frac{1}{\pi\sqrt{x(1-x)}} \text{ as } n \to \infty,$$

and since the graph of $1/\sqrt{x(1-x)}$ has the shape of the letter "U" with arms at $x = 0$ and $x = 1$, this is the reason for the observed behaviour of the histograms in Fig. 8.2. To summarize, although we are unable to describe *individual* orbits of the map $f(x) = 4x - 4x^2$, we can still describe the behaviour of *large assemblies* of its orbits. We will now show how to carry over this approach to orbits of cellular automata.

8.2 Ruelle-Frobenius-Perron Equation for Cellular Automata

Instead of orbits of maps $f : [0,1] \to [0,1]$, our goal is to describe behaviour of orbits of maps $F : \mathcal{A}^{\mathbb{Z}} \to \mathcal{A}^{\mathbb{Z}}$. What is the object corresponding to interval? The cylinder set is a natural candidate. Let $\mathbf{a} = a_1 a_2 \ldots a_m$, $a_i \in \mathcal{A}$ be a block of length m. Recall that the cylinder set generated by $\mathbf{a}$ and and anchored at i is defined as

$$[\mathbf{a}]_i = \{x \in X : x_i = a_1, x_{i+1} = a_2, \ldots x_{i+m-1} = a_m\}.$$

Suppose that we have a large collection of points of X, some of them belonging to set $E \subset X$ and others not. When we apply F to all of them, the same "conservation of points" principle as in Eq. (8.1) applies here, therefore

$$\begin{array}{ccc} \text{fraction of points in } E & = & \text{fraction of points in } F^{-1}(E) \\ \text{after application of } F & & \text{initially} \end{array} \tag{8.3}$$

We will use the term *measure* of E instead of "fraction of points in E". We will assume that the measure is probabilistic and shift-invariant. If μ is such a measure, then Eq. (8.3) becomes

$$(F\mu)(E) = \mu(F^{-1}(E)). \tag{8.4}$$

Here $F\mu$ is the measure describing the distribution of points after F has been applied to all of them. We will say that eq. (8.4) defines *pushback measure* $F\mu$. Replacing μ by $F^n\mu$ on both sides we obtain

$$(F^{n+1}\mu)(E) = (F^n\mu)(F^{-1}(E)), \tag{8.5}$$

which we will call *Ruelle-Frobenius-Perron equation for cellular automata.*

Since we know that the shift-invariant probability measure is completely characterized by probabilities of cylinder sets, we will now assume that E is a cylinder set, $E = [\mathbf{a}]_i$ for some $\mathbf{a} \in \mathcal{A}^{\mathbb{Z}}$. If μ is shift-invariant and probabilistic, then $F^n\mu$ is also shift invariant and probabilistic. The measure of $[\mathbf{a}]_i$ can be interpreted as probability of occurrence of block $\mathbf{a}$ in the configuration, or, in other words, the block probability $P(\mathbf{a})$, as defined in Eq. (7.13). Let us denote the probability of occurrence of $\mathbf{a}$ after n iterations of F by $P_n(\mathbf{a})$,

$$P_n(\mathbf{a}) = F^n\mu([\mathbf{a}]_i).$$

Note that $F^{-1}([\mathbf{a}]_i)$ is the union of cylinder sets generated by preimages of $\mathbf{a}$ under the block mapping $\mathbf{f}$,

$$F^{-1}([\mathbf{a}]_i) = \bigcup_{\mathbf{b}\in\mathbf{f}^{-1}(\mathbf{a})} [\mathbf{b}]_{i-n}.$$

Using this together with the fact that probability of union of disjoint sets is the sum of their probabilities we can rewrite Eq. (8.5) as

$$P_{n+1}(\mathbf{a}) = \sum_{\mathbf{b}\in\mathbf{f}^{-1}(\mathbf{a})} P_n(\mathbf{b}). \tag{8.6}$$

This is the Ruelle-Frobenius-Perron equation for cellular automata expressed in terms of block probabilities.

Example 8.1 For rule 14 we have $\mathbf{f}_{14}^{-1}(0) = \{000, 100, 101, 110, 111\}$ and $\mathbf{f}_{14}^{-1}(1) = \{001, 010, 011\}$, therefore

$$\begin{aligned} P_{n+1}(0) &= P_n(000) + P_n(100) + P_n(101) + P_n(110) + P_n(111), \\ P_{n+1}(1) &= P_n(001) + P_n(010) + P_n(011). \end{aligned} \tag{8.7}$$

We can similarly construct the Ruelle-Frobenius-Perron equation for blocks of length 2,

$$\begin{aligned}
P_{n+1}(11) &= P_n(0010) + P_n(0011),\\
P_{n+1}(00) &= P_n(0000) + P_n(1000) + P_n(1100) + P_n(1101) + P_n(1110) + P_n(1111),\\
P_{n+1}(01) &= P_n(0001) + P_n(1001) + P_n(1010) + P_n(1011),\\
P_{n+1}(10) &= P_n(0100) + P_n(0101) + P_n(0110) + P_n(0111),
\end{aligned} \tag{8.8}$$

and length 3,

$$\begin{aligned}
P_{n+1}(000) &= P_n(00000) + P_n(10000) + P_n(11000) + P_n(11100) + P_n(11101)\\
&\quad + P_n(11110) + P_n(11111),\\
P_{n+1}(001) &= P_n(00001) + P_n(10001) + P_n(11001) + P_n(11010) + P_n(11011),\\
P_{n+1}(010) &= P_n(10100) + P_n(10101) + P_n(10110) + P_n(10111),\\
P_{n+1}(011) &= P_n(00010) + P_n(00011) + P_n(10010) + P_n(10011),\\
P_{n+1}(100) &= P_n(01000) + P_n(01100) + P_n(01101) + P_n(01110) + P_n(01111),\\
P_{n+1}(101) &= P_n(01001) + P_n(01010) + P_n(01011),\\
P_{n+1}(110) &= P_n(00100) + P_n(00101) + P_n(00110) + P_n(00111),\\
P_{n+1}(111) &= 0.
\end{aligned} \tag{8.9}$$

Note that the last equation reflects the fact that block 111 belongs to the garden of Eden for rule 14, as discussed in Example 1.8.

Equations (8.7–8.9) describe evolution of probabilities of all 14 blocks of length of up to 3 under the action of rule 14. We know, however, that these 14 probabilities are not independent, and they can all be expressed by only 4 probabilities $P_n(0)$, $P_n(00)$, $P_n(000)$ and $P_n(010)$ using the short block representation given in Eq. (7.41). Using Eq. (7.41), our 14 equations reduce to just four,

$$\begin{aligned}
P_{n+1}(0) &= 1 - P_n(0) + P_n(000), \qquad (8.10)\\
P_{n+1}(00) &= 1 - 2P_n(0) + P_n(00) + P_n(000),\\
P_{n+1}(000) &= 1 - 3P_n(0) + 2P_n(00) + P_n(000) + P_n(010) - P_n(01000),\\
P_{n+1}(010) &= P_n(0) - 2P_n(00) + P_n(000).
\end{aligned}$$

Note that the above system of difference equations cannot be solved, because it includes $P_n(01000)$ for which P_{n+1} is not given. Of course we could find preimages of 01000 and write additional equation for $P_{n+1}(01000)$, yet other new probabilities would appear in it. This shows that, in general, block probabilities cannot be computed by iteration of a finite-dimensional system of equations.

The system of Ruelle-Frobenius-Perron equations (abbr. RFB) written for blocks of length k and then reduced to fundamental block using short block representation, as in the above example, will be subsequently called the *Ruelle-Frobenius-Perron equations of order k*.

In the Appendix D we show RFB equations of order 3 for all minimal elementary CA. Most of them cannot be solved by reduction to a finite-dimensional system, but there are some exceptions, as the following example illustrates.

Example 8.2 For rule 2 the procedure outlined in the previous example yields

$$\begin{aligned} P_{n+1}(0) &= P_n(000) - P_n(00) + 1, \\ P_{n+1}(00) &= 2P_n(000) - 2P_n(00) + 1, \\ P_{n+1}(000) &= 1 + 3P_n(000) - 3P_n(00), \\ P_{n+1}(010) &= -P_n(000) + P_n(00). \end{aligned}$$

We can see that this is a system of four linear difference equations, thus it can be easily solved by standard methods [2], yielding expressions for P_n in terms of P_0,

$$\begin{aligned} P_n(0) &= -P_0(00) + P_0(000) + 1, \\ P_n(00) &= -2P_0(00) + 2P_0(000) + 1 \\ P_n(000) &= -3P_0(00) + 3P_0(000) + 1, \\ P_n(010) &= P_0(00) - P_0(000). \end{aligned}$$

Note that these expressions do not depend on n, reflecting the fact that rule 2 emulates shift, $F_2^2 = F_{170}F_2$. After the first iteration it behaves as shift, and shift cannot change block probabilities (the probability measure is shift-invariant).

If the reduction to finite-dimensional system is not possible (and in most cases it is not), we can do something else. Iterative application of Eq. (8.6) leads to the following expression for block probabilities after n iterations in terms of initial block probabilities,

$$P_n(\mathbf{a}) = \sum_{\mathbf{b} \in \mathbf{f}^{-n}(\mathbf{a})} P_0(\mathbf{b}). \tag{8.11}$$

This form of the Ruelle-Frobenius-Perron equation will be useful in solving the probabilistic initial value problem which we will shortly formulate.

8.3 Orbits of Measures Under the Action of Cellular Automata

Definition 8.1 Let μ be a shift invariant probability measure. The *orbit of μ under the action of cellular automaton F* is the sequence

$$\{\mu, F\mu, F^2\mu \ldots\}$$

where, for a measurable set E, the measure $F^n\mu$ is defined as

$$(F^n\mu)(E) = \mu(F^{-n}(E)).$$

By the virtue of Theorem 7.2, in order to fully characterize consecutive iterates of μ under F, all we need to know are the block probabilities $P_n(\mathbf{a})$. We can, therefore, pose a general problem as follows.

Definition 8.2 Let $F : \mathcal{A}^{\mathbb{Z}} \to \mathcal{A}^{\mathbb{Z}}$ be the global function of a cellular automaton. The *probabilistic initial value problem* for F is the problem of finding block probabilities $P_n(\mathbf{a})$ for all $\mathbf{a} \in \mathcal{A}^\star$ and all $n > 0$, assuming that all initial block probabilities $P_0(\mathbf{a})$ are given. The *k-th order probabilistic initial value problem* for F is the problem of finding block probabilities $P_n(\mathbf{a})$ for all $\mathbf{a} \in \mathcal{A}^k$ and all $n > 0$, assuming that all initial block probabilities $P_0(\mathbf{a})$ are given.

We will be interested mostly in binary rules and orbits of the Bernoulli measure $\nu_p,\ p \in [0, 1]$, defined in terms of block probabilities

$$P(\mathbf{a}) = \nu_p([\mathbf{a})]) = p^{\#_1(\mathbf{a})}(1-p)^{\#_0(\mathbf{a})},$$

where $\#_s(\mathbf{a})$ denotes the number of symbols s in the string $\mathbf{a}$. The reason why this measure is of particular interest is because it represents what is often (imprecisely) simply called "a random configuration", the type of distribution of initial configurations commonly used in models of complex systems. When $p = 1/2$, we will say that the measure $\mu_{1/2}$ is *symmetric* or *uniform*.

For a Bernoulli measure,

$$P_0(\mathbf{a}) = p^{\#_1(\mathbf{a})}(1-p)^{\#_0(\mathbf{a})},$$

therefore Eq. (8.11) becomes

$$P_n(\mathbf{a}) = \sum_{\mathbf{b}\in\mathbf{f}^{-n}(\mathbf{a})} p^{\#_1(\mathbf{b})}(1-p)^{\#_0(\mathbf{b})}. \tag{8.12}$$

The above expression in principle provides the solution of the probabilistic initial value problem for the Bernoulli measure, assuming that we know all the relevant preimages. This is unfortunately rarely practical because the set $\mathbf{f}^{-n}(\mathbf{a})$ is typically very hard to characterize for general $\mathbf{a}$ and arbitrary n. Nevertheless, for rules for which the deterministic initial value problem is solvable, we can obtain "closed form" expressions for $P_n(1)$ and $P_n(0)$, and often also for probabilities of longer blocks, ans we will see in the subsequent sections.

8.4 The First Order Probabilistic Initial Value Problem

One way to think of the space $\{0, 1\}^{\mathbb{Z}}$ equipped with Bernoulli measure ν_p is to consider a bisequence of random variables X_i, identical and identically distributed, such that $Pr(X_i = 1) = p$ and $Pr(X_i = 0) = 1 - p$. For a given $\mathbf{a} \in \{0, a\}^m$ we have

$$
\begin{aligned}
&Pr(X_i X_{i+1} \dots X_{i+m-1} = \mathbf{a}) \\
&\qquad = Pr(X_i = a_1) \cdot Pr(X_{i+1} = a_2) \cdot \dots \cdot Pr(X_{i+m-1} = a_m) \\
&\qquad\qquad = p^{\#_1(\mathbf{a})}(1-p)^{\#_0(\mathbf{a})}. \qquad (8.13)
\end{aligned}
$$

Furthermore, if we denote by $\langle Y \rangle$ the expected value of the binary random variable with respect to ν_p, we have

$$\langle Y \rangle = Pr(Y = 1) \cdot 1 + Pr(Y = 0) \cdot 0,$$

hence

$$\langle Y \rangle = Pr(Y = 1).$$

If we have a solvable cellular automaton and we know the expression for $[F^n(x)]_j$, we can replace all x_i's in the solution formula by random variables X_i, thus making $[F^n(X)]_j$ a binary random variable. Its expected value will then be equal to the probability of occurrence of 1 after n iterations of the rule,

$$P_n(1) = \langle [F^n(X)]_i \rangle, \qquad (8.14)$$

meaning that we can obtain this way the solution of the first order probabilistic initial value problem.

Example 8.3 For rule 200, the solution of the deterministic initial value problem is given by Eq. (2.22),

$$[F_{200}^n(X)]_j = X_{j-1}X_j + X_j X_{j+1} - X_{j-1}X_j X_{j+1}.$$

Applying Eq. (8.14) we obtain

$$P_n(1) = \langle X_{j-1}X_j \rangle + \langle X_j X_{j+1} \rangle - \langle X_{j-1}X_j X_{j+1} \rangle,$$

where we used the fact that the expected value of the sum is the sum of expected values. Since the variables X_{j-1}, X_j, X_{j+1} are all independent, the expected value of each product in the above is the product of expected values,

$$P_n(1) = \langle X_{j-1} \rangle \langle X_j \rangle + \langle X_j \rangle \langle X_{j+1} \rangle - \langle X_{j-1} \rangle \langle X_j \rangle \langle X_{j+1} \rangle,$$

Table 8.1 Elementary rules for which $P_n(0)$ is independent of n

Rule number	$P_n(0)$
0	1
2	$-p^3+2\,p^2-p+1$
4	$-p^3+2\,p^2-p+1$
10	p^2-p+1
12	p^2-p+1
34	p^2-p+1
42	p^3-p+1
76	p^3-p+1
138	$(1-p)(p^2+1)$
170	$1-p$
200	$p^3-2\,p^2+1$
204	$1-p$

and because $\langle X_i\rangle = p$ for every i, this yields

$$P_n(1) = 2p^2 - p^3.$$

Since $P(0) = 1 - P(1)$, we have $P_n(0)$ as well,

$$P_n(0) = 1 - 2p^2 + p^3.$$

As we can see, the probability of 1 for rule 200 is independent of n. There are many other rules with the same property, and the expressions for $P_n(1)$ can be obtained in the same way. These rules are shown in Table 8.1. Note that the table shows only $P_n(0)$. This is because $P_n(0)$ is included in the short block representation, and $P_n(1)$ can easily be computed by $P_n(1) = 1 - P_n(0)$.

Example 8.4 As a slightly more complicated example consider rule 77,

$$[F_{77}^n(X)]_j = x_j + \sum_{r=1}^{n}(-1)^r \prod_{i=-r}^{r} X_{j+i} + \sum_{r=1}^{n}(-1)^{r+1} \prod_{i=-r}^{r} \overline{X}_{j+i}.$$

Taking the expected value we obtain

$$\begin{aligned}
P_n(1) &= \langle X_j\rangle + \sum_{r=1}^{n}(-1)^r \prod_{i=-r}^{r} \langle X_{j+i}\rangle + \sum_{r=1}^{n}(-1)^{r+1} \prod_{i=-r}^{r} \langle \overline{X}_{j+i}\rangle \\
&= p + \sum_{r=1}^{n}(-1)^r\, p^{2r+1} + \sum_{r=1}^{n}(-1)^{r+1}\,(1-p)^{2r+1} \\
&= p - \frac{(-1)^{n+1}\,p^{2n+3} + p^3}{p^2+1} - \frac{(-1)^n\,(1-p)^{2n+3} - (1-p)^3}{p^2-2\,p+2},
\end{aligned}$$

and

$$P_n(0) = 1 - p + \frac{(-1)^{n+1}\, p^{2n+3} + p^3}{p^2+1} + \frac{(-1)^n\,(1-p)^{2n+3} - (1-p)^3}{p^2 - 2p + 2}. \tag{8.15}$$

In a similar fashion we can obtain expressions for $P_n(1)$ and $P_n(0)$ for many rules for which we obtained deterministic solutions in earlier chapters. This includes rules 0, 1, 2, 3, 4, 5, 8, 10, 12, 13, 15, 19, 24, 29, 36, 38, 32, 46, 50, 51, 108, 128, 132, 136, 138, 140, 162, 170, 178 and 204. The resulting expressions, obtained and verified with the help of symbolic algebra software, are shown in the Appendix C. All of the aforementioned rules have deterministic solutions in which only sums and products of independent variables appear. Sometimes, however, the products do not involve only independent variables, and in such case other methods must be used, as we will see in the next section.

8.5 Solutions with Dependencies in Products: Rule 172

As an example of the rule where the expected value of a product cannot be evaluated as a product of expected values, we will consider rule 172 with solution given by Eq. (4.11), which we will reproduce here for $j = 0$,

$$[F_{172}^n(X)]_0 = \overline{X}_{-2}\overline{X}_{-1}X_0 + \left(\overline{X}_{n-2}X_{n-1} + X_{n-2}X_n\right)\prod_{i=-2}^{n-3}(1 - \overline{X}_i\overline{X}_{i+1}).$$

Before we take the expected value of both sides of the above, let us rearrange the first factor,

$$\begin{aligned}&\bar{X}_{n-2}X_{n-1} + X_{n-2}X_n \\ &\quad = (1 - X_{n-2})X_{n-1} + X_{n-2}X_n = X_{n-1} + X_{n-2}(X_n - X_{n-1}).\end{aligned} \tag{8.16}$$

Using this, we obtain

$$\begin{aligned}P_n(1) = \langle \bar{X}_{-2}\bar{X}_{-1}X_0\rangle &+ \left\langle\left(\prod_{i=-2}^{n-3}(1 - \bar{X}_i\bar{X}_{i+1})\right)X_{n-1}\right\rangle \\ &+ \left\langle\left(\prod_{i=-2}^{n-3}(1 - \bar{X}_i\bar{X}_{i+1})\right)X_{n-2}(X_n - X_{n-1})\right\rangle.\end{aligned} \tag{8.17}$$

We now apply the property of the product of independent random variables being equal to the product of their expected values, therefore

$$P_n(1) = \langle \bar{X}_{-2} \rangle \langle \bar{X}_{-1} \rangle \langle X_0 \rangle + \left\langle \prod_{i=-2}^{n-3} (1 - \bar{X}_i \bar{X}_{i+1}) \right\rangle \langle X_{n-1} \rangle$$
$$+ \left\langle \left(\prod_{i=-2}^{n-3} (1 - \bar{X}_i \bar{X}_{i+1}) \right) X_{n-2} \right\rangle \langle X_n - X_{n-1} \rangle. \quad (8.18)$$

Because $\langle X_n - X_{n-1} \rangle = 0$, the second line vanishes, and by substituting $\langle X_i \rangle = p$ we obtain

$$P_n(1) = (1-p)^2 p + p \left\langle \prod_{i=-2}^{n-3} (1 - \bar{X}_i \bar{X}_{i+1}) \right\rangle. \quad (8.19)$$

Note that

$$\left\langle \prod_{i=-2}^{n-3} (1 - \bar{X}_i \bar{X}_{i+1}) \right\rangle \neq \prod_{i=-2}^{n-3} \langle 1 - \bar{X}_i \bar{X}_{i+1} \rangle,$$

because, for example, the factor $(1 - X_0 \overline{X}_1)$ is not independent of $(1 - X_1 \overline{X}_2)$, as the variable X_1 appears in both. We have to devise another method of computing $\left\langle \prod_{i=-2}^{n-3} (1 - \bar{X}_i \bar{X}_{i+1}) \right\rangle$, and we will do it by setting up a recursion and solving it, similarly as done in [3].

Lemma 8.1 *Let $q \in (0, 1)$ and let X_i be independent and identically distributed Bernoulli random variables for $i \in \{0, 1, \ldots, n\}$ such that $Pr(X_i = 1) = p$, $Pr(X_i = 0) = 1 - p$. Then*

$$\left\langle \prod_{i=1}^{n-1} \left(1 - \bar{X}_i \bar{X}_{i+1}\right) \right\rangle = \frac{p}{\lambda_2 - \lambda_1} \left(a_1 \lambda_1^{n-2} + a_2 \lambda_2^{n-2}\right), \quad (8.20)$$

where

$$\lambda_{1,2} = \frac{1}{2} p \pm \frac{1}{2} \sqrt{p(4-3p)}, \quad (8.21)$$

$$a_{1,2} = \left(\frac{p}{2} - 1\right) \sqrt{p\,(4 - 3\,p)} \pm \left(\frac{p^2}{2} - 1\right). \quad (8.22)$$

Proof To prove this lemma, we will setup recursion for two variables defined as

$$U_n = \prod_{i=1}^{n-1} \left(1 - \bar{X}_i \bar{X}_{i+1}\right) \text{ and } V_n = \bar{X}_n \prod_{i=1}^{n-1} \left(1 - \bar{X}_i \bar{X}_{i+1}\right). \quad (8.23)$$

We observe that

$$U_n = U_{n-1}(1 - \bar{X}_{n-1}\bar{X}_n) = U_{n-1} - \bar{X}_n V_{n-1}, \tag{8.24}$$

and

$$V_n = \bar{X}_n U_n = \bar{X}_n U_{n-1}(1 - \bar{X}_{n-1}\bar{X}_n) = \bar{X}_n U_{n-1} - \bar{X}_n V_{n-1}, \tag{8.25}$$

where we used the fact X_n is a Boolean variable for which $\bar{X}_n^2 = \bar{X}_n$. This yields the system of recurrence equations for U_n and V_n,

$$\begin{aligned} U_n &= U_{n-1} - \bar{X}_n V_{n-1}, \\ V_n &= \bar{X}_n U_{n-1} - \bar{X}_n V_{n-1}. \end{aligned} \tag{8.26}$$

We are interested in expected values, and since $\bar{X}_n$ is independent of both U_{n-1} and V_{n-1}, we can write

$$\langle U_n \rangle = \langle U_{n-1} \rangle - \langle \bar{X}_n \rangle \langle V_{n-1} \rangle, \tag{8.27}$$

$$\langle V_n \rangle = \langle \bar{X}_n \rangle \langle U_{n-1} \rangle - \langle \bar{X}_n \rangle \langle V_{n-1} \rangle. \tag{8.28}$$

Using $\langle \bar{X}_n \rangle = 1 - p$ this leads to

$$\begin{bmatrix} \langle U_n \rangle \\ \langle V_n \rangle \end{bmatrix} = M \begin{bmatrix} \langle U_{n-1} \rangle \\ \langle V_{n-1} \rangle \end{bmatrix}, \tag{8.29}$$

where

$$M = \begin{bmatrix} 1 & q - 1 \\ 1 - q & q - 1 \end{bmatrix}. \tag{8.30}$$

This is a system of linear difference equation with solution

$$\begin{bmatrix} \langle U_n \rangle \\ \langle V_n \rangle \end{bmatrix} = M^{n-2} \begin{bmatrix} \langle U_2 \rangle \\ \langle V_2 \rangle \end{bmatrix}. \tag{8.31}$$

We used U_2, V_2 as initial values because U_1, V_1 are undefined. Both $\langle U_2 \rangle$ and $\langle V_2 \rangle$ can be explicitly computed,

$$\langle U_2 \rangle = \langle 1 - \bar{X}_1 \bar{X}_2 \rangle = 1 - (1-p)^2 = 2p - p^2, \tag{8.32}$$

$$\langle V_2 \rangle = \langle \bar{X}_2 (1 - \bar{X}_1 \bar{X}_2) \rangle = \langle \bar{X}_2 - \bar{X}_1 \bar{X}_2 \rangle = 1 - p - (1-p)^2 = p - p^2. \tag{8.33}$$

The solution now becomes

$$\begin{bmatrix} \langle U_n \rangle \\ \langle V_n \rangle \end{bmatrix} = M^{n-2} \begin{bmatrix} 2p - p^2 \\ p - p^2 \end{bmatrix}, \tag{8.34}$$

and the only thing we need to do is to compute M^{n-2}. This can be done by diagonalizing M,

$$M^{n-2} = P \begin{bmatrix} \lambda_1^{n-2} & 0 \\ 0 & \lambda_2^{n-2} \end{bmatrix} P^{-1}, \tag{8.35}$$

where $\lambda_{1,2}$ are eigenvalues of M, as defined in Eq. (8.21), and P is the matrix of eigenvectors of P,

$$P = \begin{bmatrix} \frac{1-q}{1-\lambda_1} & \frac{1-q}{1-\lambda_2} \\ 1 & 1 \end{bmatrix}, P^{-1} = \frac{1}{\lambda_1 - \lambda_2} \begin{bmatrix} -\frac{(\lambda_1-1)(\lambda_2-1)}{-1+q} & \lambda_1 - 1 \\ \frac{(\lambda_1-1)(\lambda_2-1)}{-1+q} & -\lambda_2 + 1 \end{bmatrix}. \tag{8.36}$$

The final formula for the expected values of U_n and V_n is

$$\begin{bmatrix} \langle U_n \rangle \\ \langle V_n \rangle \end{bmatrix} = P \begin{bmatrix} \lambda_1^{n-2} & 0 \\ 0 & \lambda_2^{n-2} \end{bmatrix} P^{-1} \begin{bmatrix} 2p - p^2 \\ p - p^2 \end{bmatrix}. \tag{8.37}$$

By carrying out the multiplications of matrices and simplification we obtain

$$\begin{aligned} \langle U_n \rangle = \frac{p}{\lambda_2 - \lambda_1} \Big(\left(-1 - p + p^2 + 2\lambda_2 - p\lambda_2\right) \lambda_1^{n-2} \\ + \left(1 + p - p^2 - 2\lambda_1 + p\lambda_1\right) \lambda_2^{n-2} \Big), \end{aligned} \tag{8.38}$$

which further simplifies to become Eq. (8.20) if we use expressions for $a_{1,2}$ defined there. □

We can now continue with Eq. (8.19). Using Lemma 8.1 we obtain

$$\left\langle \prod_{i=-2}^{n-3} (1 - \bar{X}_i \bar{X}_{i+1}) \right\rangle = \frac{p}{\lambda_2 - \lambda_1} \left(a_1 \lambda_1^{n-1} + a_2 \lambda_2^{n-1} \right),$$

where the exponent $n-1$ appears in place of $n-2$ because we have one more factor in the product $\prod_{i=-2}^{n-3} (1 - \bar{X}_i \bar{X}_{i+1})$ compared to $\prod_{i=1}^{n-1} (1 - \bar{X}_i \bar{X}_{i+1})$ in the lemma. The final expression for $P_n(1)$ for rule 172 is, therefore,

$$P_n(1) = (1-p)^2 p + \frac{p^2}{\lambda_2 - \lambda_1} \left(a_1 \lambda_1^{n-1} + a_2 \lambda_2^{n-1} \right), \tag{8.39}$$

where

$$\begin{aligned} \lambda_{1,2} &= \frac{1}{2} p \pm \frac{1}{2} \sqrt{p(4-3p)}, \\ a_{1,2} &= \left(\frac{p}{2} - 1 \right) \sqrt{p(4-3p)} \pm \left(\frac{p^2}{2} - 1 \right). \end{aligned}$$

We can put this in a more compact form if we define

$$\alpha_1 = \frac{-p^2 a_1}{\lambda_1(\lambda_2 - \lambda_1)},$$
$$\alpha_2 = \frac{-p^2 a_2}{\lambda_2(\lambda_2 - \lambda_1)},$$

which simplifies to

$$\alpha_{1,2} = -\frac{1}{2}\,p \mp \frac{1}{2}\,\frac{\sqrt{p\,(4-3\,p)}\,p\,(2\,p-3)}{3p-4}.$$

With this notation,

$$P_n(1) = p\,(1-p)^2 - \alpha_1 {\lambda_1}^n - \alpha_2 {\lambda_2}^n, \tag{8.40}$$

and

$$P_n(0) = 1 - p\,(1-p)^2 + \alpha_1 {\lambda_1}^n + \alpha_2 {\lambda_2}^n. \tag{8.41}$$

Note that for $p < 1$ we have $|\lambda_{1,2}| < 1$, therefore

$$\lim_{n\to\infty} P_n(0) = 1 - (1-p)^2 p.$$

Furthermore, for symmetric Bernoulli measure, when $p = 1/2$, Eq. (8.41) becomes

$$P_n(0) = \frac{7}{8} - \left(\frac{1}{4} + \frac{\sqrt{5}}{10}\right)\left(\frac{1}{4} + \frac{\sqrt{5}}{4}\right)^n - \left(\frac{1}{4} - \frac{\sqrt{5}}{10}\right)\left(\frac{1}{4} - \frac{\sqrt{5}}{4}\right)^n. \tag{8.42}$$

Since $\frac{1}{4} + \frac{\sqrt{5}}{4}$ is half of the golden ratio, one can express this in terms of Fibonacci numbers,

$$P_n(0) = \frac{7}{8} - \frac{\mathcal{F}_{n+3}}{2^{n+2}},$$

where $\mathcal{F}_n$ is the n-th Fibonacci number. For the symmetric Bernoulli measure,

$$\lim_{n\to\infty} P_n(0) = \frac{7}{8}.$$

8.6 Additive Rules

For rules 60, 90 and 150 we can obtain the probabilistic solution by a slightly different method. Recall that for rule 60 we have obtained the following solution of the initial value problem in Eq. (3.10),

$$[F_{60}^n(x)]_j = \sum_{\substack{0 \le i \le n \\ \binom{n}{i} \text{ odd}}} x_{i-n+j} \mod 2. \tag{8.43}$$

The number of terms in the above sum can be written as

$$G(n) = \sum_{k=0}^{n} \left(\binom{n}{k} \mod 2 \right).$$

The sequence $G(n)$ is called Gould's sequence,

$$\{G(n)\}_{n=1}^{\infty} = \{2, 2, 4, 2, 4, 4, 8, 2, 4, 4, 8, 4, 8, 8, 16, 2, \ldots\}.$$

The sum on the right hand side of Eq. (8.43) is, therefore, a sum of $G(n)$ independent Bernoulli variables. Since we taking mod 2, the result will be equal to 0 only and only if the number of 1's among x_i's is even. It is well known that the probability of having even number of successes in k Bernoulli trials is $\frac{1}{2} + \frac{1}{2}(1-2p)^n$, therefore

$$P_n(0) = \frac{1}{2} + \frac{1}{2}(1 - 2p)^{G(n)}. \tag{8.44}$$

For rule 90 the solution formula (Eq. 3.11) has the same number of terms as for rule 60,

$$[F_{90}^n(x)]_j = \sum_{\substack{0 \le i \le n \\ \binom{n}{i} \text{ odd}}} x_{2i-n+j} \mod 2,$$

thus the reasoning is identical and the probabilistic solution given in Eq. (8.44) remains valid for rule 90 as well.

For rule 150 we obtained

$$[F_{150}^n(x)]_j = \sum_{\substack{0 \le i \le 2n \\ \binom{n}{i-n}_2 \text{ odd}}} x_{i+j-n} \mod 2.$$

The only difference compared to rules 60 and 90 is that the number of terms in the sum is now equal to

$$R(n) = \sum_{i=0}^{2n} \left(\binom{n}{i-n}_2 \mod 2 \right),$$

$$\{R(n)\}_{n=1}^{\infty} = \{3, 3, 5, 3, 9, 5, 11, 3, 9, 9, 15 \ldots\}.$$

The probabilistic solution formula for rule 150 is, therefore,

$$P_n(0) = \frac{1}{2} + \frac{1}{2} (1 - 2p)^{R(n)}. \tag{8.45}$$

We shall add that the solution formula for rule 164 obtained in Sect. 4.3 also exhibits summation terms with binomial coefficients and modulo 2 operation. The expected value of these terms can be obtained in the same way as for the additive rules discussed here. Without going into details, we will only present the final expression for the probabilistic solution for rule 164,

$$P_n(0) = -\frac{1}{2} \frac{2p^4 - 3p^3 - p - 2}{1 + p} - \frac{3}{2} p^{1+n} + 2 \frac{p^{2+2n}}{1 + p} - \frac{1}{2} p^n (1 - 2p)^{G(n)}. \tag{8.46}$$

The origin of the last term with $(1 - 2p)^{G(n)}$ should become clear to the reader upon close inspection of Eq. (4.27).

8.7 Higher Order Probabilistic Solutions

If we want to find $P_n(\mathbf{a})$ for $|\mathbf{a}| > 1$, the fist thing we should check is whether the reduction of the RFP system of equations to finite-dimensional system is possible. This was done in Example 8.2 for rule 2, for which we obtained expressions for the first four fundamental blocks in therms of their initial values,

$$\begin{aligned}
P_n(0) &= -P_0(00) + P_0(000) + 1, \\
P_n(00) &= -2P_0(00) + 2P_0(000) + 1 \\
P_n(000) &= -3P_0(00) + 3P_0(000) + 1, \\
P_n(010) &= P_0(00) - P_0(000).
\end{aligned}$$

For Bernoulli measure, $P(0) = 1 - p$, $P(00) = (1 - p)^2$, $P(000) = (1 - p)^3$ and $P(010) = p(1 - p)^2$, thus the aforementioned expressions become

$$\begin{aligned}
P_n(0) &= -p^3 + 2\,p^2 - p + 1,\\
P_n(00) &= -2\,p^3 + 4\,p^2 - 2\,p + 1,\\
P_n(000) &= -3\,p^3 + 6\,p^2 - 3\,p + 1,\\
P_n(010) &= p\,(p-1)^2\,.
\end{aligned}$$

This can be done for many other rules possessing finite-dimensional subsystem in their RFP equations. Close inspection of the Appendix D reveals that this is the case for rules 2, 3, 4, 12, 15, 34, 42, 51, 170, 200 and 204.

If the reduction to the finite-dimensional system is not possible, another method may be used. In Sect. 8.4, we presented a method of computing $P_n(1)$ by using the expected value. This can be easily extended to probabilities of longer blocks. As before, let X be a bisequence of random variables X_i, identical and identically distributed, such that $Pr(X_i = 1) = p$ and $Pr(X_i = 0) = 1 - p$. To find the probability of occurrence of block $\mathbf{a}$ of length k in $F^n(X)$, we first find the density polynomial $\Psi_{\mathbf{a}}$. Then we construct the random variable

$$Y = \Psi_{\mathbf{a}}\left([F^n(X)]_j, [F^n(X)]_{j+1}, \ldots [F^n(X)]_{j+k-1}\right). \tag{8.47}$$

This random variable will be equal to 1 if and only if block $\mathbf{a}$ occurs in $F^n(X)$ at the position starting from i, so that $[F^n(X)]_j = \mathbf{a}_1$, $[F^n(X)]_{j+1} = \mathbf{a}_2$, etc. Since Y is a Boolean variable, as already explained in Sect. 8.4, we have $P_n(\mathbf{a}) = \langle Y \rangle$, or more explicitly,

$$P_n(\mathbf{a}) = \langle \Psi_{\mathbf{a}}\left([F^n(X)]_j, [F^n(X)]_{j+1}, \ldots [F^n(X)]_{j+k-1}\right)\rangle. \tag{8.48}$$

Obviously, since everything is shift-invariant, we can use $j = 0$ without the loss of generality. Moreover, because of Corollary 7.2, we only need to know fundamental block probabilities of the short block representation, and subsequently we will mainly show examples of computing such probabilities.

Example 8.5 Let us find $P_n(010)$ for rule 10, for which

$$[F_{10}^n(X)]_j = X_{j+n} - X_{j+n}X_{j+n-2}.$$

The density polynomial for 010 is

$$\Psi_{010}(x_1, x_2, x_3) = (1 - x_1)x_2(1 - x_3),$$

therefore Eq. (8.48) for $j = 0$ becomes

$$\begin{aligned}
P_n(010) = &\left\langle \Psi_{\mathbf{a}}\left([F_{200}^n(X)]_0, [F_{10}^n(X)]_1, [F_{10}^n(X)]_2\right)\right\rangle\\
&= \left\langle (1 - [F_{10}^n(X)]_0)[F_{10}^n(X)]_1(1 - [F_{10}^n(X)]_2)\right\rangle.
\end{aligned}$$

After substituting the expressions for $[F_{10}^n(X)]_j$ we obtain

$$P_n(010) = \big\langle (X_n X_{n-2} - X_n + 1)\,(X_{n+1} - X_{n-1} X_{n+1}) \times (X_n X_{n+2} - X_{n+2} + 1) \big\rangle.$$

Multiplication and simplification using $X_i^2 = X_i$ yields

$$\begin{aligned} P_n(010) = \big\langle & - X_n X_{n-2} X_{n-1} X_{n+1} - X_n X_{n-1} X_{n+1} X_{2+n} \\ & + X_n X_{n-2} X_{n+1} + X_n X_{n-1} X_{n+1} + X_n X_{n+1} X_{n+2} + X_{n-1} X_{n+1} X_{n+2} \\ & - X_n X_{n+1} - X_{n-1} X_{n+1} - X_{n+1} X_{n+2} + X_{n+1} \big\rangle. \end{aligned}$$

The expected value of each term with m variables will be p^m, therefore

$$\begin{aligned} P_n(010) &= -p^4 - p^4 + p^3 + p^3 + p^3 + p^3 - p^2 - p^2 - p^2 + p \\ &= -2p^4 + 4p^3 - 3p^2 + p = p(1-p)(2p^2 - 2p + 1). \end{aligned}$$

Example 8.6 As a second example we will consider $P_n(00)$ for rule 140,

$$[F_{140}^n(X)]_j = \overline{X}_{j-1} X_j + \prod_{i=n-1}^{2n} X_{i-n+j}.$$

The same reasoning as before leads to

$$P_n(00) = \big\langle (1 - [F_{140}^n(X)]_0)(1 - [F_{140}^n(X)]_1) \big\rangle.$$

hence

$$P_n(00) = \left\langle \left(1 - \overline{X}_{-1} X_0 - \prod_{i=n-1}^{2n} X_{i-n}\right)\left(1 - \overline{X}_0 X_1 - \prod_{i=n-1}^{2n} X_{i-n+1}\right) \right\rangle.$$

Before we carry out multiplication, it will be easier to rearrange dummy indices so that the terms in each product have the same index,

$$P_n(00) = \left\langle \left(1 - \overline{X}_{-1} X_0 - \prod_{k=-1}^{n} X_k\right)\left(1 - \overline{X}_0 X_1 - \prod_{k=0}^{n+1} X_k\right) \right\rangle.$$

Let us denote the products by $\Pi_1 = \prod_{k=-1}^{n} X_k$ and $\Pi_2 = \prod_{k=0}^{n+1} X_k$. Now it is clear that some terms will vanish because $\overline{X}_0 X_0 = 0$, and, furthermore,

$$\Pi_1 \Pi_2 = \prod_{k=-1}^{n+1} X_k = \Pi_1 X_{n+1}.$$

Carrying out the multiplication and simplification with the above in mind we obtain

$$P_n(00) = \left\langle \Pi_1 X_{n+1} + \Pi_2 \overline{X}_{-1} - \overline{X}_{-1} X_0 - \overline{X}_0 X_1 - \Pi_1 - \Pi_2 + 1 \right\rangle.$$

Now we can use the fact that $\langle \Pi_{1,2} \rangle = p^{n+2}$, therefore

$$\begin{aligned} P_n(00) =& p^{n+3} + p^{n+2}(1-p) - 2(1-p)p - 2p^{n+2} + 1 \\ =& 2p^2 - 2p + 1 - p^{n+2}. \end{aligned}$$

The method described in the preceding two examples works for many cellular automata rules. Whenever this was possible, the solution formulae obtained this way for $P_n(0)$, $P_n(00)$, $P_n(000)$ and $P_n(010)$ are shown in the Appendix C. In some cases direct computation of the expected values was not possible, and similar recursion like in Sect. 8.5 had to be setup and solved. Again, all cases for which obtaining probabilistic solutions this way is possible are included in the Appendix C.

8.8 Rule 184

For rule 184, the formula which we obtained in Eq. (6.26) is not suitable for direct computation of the expected value of $[F^n_{184}(X)]_j$. We will, therefore, resort to a different method.

Let us consider a sequence of $2n + 1$ independent and identically distributed Bernoulli random variables $X_0, X_1, \ldots, X_{2n}$, such that $Pr(X_i = 1) = p$, $Pr(X_i = 0) = 1 - p$ for all $0 \le i \le 2n$. Construct a bi-infinite periodic sequence such that $X_0, X_1, \ldots, X_{2n}$ is repeated to the right and to the left, so that X_i and X_j are the same random variables (producing the same output) whenever $i = j \mod 2n + 1$. We apply rule 184 n times to X obtaining Y,

$$Y_j = [F^n_{184}(X)]_j.$$

Since rule 184 conserves the number of 1's,

$$\sum_{j=0}^{2n} X_j = \sum_{j=0}^{2n} Y_j,$$

and furthermore

$$\left\langle \sum_{j=0}^{2n} X_j \right\rangle = \left\langle \sum_{j=0}^{2n} Y_j \right\rangle.$$

Each Y_i has the same expected value, because each depends on $2n + 1$ random variables, all independent. The right hand side, therefore, is equal to $(2n + 1)\langle Y_j \rangle = (2n + 1)P_n(1)$. The left hand side is the expected value of the sum of $2n + 1$ Bernoulli variables, thus it is equal to $(2n + 1)p$. We conclude that for rule 184,

$$P_n(1) = p,$$
$$P_n(0) = 1 - p.$$

We will now compute $P_n(11)$ for this rule. Although $P_n(11)$ does not belong to the short block representation of block probabilities, it is more convenient to compute than $P_n(00)$ using the FSM for preimages of 1 which we discussed in Sect. 6.3.

We start by observing that $\mathbf{f}_{184}^{-1}(11) = \{0111, 1011, 1111\}$. This, by inductive reasoning, leads to the conclusion that every preimage of 11 must end with 11. Suppose now that $\mathbf{b} \in \mathbf{f}_{184}^{-n}(11)$. This means that $\mathbf{b}$ must belong to the set of strings of the form $\mathbf{a}1$ where $\mathbf{a} \in \mathbf{f}_{184}^{-n}(1)$, although of course not every string of that form is a preimage of 1. Preimages of $\mathbf{a}$ are represented as paths on the FSM graph shown in Figure 6.4b. Let us rearrange this graph by folding it in half, obtaining Fig. 8.3a. FSM of $\mathbf{a}1$ where $\mathbf{a} \in \mathbf{f}_{184}^{-n}(1)$ is shown in Fig. 8.3b, where we added additional edge to state ㉚ to represent 1 at the end of $\mathbf{a}1$.

Every $\mathbf{b} \in \mathbf{f}_{184}^{-4}(11)$ must be represented by a path on the graph of Fig. 8.3b, but not all paths correspond to valid preimages of 11. We need to make some further modifications to this graph to remove unwanted paths. We are now guaranteed that every block $\mathbf{b}$ accepted by the FSM of Fig. 8.3b yields $\mathbf{f}_{184}^{4}(\mathbf{b}) = 1c$, with some unknown $c \in \{0, 1\}$. This is because $\mathbf{b}$ starts with $\mathbf{a}$, and $\mathbf{f}_{184}^{4}(\mathbf{a}) = 1$, so $\mathbf{f}_{184}^{4}(\mathbf{b})$ must start with 1. How do we ensure that $c = 1$? This will guaranteed if $b_2 b_3 \dots b_{10}$ is a preimage of 1 as well. In terms of paths on the FSM graph, we have to make sure that every path starting from state ① or ② and ending with ㉚ in the graph of Fig. 8.3b is represented by a string which is also accepted by the FSM of Fig. 8.3a. The reader can convince himself that this will happen if we remove double edges from the FSM, resulting in the graph shown in Fig. 8.3c. We already know from Sect. 6.3 that the FSM without double edges represents tail-dense strings (c.f. definition 6.2), therefore we obtain the following result.

Proposition 8.1 *The set of n-step preimages of 11 under the rule 184 consists of tail-dense strings followed by 1.*

We can state the above more formally, namely $\mathbf{a} \in \mathbf{f}_{184}^{-n}(11)$ if and only if $a_{2n+2} = 1$ and

$$\sum_{i=1}^{k} a_{2n+2-i} \geq k/2,$$

for all $k \in \{1, \dots 2n + 1\}$. Note that this inequality can be combined with the requirement that $a_{2n+2} = 1$ if we start the summation with $i = 0$ and and add 1 to the right hand side, yielding

$$\sum_{i=0}^{k} a_{2n+2-i} \geq \frac{k}{2} + 1.$$

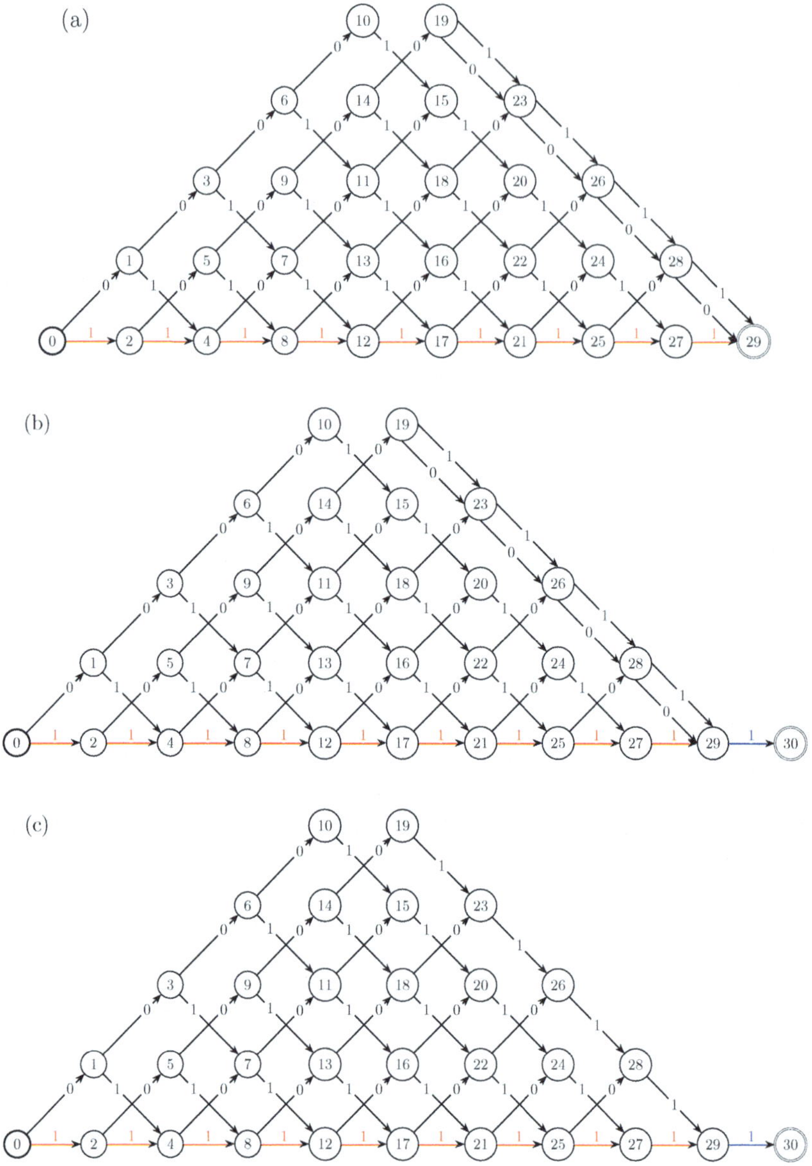

Fig. 8.3 FSM of $\mathbf{f}_{184}^{-4}(1)$ (top), $\mathbf{f}_{184}^{-4}(1)$ followed by 1 (center), and $\mathbf{f}_{184}^{-4}(11)$ (bottom)

We can make the inequality strong by subtracting 1/2,

$$\sum_{i=0}^{k} a_{2n+2-i} > \frac{k+1}{2}.$$

Replacing the dummy index i by $2n+2-i$ and shifting k by 1 (k replaced by $k+1$), we obtain a simplified form of the condition.

Proposition 8.2 *The string* $\mathbf{a} \in \{0,1\}^{2n+2}$ *belongs to* $\mathbf{f}_{184}^{-n}(11)$ *if and only if*

$$\sum_{i=2n+3-k}^{2n+2} a_i > k/2, \tag{8.49}$$

for all $k \in \{1, \dots, 2n+2\}$.

Our derivation of the above was somewhat informal, using the example of the FSM graph for $m = 4$. An alternative proof, not using FSM, can be found in [4].

We could compute $P_n(11)$ now using Proposition 8.2, but block 11 does not belong to the short block representation. We will now show, therefore, how to obtain condition for being a preimage of 00, exploiting rule symmetries. Rule 184 is the same as the conjugated and reflected version of itself. This means that if $a_1 a_2 \dots a_{2n+2}$ is a preimage of 11, then $b_1 b_2 \dots b_{2n+2} = \overline{a}_{2n+2}\overline{a}_{2n+1} \dots \overline{a}_1$ is a preimage of 00, so that $a_i = \overline{b}_{2n+3-i}$. Condition given by inequality (8.49) becomes

$$\sum_{i=2n+3-k}^{2n+2} 1 - b_{2n+3-i} > k/2,$$

which, by changing the dummy index i to $j = 2n+3-i$, becomes

$$\sum_{j=1}^{k} 1 - b_j > k/2.$$

This leads to the condition for being a preimage of 00.

Proposition 8.3 *The string* $\mathbf{b} \in \{0,1\}^{2n+2}$ *belongs to* $\mathbf{f}_{184}^{-n}(00)$ *if and only if*

$$\sum_{i=1}^{k} b_i < k/2, \tag{8.50}$$

for all $k \in \{1, \dots, 2n+2\}$.

Consider now a string $\mathbf{b}$ which contains n_0 zeros and n_1 ones. The number of such strings can be immediately obtained if we realize that it is equal to the number of lattice paths from the origin to (n_0, n_1) which do not touch nor cross the line $x = y$. This is a well known combinatorial problem [5], and the number of such paths equals

$$\frac{n_0 - n_1}{n_0 + n_1}\binom{n_0 + n_1}{n_1}. \tag{8.51}$$

Low let us suppose that $b_1 b_2 \ldots b_{2n+2} = X_1 X_2 \ldots X_{2n+2}$, where $X_0, X_1, \ldots, X_{2n+1}$ are Bernoulli random variables such that $Pr(X_i = 1) = p$, $Pr(X_i = 0) = 1 - p$ for all $0 \le i \le 2n + 2$. Probability that $b_1 b_2 \ldots b_{2n+2}$ satisfies condition (8.50) and has n_0 zeros and n_1 ones is then equal to

$$\frac{n_0 - n_1}{n_0 + n_1}\binom{n_0 + n_1}{n_1} p^{n_1}(1 - p)^{n_0}. \tag{8.52}$$

In an n-step preimage of 00 the minimum number of zeros is $n + 2$, while the maximum is $2n + 2$, corresponding to all zeros. The probability $P_n(00)$ is equal to the sum of probabilities of all strings $b_1 b_2 \ldots b_{2n+2}$ satisfying (8.50), thus

$$P_n(00) = \sum_{i=n+2}^{2n+2} \frac{i - (2n + 2 - i)}{2n + 2}\binom{2n + 2}{2n + 2 - i} \times p^{2n+2-i}(1 - p)^i.$$

Changing the summation index $j = i - (n + 1)$ we obtain

$$P_n(00) = \sum_{j=1}^{n+1} \frac{j}{n + 1}\binom{2n + 2}{n + 1 - j} p^{n+1-j}(1 - p)^{n+1+j}. \tag{8.53}$$

Probability $P_n(000)$ can be obtained from the above as well, because preimages of 000 can be constructed from preimages of 00 by adding extra 0 at the beginning, hence

$$P_n(000) = \sum_{j=1}^{n+1} \frac{j}{n + 1}\binom{2n + 2}{n + 1 - j} p^{n+1-j}(1 - p)^{n+2+j}. \tag{8.54}$$

The probability $P_n(010)$ can also be obtained by a similar method, and it is listed in the Appendix C.

8.9 Rule 14 and Other Rules with Second Order Invariants

For some rules, it is not possible to obtain probabilistic solutions for arbitrary p. In such cases, however, it is sometimes still possible to obtain solutions for the symmetric initial measure, when $p = 1/2$. Recall that at the end of Sect. 8.2 we derived the following version of the Ruelle-Frobenius-Perron equation,

$$P_n(\mathbf{a}) = \sum_{\mathbf{b} \in \mathbf{f}^{-n}(\mathbf{a})} P_0(\mathbf{b}). \tag{8.55}$$

When $p = 1/2$, for any block $\mathbf{b}$,

$$P_0(\mathbf{b}) = \left(\frac{1}{2}\right)^{\#_0(\mathbf{b})} \left(\frac{1}{2}\right)^{\#_1(\mathbf{b})} = \frac{1}{2^{|\mathbf{b}|}},$$

therefore

$$P_n(\mathbf{a}) = \sum_{\mathbf{b} \in \mathbf{f}^{-n}(\mathbf{a})} \frac{1}{2^{|\mathbf{b}|}} = \frac{\operatorname{card} \mathbf{f}^{-n}(\mathbf{a})}{2^{|\mathbf{b}|}}. \tag{8.56}$$

For elementary rules $|\mathbf{b}| = 2n + |\mathbf{a}|$, and thus

$$P_n(\mathbf{a}) = \frac{\operatorname{card} \mathbf{f}^{-n}(\mathbf{a})}{2^{2n+|\mathbf{a}|}}. \tag{8.57}$$

We will now use this expression to find $P_n(0)$, $P_n(00)$, $P_n(000)$ and $P_n(010)$ for elementary rule 14. We start from computing $P_n(1)$. Inspection of the FSM graph representing preimages of 1 in Fig. 6.8 on p. 87 reveals that the number of preimages of 1 is simply the number of paths from the starting node (0,0) to the final node (5,4).

We can reach node (5,4) via nodes:

- (5,0) in 2^4 ways;
- (5,1) in $\binom{5}{1} \cdot 2^3$ ways;
- (5,2) in $\binom{6}{2} \cdot 2^2$ ways;
- (5,3) in $\binom{7}{3} \cdot 2^1$ ways;
- (4,3) and (4,4) in $\binom{7}{3}$ ways.

The total number of paths is, therefore,

$$\operatorname{card} \mathbf{f}^{-4}(\mathbf{a}) = \sum_{i=0}^{3} \binom{4+i}{i} 2^{4-i} + \binom{7}{3}.$$

For a general n, the total number of paths will be

$$\begin{aligned} \operatorname{card} \mathbf{f}^{-n}(\mathbf{a}) &= \sum_{i=0}^{n-1} \binom{n+i}{i} 2^{n-i} + \binom{2n-1}{n-1} \\ &= \sum_{i=0}^{n} \binom{n+i}{i} 2^{n-i} - \binom{2n}{n} + \binom{2n-1}{n-1} = 4^n - \binom{2n}{n} + \binom{2n-1}{n-1}. \end{aligned}$$

Since $\binom{2n-1}{n-1} = \frac{1}{2}\binom{2n}{n}$, this simplifies to

$$\operatorname{card} \mathbf{f}^{-n}(\mathbf{a}) = 4^n - \frac{1}{2}\binom{2n}{n}.$$

Using this count of preimages, Eq. (8.57) yields

$$P_n(1) = 2^{-2n-1} \operatorname{card} \mathbf{f}^{-n}(\mathbf{a}) = \frac{1}{2} - 2^{-2n-2}\binom{2n}{n},$$

and hence

$$P_n(0) = \frac{1}{2} + 2^{-2n-2}\binom{2n}{n}.$$

In order to obtain $P_n(00)$, we will take advantage of the fact that rule 14 conserves the number of blocks 10. Using similar argument as what we used for rule 184 at the beginning of Sect. 8.8, one can show that for general Bernoulli measure with parameter p,

$$P_n(10) = p(1-p),$$

thus $P_n(10) = 1/4$ for $p = 1/2$. Consistency conditions require that

$$P_n(00) + P_n(01) = P_n(0),$$

therefore

$$P_n(00) = \frac{1}{4} + 2^{-2n-2}\binom{2n}{n}.$$

From relationships listed in Eq. (7.41) we know that

$$P_n(111) = 1 - 3P_n(0) + 2P_n(00) + P_n(010),$$

and from Example 1.8 we know that 111 belongs to the garden of Eden of rule 14. This means that $P_n(111)$ must be zero for $n > 0$, therefore

$$P_n(010) = -1 + 3P_n(0) - 2P_n(00),$$

yielding, after substitution of expressions for $P_n(0)$ and $P_n(00)$ obtained so far,

$$P_n(010) = 2^{-2n-2}\binom{2n}{n}.$$

The final probability, $P_n(000)$, can be now computed using the last line of the set of Ruelle-Frobenius-Perron equations (Eq. 8.10) for rule 14,

$$P_{n+1}(010) = P_n(0) - 2P_n(00) + P_n(000).$$

Since we already know $P_n(010)$, we can express $P_n(000)$ as

$$\begin{aligned}
P_n(000) &= P_{n+1}(010) - P_n(0) + 2P_n(00) \\
&= 2^{-2n-1}\binom{2n+2}{n+1} - \frac{1}{2} - 2^{-2n-2}\binom{2n}{n} + 2\left(\frac{1}{4} + 2^{-2n-2}\binom{2n}{n}\right) \\
&= \frac{2^{-2n-3}(4n+3)}{n+1}\binom{2n}{n}. \qquad (8.58)
\end{aligned}$$

Let us summarize our results for rule 14, listing all fundamental probabilities for blocks of length up to 3,

$$\begin{aligned}
P_n(0) &= \frac{1}{2} + 2^{-2n-2}\binom{2n}{n}, \\
P_n(00) &= \frac{1}{4} + 2^{-2n-2}\binom{2n}{n}, \\
P_n(000) &= \frac{2^{-2n-3}(4n+3)}{n+1}\binom{2n}{n}, \\
P_n(010) &= 2^{-2n-2}\binom{2n}{n}.
\end{aligned}$$

In a very similar way one can obtain fundamental probabilities $P_n(0)$, $P_n(00)$, $P_n(000)$ and $P_n(010)$ for two other rules conserving 10, namely for rules 43 and 142. These formulas, as usual, are listed in the Appendix C.

8.10 Surjective Rules

One important class of CA rules for which the probabilistic IVP can be solved are surjective rules. We say that a CA rule with global function F is *surjective* if for all $y \in \mathcal{A}^{\mathbb{Z}}$ there exists $x \in \mathcal{A}^{\mathbb{Z}}$ such that $y = F(x)$. We will first describe a method for finding surjective rules, state and prove some of their important properties, and then use these properties to solve the IVP for elementary surjective rules.

8.10.1 The Algorithm for Determining Surjectivity

The method for determining if a given rule is surjective to be introduced here has been first proposed by Amoroso and Patt [6]. It is based on the following result from the classic work of Hedlund [7]. As usual, we assume that the global function F corresponds to the local function f and to the block mapping $\mathbf{f}$.

Theorem 8.1 *A cellular automaton with global function $F : \mathcal{A}^{\mathbb{Z}} \to \mathcal{A}^{\mathbb{Z}}$ is surjective if and only if every finite block has at least one preimage, that is, if for all $\mathbf{b} \in \mathcal{A}^{\star}$.*

$$\operatorname{card} \mathbf{f}^{-1}(\mathbf{b}) > 0.$$

Proof ($\Rightarrow$) Assume that F is surjective and let $\mathbf{b} \in \mathcal{A}^m$. Define $x \in \mathcal{A}^{\mathbb{Z}}$ by setting $x_1 x_2 \ldots x_m = b_1 b_2 \ldots b_m$ and $x_i = 0$ for $i < 1$ or $i > m$. Since F is surjective, there exist $y \in \mathcal{A}^{\mathbb{Z}}$ such that $F(y) = x$. Define $\mathbf{a} = y_{-r+1} \ldots y_{m+r}$. Then $\mathbf{f}(\mathbf{a}) = \mathbf{b}$, meaning that $\mathbf{a}$ is a preimage of $\mathbf{b}$.

($\Leftarrow$) Now suppose that every blocks has a preimage. Let $x \in \mathcal{A}^{\mathbb{Z}}$ and let $k > 0$ be an arbitrarily selected integer. Define $\mathbf{b} = x_{-k} \ldots x_k$. By hypothesis the block $\mathbf{b}$ must have at least one preimage $\mathbf{a} = a_{-k-r} \ldots a_{k+r}$ such that $\mathbf{f}(\mathbf{a}) = \mathbf{b}$. Now let us construct $y \in \mathcal{A}^{\mathbb{Z}}$ such that it agrees with $\mathbf{a}$ on indices $-k-r$ to $k+r$ and $y_i = 0$ elsewhere,

$$y = \ldots 000 a_{-k-r} \ldots a_{k+r} 000 \ldots$$

Let $u = F(y)$. Then $u_{-k} \ldots u_k = x_{-k} \ldots x_k$, meaning that the first position on which x and u are potentially different is further then k from the origin. By the definition of the Cantor metric given in the first chapter,

$$d(x, u) = 2^{\min\{|i|:x_i \neq u_i\}} < 2^{-k}.$$

Since k was arbitrarily selected, this means that u can be arbitrarily close to x, or in other words, in every neighbourhood of x there is at least one u such that $u = F(y)$ for some $y \in \mathcal{A}^{\mathbb{Z}}$. This implies that the set $F(\mathcal{A}^{\mathbb{Z}})$ is dense in $\mathcal{A}^{\mathbb{Z}}$. On the other hand, F is a continuous function and its domain $\mathcal{A}^{\mathbb{Z}}$ is compact. Continuous image of a compact set is compact, therefore $F(\mathcal{A}^{\mathbb{Z}})$ is compact too, and as such it must be closed.

By basic topological properties of dense sets, $F(\mathcal{A}^{\mathbb{Z}})$ is dense in $\mathcal{A}^{\mathbb{Z}}$ if the smallest closed subset of $\mathcal{A}^{\mathbb{Z}}$ containing $F(\mathcal{A}^{\mathbb{Z}})$ is $\mathcal{A}^{\mathbb{Z}}$ itself. Yet $F(\mathcal{A}^{\mathbb{Z}})$ is closed, thus we must have $F(\mathcal{A}^{\mathbb{Z}}) = \mathcal{A}^{\mathbb{Z}}$, meaning that F is surjective. □

According to Theorem 8.1, if a given rule is nonsurjective, then there must exist a finite block $\mathbf{b}$ which has no preimage. For rule F, the set of blocks without preimages will be called *Garden of Eden set of* F, defined as

$$\mathrm{GOE}_F = \{\mathbf{b} \in \mathcal{A}^{\star} : \mathbf{f}^{-1}(\mathbf{b}) = \emptyset\}.$$

The method for deciding if a given rule F is surjective or not works as follows. If we can find at least one block belonging to GOE_F, then the rule F is not surjective. If we can show that no such block exists, the rule F is surjective.

Consider first the problem of finding preimages of a given block $\mathbf{b} = b_1 b_2 \dots b_n$ assuming that we know $\mathbf{f}^{-1}(a)$ for all $a \in \mathcal{A}$. If the radius of F is r, then the preimages must be blocks of length $n + 2r$, having the form $\mathbf{a} = a_1 a_2 \dots a_{n+2r}$, where $b_1 = f(a_1, a_2, \dots a_{2r+1})$, $b_2 = f(a_2, a_3, \dots a_{2r+2})$, etc. We can say, therefore, that the block $a_1 a_2 \dots a_{2r+1}$ must belong to preimages of b_1, the block $a_2 a_3 \dots a_{2r+2}$ must belong to preimages of b_2, etc. Note that the last $2r$ symbols of $a_1 a_2 \dots a_{2r+1}$ and the first $2r$ symbols of $a_2 a_3 \dots a_{2r+2}$ are overlapping. In general, we will say that the block $\mathbf{d} \in \mathcal{A}^m$ is *a follower* of $\mathbf{c} \in \mathcal{A}^m$ if the first $m - 1$ symbols of $\mathbf{d}$ and the last $m - 1$ symbols of $\mathbf{c}$ are overlapping, that is, $c_2 = d_1, c_3 = d_2, \dots c_m = d_{m-1}$. If $\mathbf{d}$ is a follower of $\mathbf{c}$, we will write $\mathbf{c} \rightsquigarrow \mathbf{d}$.

To find preimages of $b_1 b_2 \dots b_n$, we start by constructing the set $A_1 = \mathbf{f}^{-1}(b_1)$. Then we construct $A_2, A_3, \dots A_n$ by recursion such that A_{k+1} is the set of those elements of $\mathbf{f}^{-1}(b_{k+1})$ which are followers of at least one element of A_k,

$$A_{k+1} = \{\mathbf{a} \in \mathbf{f}^{-1}(b_{k+1}) : \exists_{\mathbf{c} \in A_k} \mathbf{c} \rightsquigarrow \mathbf{a}\}.$$

Once the sets $A_1, A_2 \dots A_n$ are determined, we can take any path of followers

$$\mathbf{a}^{(1)} \rightsquigarrow \mathbf{a}^{(2)} \dots \rightsquigarrow \mathbf{a}^{(n)},$$

such that $\mathbf{a}^{(k)} \in A_k$. The block constructed by concatenation of $\mathbf{a}^{(1)}$ and last symbols of $\mathbf{a}^{(k)}$ for $k = 2, 3, \dots, n$, that is, the block

$$\mathbf{a}^{(1)} a_n^{(2)} a_n^{(3)} \dots a_n^{(n)}$$

will be the preimage of $\mathbf{b}$.

Example 8.7 For ECA rule 18, $\mathbf{f}^{-1}(0) = \{000, 010, 011, 101, 110, 111\}$ and $\mathbf{f}^{-1}(1) = \{001, 100\}$. Suppose we want to determine preimages of $\mathbf{b} = 1010$ under this rule. The first symbol is $b_1 = 1$, thus we have

$$A_1 = \mathbf{f}^{-1}(1) = \{001, 100\}.$$

The second symbol is $b_2 = 0$, so we construct A_2 from those elements of $\mathbf{f}^{-1}(0)$ which are followers of some element of A_1. This means that from 6 elements of $\mathbf{f}^{-1}(0)$ we select only those which start with 01 or 00,

$$A_2 = \{000, 010, 011\}.$$

Similarly, $b_3 = 1$, so we select as members of A_3 only those elements of $\mathbf{f}^{-1}(1)$ which start from 00, 10 or 11, and these are,

$$A_3 = \{001, 100\}.$$

The last symbol is 0, and we again selected from $\mathbf{f}^{-1}(0)$ only elements which are followers of 001 or 100, meaning that they start from 01 or 00,

$$A_4 = \{000, 010, 011\}.$$

We can schematically represent our results as a directed graph with nodes labeled by sets $A_1 \dots A_4$,

$$\boxed{\begin{array}{c}001\\100\end{array}} \longrightarrow \boxed{\begin{array}{c}000\\010\\011\end{array}} \longrightarrow \boxed{\begin{array}{c}001\\100\end{array}} \longrightarrow \boxed{\begin{array}{c}000\\010\\011\end{array}}.$$

On this graph there are three possible paths of followers chosen consecutively from $A_1 \dots A_4$,

$$\begin{aligned}
&001 \rightsquigarrow 010 \rightsquigarrow 100 \rightsquigarrow 000,\\
&100 \rightsquigarrow 000 \rightsquigarrow 001 \rightsquigarrow 010,\\
&100 \rightsquigarrow 000 \rightsquigarrow 001 \rightsquigarrow 011.
\end{aligned}$$

Taking the first block and last elements of the remaining blocks from each path we obtain three preimages of 1010,

$$\mathbf{f}^{-1}(1010) = 001000,\ 100010,\ 100011\}.$$

Example 8.8 As another example, consider what happens if we want to find preimages of a block which belongs to GOE_F. For the same ECA rule 18, let $\mathbf{b} = 111$. We have

$$A_1 = \mathbf{f}^{-1}(1) = \{001, 100\}.$$

The second symbol is $b_2 = 1$, so we need to select as members of A_2 the elements of $\mathbf{f}^{-1}(1)$ which start from 01 or 00, and there is only one,

$$A_2 = \{001\}.$$

The third symbol is $b_3 = 1$, so we need to select as members of A_3 the elements of $\mathbf{f}^{-1}(1)$ which start from 01, yet such elements do not exist, so

$$A_3 = \varnothing.$$

Again, we summarize our findings in a form of graph,

$$\boxed{\begin{array}{c}001\\100\end{array}} \longrightarrow \boxed{001} \longrightarrow \boxed{\varnothing}.$$

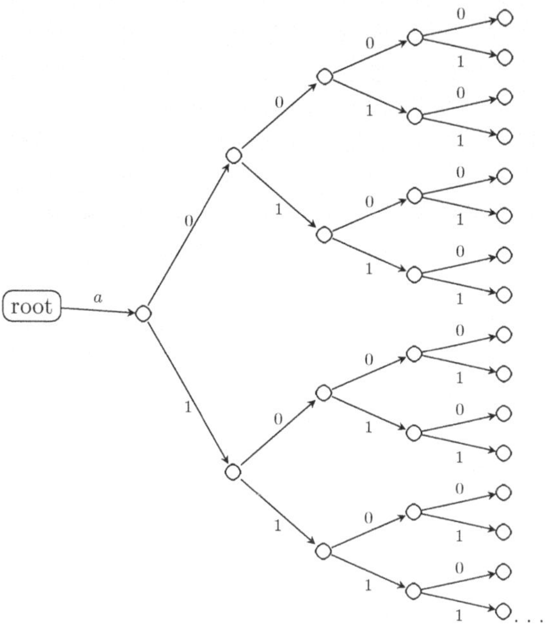

Fig. 8.4 First five levels of an infinite tree used in the Patt and Amoroso algorithm for $N = 2$. The symbol a can be arbitrarily selected from $\{0, 1\}$

Since A_3 is empty, we won't be able to construct a path of consecutive followers from elements of A_1, A_2 and A_3. The block 111 thus has no preimages under the ECA rule 18.

The idea of Amoroso and Patt algorithm [6] is to construct a graph similar to the graphs of previous two examples but describing preimages of *all* blocks, not just a single one. It is done as follows.

- For an alphabet $\mathcal{A}$ with N symbols, we start with an infinite N-ary tree, horizontally oriented, stemming from a "trunk" with a single edge carrying an arbitrarily selected symbol $a \in \mathcal{A}$. Paths on this tree beginning at the root represent all blocks of $\mathcal{A}^\star$ starting with symbol a. Example of such a tree for $N = 2$ is shown in Fig. 8.4.
- The first node to the right of the root receives label $\mathbf{f}^{-1}(a)$.
- Then we label nodes recursively, moving from one level of the tree to the next one, and within each level from the top to bottom. If an edge carrying symbol b originates from already labeled node A and arrives to an unlabeled node, the unlabeled node receives label B consisting of all blocks of $\mathbf{f}^{-1}(b)$ which are followers of at least one element of the label A.
- If no such block exists, we label the node with symbol $\varnothing$ calling it *null* and we remove all further branches stemming form it.
- If the new label B has already been encountered earlier, we call it *duplicate*, mark it with double frame and remove all further branches stemming form it.

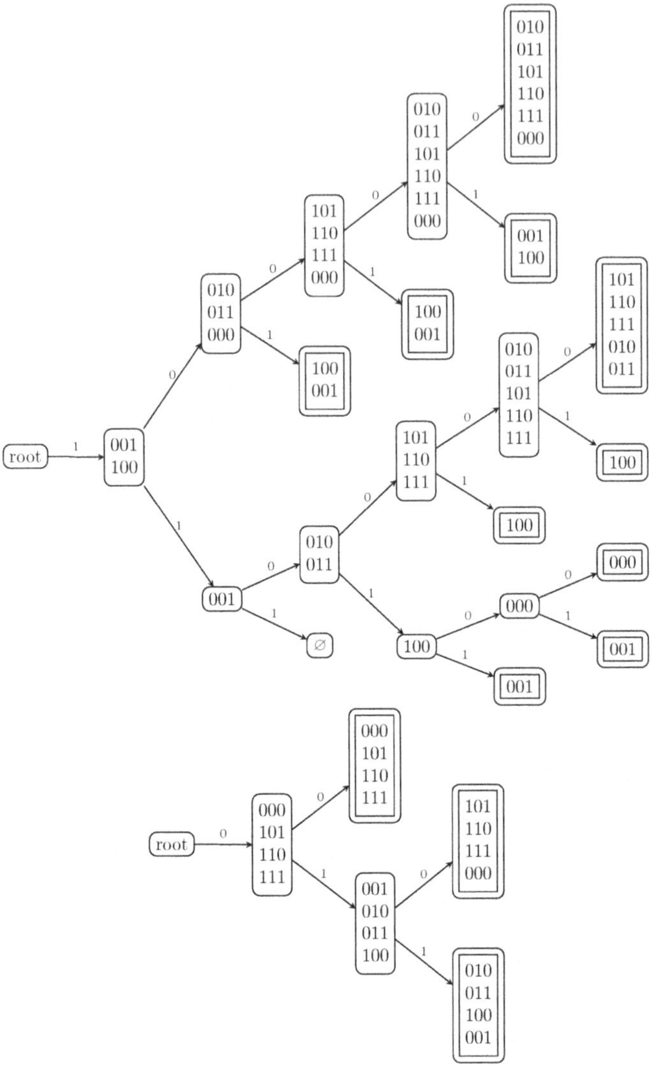

Fig. 8.5 Graphs obtained by applying the algorithm of Amoroso and Patt to ECA rules 18 (top) and 30 (bottom)

It is clear that since the number of possible labels is finite, this procedure must end sooner or later, resulting in a finite labeled tree in which all terminal nodes are either duplicates or null. If the tree contains a null node, the rule is not surjective, otherwise it is surjective.

Example 8.9 For the ECA rule 18 discussed earlier the resulting tree is shown in the top of Fig. 8.5. It includes one node with label ∅, thus the rule is not surjective. Note that the path from the root to the null node carries edge labels 111, meaning that the block 111 has no preimages under this rule, as already demonstrated in Example 8.8. The tree shown in Fig. 8.5 is complete and has 6 levels, but it was not necessary to continue its labeling once the first node with $\varnothing$ label appeared. We should also add that if the leftmost edge of the tree carried symbol 0 instead of 1, we would still reach the same result (non-surjectivity of rule 18). The only difference would be that the block corresponding to label $\varnothing$ would be 0111 instead of 111.

Example 8.10 For the ECA rule 30 we have $\mathbf{f}^{-1}(0) = \{000, 101, 110, 111\}$ and $\mathbf{f}^{-1}(1) = \{001, 010, 011, 100\}$. The resulting graph is shown in the bottom of Fig. 8.5. All terminal nodes of the tree are duplicates, and none is labeled $\varnothing$. The rule is surjective.

Application of the Amoroso and Patt algorithm to all minimal elementary CA rules reveals that only the following minimal rules are surjective: 15, 30, 45, 51, 60, 90, 105, 106, 150, 154, 170 and 204. Of course, other members of equivalence classes of these rules are surjective as well.

8.10.2 The Balance Theorem

We already established that for surjective rules every block must have at least one preimage. It turns out that even stronger property holds, namely preimage sets of all blocks have the same cardinality. The following theorem was first proved by Hedlund [8] who acknowledged earlier contributions from W. A. Blankenship and O. S. Rothaus. Another version of the proof can be found in [9] or, for binary CA, in [10].

Theorem 8.2 (Balance Theorem) *Let $\mathcal{A} = \{0, 1, \dots N-1\}$ and let $F : \mathcal{A}^{\mathbb{Z}} \to \mathcal{A}^{\mathbb{Z}}$ be a cellular automaton with radius r. Then F is surjective if and only if every block $\mathbf{b} \in \mathcal{A}^\star$ has exactly N^{2r} preimages, i.e.,*

$$\operatorname{card} \mathbf{f}^{-1}(\mathbf{b}) = N^{2r}.$$

We will closely follow the version of the proof given by Hedlund [7]. It depends on two lemmas which will be proved first. In both proofs we will use the notation for concatenation of sets of block which is a natural extension of concatenation of individual blocks. If $B \subset \mathcal{A}^p$ and $C \subset \mathcal{A}^q$ then

$$BC = \{\mathbf{bc} \in \mathcal{A}^{p+q} : \mathbf{b} \in B, \mathbf{c} \in C\},$$

where $\mathbf{bc} = b_1 b_2 \dots b_p c_1 c_2 \dots c_q$.

Lemma 8.2 *Let $F : \mathcal{A}^{\mathbb{Z}} \to \mathcal{A}^{\mathbb{Z}}$ be a cellular automaton rule and suppose that there exist a block $\mathbf{b} \in \mathcal{A}^\star$ such that* $\operatorname{card} \mathbf{f}^{-1}(\mathbf{b}) > 0$. *If*

$$\operatorname{card} \mathbf{f}^{-1}(\mathbf{b}a) \geq \operatorname{card} \mathbf{f}^{-1}(\mathbf{b}) \ \ \forall_{a \in \mathcal{A}},$$

then

$$\text{card}\,\mathbf{f}^{-1}(\mathbf{b}a) = \text{card}\,\mathbf{f}^{-1}(\mathbf{b}) \;\; \forall_{a\in\mathcal{A}}.$$

Proof Let $\mathbf{b} \in \mathcal{A}^m$ and $\text{card}\,\mathbf{f}^{-1}(\mathbf{b}) > 0$. We claim that

$$\mathbf{f}^{-1}(b\mathcal{A}) = \mathbf{f}^{-1}(\mathbf{b})\mathcal{A}.$$

This is because $\mathbf{f}(\mathbf{c}a) \in \mathbf{b}\mathcal{A}$ if and only if $\mathbf{f}(\mathbf{c}) = \mathbf{b}$, or $\mathbf{c} \in \mathbf{f}^{-1}(\mathbf{b})$. The above equation implies that

$$\text{card}\,\mathbf{f}^{-1}(\mathbf{b}\mathcal{A}) = N\,\text{card}\,\mathbf{f}^{-1}(\mathbf{b}) = \sum_{a\in\mathcal{A}} \text{card}\,\mathbf{f}^{-1}(\mathbf{b}a). \tag{8.59}$$

Now assume that

$$\text{card}\,\mathbf{f}^{-1}(\mathbf{b}a) \geq \text{card}\,\mathbf{f}^{-1}(\mathbf{b}) \;\; \forall_{a\in\mathcal{A}}.$$

If, for any $a \in \mathcal{A}$, we had $\text{card}\,\mathbf{f}^{-1}(\mathbf{b}a) > \text{card}\,\mathbf{f}^{-1}(\mathbf{b})$, this would also mean that $\text{card}\,\mathbf{f}^{-1}(\mathbf{b}\mathcal{A}) > N\,\text{card}\,\mathbf{f}^{-1}(\mathbf{b})$, contradicting Eq. (8.59). Therefore, $\text{card}\,\mathbf{f}^{-1}(\mathbf{b}a) = \text{card}\,\mathbf{f}^{-1}(\mathbf{b})$ for all $a \in \mathcal{A}$, concluding the proof. □

Lemma 8.3 *Let* $\mathcal{A} = \{0, 1, \ldots N-1\}$ *and let* $F : \mathcal{A}^{\mathbb{Z}} \to \mathcal{A}^{\mathbb{Z}}$ *be a cellular automaton rule with radius* $r > 0$. *Suppose that there exist a block* $\mathbf{b} \in \mathcal{A}^m$ *such that* $\text{card}\,\mathbf{f}^{-1}(\mathbf{b}) > 0$ *satisfying*

$$\text{card}\,\mathbf{f}^{-1}(\mathbf{bdb}) = \text{card}\,\mathbf{f}^{-1}(\mathbf{b}) \;\; \forall_{\mathbf{d}\in\mathcal{A}^{2r}}.$$

Then

$$\text{card}\,\mathbf{f}^{-1}(\mathbf{b}) = N^{2r}.$$

Proof We will first show that under the hypothesis of the lemma and using our notation for concatenation of sets of blocks we have

$$\mathbf{f}^{-1}(\mathbf{b}\mathcal{A}^{2r}\mathbf{b}) = \mathbf{f}^{-1}(\mathbf{b})\mathbf{f}^{-1}(\mathbf{b}). \tag{8.60}$$

To show this, let $\mathbf{a} \in \mathbf{f}^{-1}(\mathbf{b})\mathbf{f}^{-1}(\mathbf{b})$, meaning that $\mathbf{a} = \mathbf{a}_1\mathbf{a}_2$ and $\mathbf{a}_1 \in \mathbf{f}^{-1}(\mathbf{b})$, $\mathbf{a}_2 \in \mathbf{f}^{-1}(\mathbf{b})$. This implies

$$|\mathbf{a}| = |\mathbf{a}_1| + |\mathbf{a}_2| = 2(m+2r) = 4r + 2m,$$
$$|\mathbf{f}(\mathbf{a})| = 2m + 2r.$$

Since $\mathbf{f}(\mathbf{a}_1) = \mathbf{f}(\mathbf{a}_2) = \mathbf{b}$, $\mathbf{f}(\mathbf{a})$ is of the form $\mathbf{bdb}$, where $|\mathbf{d}| = 2r$, so that $\mathbf{f}(\mathbf{a}) \in \mathbf{b}\mathcal{A}^{2r}\mathbf{b}$.

Now suppose that $\mathbf{f}(\mathbf{a}) = \mathbf{bdb}$ where $\mathbf{d} \in \mathcal{A}^{2r}$. Then $|\mathbf{a}| = 4r + 2m$ and $\mathbf{a} = \mathbf{a}_1\mathbf{a}_2$, where $|\mathbf{a}_1| = |\mathbf{a}_2| = m + 2r$. Moreover, $\mathbf{f}(\mathbf{a}_1) = \mathbf{b} = \mathbf{f}(\mathbf{a}_2)$, so that $\mathbf{a}_1, \mathbf{a}_2 \in \mathbf{f}^{-1}(\mathbf{b})$ and $\mathbf{a}_1\mathbf{a}_2 \in f^{-1}(\mathbf{b})f^{-1}(\mathbf{b})$. This completes the proof of Eq. (8.60).

Let $k = \operatorname{card} \mathbf{f}^{-1}(\mathbf{b})$. The cardinality of the left hand side of Eq. (8.60) is

$$\operatorname{card} \mathbf{f}^{-1}(\mathbf{b}\mathcal{A}^{2r}\mathbf{b}) = \sum_{\mathbf{d}\in\mathcal{A}^{2r}} \mathbf{f}^{-1}(\mathbf{bdb}) = \sum_{\mathbf{d}\in\mathcal{A}^{2r}} k = N^{2r}k.$$

The right hand side of Eq. (8.60) has cardinality

$$\operatorname{card} \left(\mathbf{f}^{-1}(\mathbf{b})\mathbf{f}^{-1}(\mathbf{b})\right) = k^2.$$

These two numbers must be equal, hence $k = N^{2r}$, as the lemma claims. □

Proof (the Balance Theorem)

($\Leftarrow$) If preimages of all block have the same non-zero cardinality, surjectivity follows from Theorem 8.1.

($\Rightarrow$) Suppose F is surjective. By Theorem 8.1 every block has at least one preimage, thus there must exist a block $\mathbf{b} \in \mathcal{A}^m$ which has the smallest number of preimages, to be denoted by $k_{\min}$. This implies that for each $a \in \mathcal{A}$ we have $\operatorname{card} \mathbf{f}^{-1}(\mathbf{b}a) \geq k_{\min}$, and by Lemma 8.2 $\operatorname{card} \mathbf{f}^{-1}(\mathbf{b}a) = k_{\min}$. By induction this leads to the conclusion that

$$\operatorname{card} \mathbf{f}^{-1}(\mathbf{ba}) = k_{\min}$$

for any block $\mathbf{a} \in \mathcal{A}^{\star}$. In particular, we must have

$$\operatorname{card} \mathbf{f}^{-1}(\mathbf{bdb}) = k_{\min}$$

for any $\mathbf{d} \in \mathcal{A}^{2r}$. Applying Lemma 8.3 this leads to conclusion that $k_{\min} = N^{2r}$, therefore

$$\operatorname{card} \mathbf{f}^{-1}(\mathbf{c}) \geq N^{2r}$$

for any $\mathbf{c} \in \mathcal{A}^{\star}$.

Suppose that there exists an integer p and a block $\mathbf{a} \in \mathcal{A}^p$ for which the above inequality is strong,

$$\operatorname{card} \mathbf{f}^{-1}(\mathbf{a}) > N^{2r}.$$

Then we must have

$$\sum_{\mathbf{c}\in\mathcal{A}^p} \operatorname{card} \mathbf{f}^{-1}(\mathbf{c}) > N^{p+2r},$$

and this is because $\operatorname{card} \mathcal{A}^p = N^p$. On the other hand,

$$\sum_{\mathbf{c}\in\mathcal{A}^p} \operatorname{card} \mathbf{f}^{-1}(\mathbf{c}) = \operatorname{card} \mathbf{f}^{-1}(\mathcal{A}^p) = \operatorname{card} \mathcal{A}^{p+2r} = N^{p+2r},$$

contradicting previously obtained $\sum_{\mathbf{c}\in\mathcal{A}^p} \operatorname{card} \mathbf{f}^{-1} > N^{p+2r}$. We conclude, therefore, that $\operatorname{card} \mathbf{f}^{-1}(\mathbf{b}) = N^{2r}$ for all $\mathbf{b} \in \mathcal{A}^{\star}$, which proves the Balance Theorem. □

8.10.3 Solution of the Probabilistic IVP for Surjective Rules

Recall that in Eq. (8.11) we obtained

$$P_n(\mathbf{a}) = \sum_{\mathbf{b} \in \mathbf{f}^{-n}(\mathbf{a})} P_0(\mathbf{b}). \tag{8.61}$$

Suppose that card $\mathcal{A} = N$ and the initial measure is the *symmetric Bernoulli measure*, also called the *uniform measure*, defined in terms of block probabilities as

$$P_0(\mathbf{b}) = 1/N^{|\mathbf{b}|}$$

for all $\mathbf{b} \in \mathcal{A}^\star$. The previous equation then yields

$$P_n(\mathbf{a}) = \sum_{\mathbf{b} \in \mathbf{f}^{-n}(\mathbf{a})} \frac{1}{N^{|\mathbf{b}|}} = \frac{\text{card}\, \mathbf{f}^{-1}(\mathbf{a})}{N^{|\mathbf{b}|}} = \frac{\text{card}\, \mathbf{f}^{-1}(\mathbf{a})}{N^{|\mathbf{a}|+2r}}, \tag{8.62}$$

where r is the radius of the rule. For surjective rules the Balance Theorem yields card $\mathbf{f}^{-1}(\mathbf{a}) = N^{2r}$, thus

$$P_n(\mathbf{a}) = \frac{N^{2r}}{N^{|\mathbf{a}|+2r}} = \frac{1}{N^{|\mathbf{a}|}} = P_0(\mathbf{a}) \tag{8.63}$$

for any $\mathbf{a} \in \mathcal{A}^\star$. The Balance theorem, therefore, can be expressed as follows.

Theorem 8.3 *Cellular automaton rule is surjective if and only if it preserves the symmetric Bernoulli measure.*

This theorem is very general and it holds also in dimensions higher than one, but we will not discuss this here. Interested readers should consult the review article [11] and references therein.

For ECA rules which are surjective, that is, for rules 15, 30, 45, 51, 60, 90, 105, 106, 150, 154, 170 and 204, if the initial Bernoulli measure is symmetric, the probabilistic initial value problems has the solution

$$P_n(\mathbf{a}) = \frac{1}{2^{|\mathbf{a}|}}$$

for any block $\mathbf{a} \in \{0, 1\}^\star$. In particular, the probabilities of the first four fundamental blocks are

$$
\begin{aligned}
P_n(0) &= \frac{1}{2}, \\
P_n(00) &= \frac{1}{4}, \\
P_n(000) &= \frac{1}{8}, \\
P_n(010) &= \frac{1}{8}.
\end{aligned}
$$

Note that for rules 15, 51, 60, 90, 150, 170 and 204 we earlier obtained more general expressions for the fundamental block probabilities, valid also for non-symmetric Bernoulli measures. For the remaining surjective rules, that is, for rules 30, 105,[1] 106 and 154, the above result is the only one known and has been included in the list of solution of the probabilistic IVP in the Appendix C.

8.11 Further Use of RFP Equations

As mentioned already, in Appendix C probabilities of the first four fundamental blocks are listed for rules for which it is possible to obtain them using methods outlined in this chapter (c.f. Table 12.1 on p. 214 for a summary). Using these solutions and the RFP equations listed in Appendix D, one can obtain probabilities of some further blocks almost "for free", as the following example illustrates.

Example 8.11 For rule 168 the expressions for fundamental block probabilities are

$$
\begin{aligned}
P_n(0) &= 1 - p^{n+1}(2-p)^n, \\
P_n(00) &= 1 - p^{n+1}(2-p)^{n+1}, \\
P_n(000) &= 1 - p^{n+1}(2-p)^n\left(p^2 - 3p + 3\right), \\
P_n(010) &= (p-1)^2 p^{n+1}(2-p)^n.
\end{aligned}
$$

The RFP system of order 3 for this rule is

$$
\begin{aligned}
P_{n+1}(0) &= P_n(0) - P_n(000) + P_n(00), \\
P_{n+1}(00) &= \underline{P_n(0010)} + 2P_n(00) - P_n(000), \\
P_{n+1}(000) &= \underline{P_n(00100)} + \underline{P_n(0010)} + P_n(00), \\
P_{n+1}(010) &= \underline{P_n(00110)} + P_n(010) - \underline{P_n(0010)}.
\end{aligned}
$$

In the above, we underlined probabilities of blocks longer than 3. There are three of them, corresponding to block 0010, 00100, and 00110. It is clear than one can solve the system of RFP equations for probabilities of these blocks, obtaining

[1] Deterministic solution is known for rule 105.

$$
\begin{aligned}
P_n(0010) &= P_{n+1}(00) - 2P_n(00) + P_n(000),\\
P_n(00100) &= P_{n+1}(000) - P_{n+1}(00) + P_n(00) - P_n(000),\\
P_n(00110) &= P_{n+1}(010) + P_n(000) - P_n(010) + P_{n+1}(00) - 2P_n(00).
\end{aligned}
$$

After substituting appropriate expressions for P_n and P_{n+1} on the right hand side and simplifying, one obtains

$$
\begin{aligned}
P_n(0010) &= p^{n+1}(2-p)^n(1-p)^3,\\
P_n(00100) &= p^{n+1}(p-1)^4(2-p)^n,\\
P_n(00110) &= p^{n+2}\,(2-p)^n\left(1-p^3+3\,p^2-3\,p\right).
\end{aligned}
$$

By a similar method one can obtain various probabilities of longer blocks for many other rules. If we want to obtain all fundamental block probabilities of a given length, increasing the order of the RPF may be helpful, as in the example below.

Example 8.12 We will obtain all fundamental probabilities of length up to 4 for rule 136. Its RPF equations of order three are given by

$$
\begin{aligned}
P_{n+1}(0) &= -P_n(00) + 2P_n(0),\\
P_{n+1}(00) &= P_n(0) + P_n(010),\\
P_{n+1}(000) &= 2P_n(010) + P_n(00),\\
P_{n+1}(010) &= P_n(0110).
\end{aligned}
$$

Clearly, knowing expressions for $P_n(0)$, $P_n(00)$, $P_n(000)$ and $P_n(010)$, we can obtain the probability of the only block of length 4 present, $P_n(0110) = P_{n+1}(010)$.

Increasing the order of RFP to 4 yields additional four equations:

$$
\begin{aligned}
P_{n+1}(0000) &= P_n(0010) + P_n(0100) + P_n(01010) + P_n(000) + P_n(010),\\
P_{n+1}(0010) &= P_n(0110),\\
P_{n+1}(0100) &= P_n(0110),\\
P_{n+1}(0110) &= P_n(01110).
\end{aligned}
$$

This immediately yields three fundamental block probabilities of length 4,

$$
\begin{aligned}
P_n(0010) &= P_n(010),\\
P_n(0100) &= P_n(010),\\
P_n(0110) &= P_n(010).
\end{aligned}
$$

The fourth one seems to be a problem, because

$$
P_{n+1}(0000)P_n(0010) + P_n(0100) + P_n(01010) + P_n(000) + P_n(010)
$$

yields

$$P_{n+1}(0000) = P_n(01010) + P_n(000) + 3 * P_n(010),$$

and we do not know $P_n(01010)$. However, if we write 5-th order RFP equations, we will find that one of them is

$$P_{n+1}(01010) = 0,$$

meaning that $P_n(01010)$ vanishes and therefore we get

$$P_{n+1}(0000) = P_n(000) + 3P_n(010),$$

hence

$$P_n(0000) = P_{n-1}(000) + 3P_{n-1}(010).$$

Using probabilistic solutions for rule 136 listed in the Appendix C, namely

$$\begin{aligned} P_n(0) &= 1 - p^{n+1}, \\ P_n(00) &= p^{2+n} - 2\,p^{n+1} + 1, \\ P_n(000) &= 2\,p^{2+n} - 3\,p^{n+1} + 1, \\ P_n(010) &= p^{n+1}\,(p-1)^2, \end{aligned}$$

we can now obtain expressions for all 4-block fundamental probabilities,

$$\begin{aligned} P_n(0000) &= P_{n-1}(000) + 3P_{n-1}(010) = 1 + 2\,p^{1+n} - 3\,p^n + 3p^n\,(p-1)^2 \\ P_n(0010) &= P_n(010) = p^{n+1}\,(p-1)^2, \\ P_n(0100) &= P_n(010) = p^{n+1}\,(p-1)^2, \\ P_n(0110) &= P_n(010) = p^{n+1}\,(p-1)^2 \end{aligned}$$

Note that these expressions are valid for $n > 1$ since in their derivation we used the formula for $P_{n-1}(000)$, not valid for $n = 0$. For $n = 0$ one simply needs to use the relevant expressions for the block probabilities in the Bernoulli measure.

References

1. Devaney, R.: An Introduction to Chaotic Dynamical Systems. CRC Press (2003)
2. Elaydi, S.: An Introduction to Difference Equations. Springer (2005)
3. Fukś, H.: Explicit solution of the Cauchy problem for cellular automaton rule 172. J. of Cellular Automata **12**(6), 423–444 (2017)
4. Fukś, H.: Exact results for deterministic cellular automata traffic models. Phys. Rev. E **60**, 197–202 (1999). https://doi.org/10.1103/PhysRevE.60.197

5. Mohanty, S.G.: Lattice Path Counting and Applications. Academic Press, New York (1979)
6. Amoroso, S., Patt, Y.N.: Decision procedures for surjectivity and injectivity of parallel maps for tesselation structures. Journal of Computer and System Sciences **6**, 448–464 (1972)
7. Hedlund, G.: Endomorphisms and automorphisms of shift dynamical systems. Mathematical Systems Theory **3**, 320–375 (1969)
8. Hedlund, G.A.: Endomorphisms and automorphisms of the shift dynamical system. Mathematical systems theory **3**, 320–375 (1968)
9. Kleveland, R.: Mixing properties of one-dimensional cellular automata. Proc. Amer. Math. Soc. **125**, 1755–1766 (1997)
10. Shirvani, M., Rogers, T.D.: On ergodic one-dimensional cellular automata. Communications in Mathematical Physics **136**, 599–605 (1991)
11. Pivato, M.: Ergodic theory of cellular automata. In: R.A. Meyers (ed.) Encyclopedia of Complexity and System Science. Springer (2009)

Chapter 9
Probabilistic Cellular Automata

9.1 Probabilistic CA as a Markov Process

In the first chapter of the book, we defined cellular automata as continuous transformations of the Cantor space $\mathcal{A}^{\mathbb{Z}}$. Such view of cellular automata is very convenient to study deterministic cellular automata (the only ones we have considered so far), but it is not easily generalized to include what is called *probabilistic cellular automata* (PCA). In what follows, therefore, we will first define PCA in a more "traditional way", as Markov process on a lattice. Afterwards we will show that we can also view PCA as transformations of the space of shift-invariant probabilistic measures.

The definition of PCA often used in the literature views them as a dynamical system, assuming that the dynamics takes place on a one-dimensional lattice of length L with periodic boundary conditions. Let $s_i(t)$ denote the state of the lattice site i at time step t, where $i \in \{0, 1, \ldots, L-1\}$, $t \in \mathbb{N}$. All operations on spatial indices i are assumed to be modulo L.

In a probabilistic cellular automaton, lattice sites simultaneously change states with probabilities depending on states of local neighbours. A common method for defining the rule for such changes is to specify a set of local transition probabilities. For example, in order to define a nearest-neighbour PCA, one has to specify the probability that the site $s_i(t)$ with nearest neighbors $s_{i-1}(t)$, $s_{i+1}(t)$ changes its state to $s_i(t+1)$ in a single time step. This probability is usually denoted by $w\big(s_i(t+1))|s_{i-1}(t), s_i(t), s_{i+1}(t)\big)$.

Example 9.1 As an illustration, consider the PCA rule investigated in [1], with $\mathcal{A} = \{0, 1\}$. Sites in state 0 are called empty, and those in state 1 are called occupied. Empty sites become occupied with a probability proportional to the number of occupied sites in the neighbourhood, and occupied sites become empty with a probability proportional to the number of empty sites in the neighbourhood. The following set of transition probabilities defines this PCA rule:

H. Fukś, *Solvable Cellular Automata*, Understanding Complex Systems,
https://doi.org/10.1007/978-3-031-38700-5_9

$$
\begin{aligned}
w(1|0,0,0) &= 0 & w(1|0,0,1) &= \alpha \\
w(1|0,1,0) &= 1-2\alpha & w(1|0,1,1) &= 1-\alpha \\
w(1|1,0,0) &= \alpha & w(1|1,0,1) &= 2\alpha \\
w(1|1,1,0) &= 1-\alpha & w(1|1,1,1) &= 1,
\end{aligned} \tag{9.1}
$$

where $\alpha \in [0, 1/2]$. The remaining eight transition probabilities can be obtained using $w(0|a, b, c) = 1 - w(1|a, b, c)$ for $a, b, c \in \{0, 1\}$.

We will now proceed to construct a general definition of PCA, with arbitrary neighbourhood radius. Let $r > 0$ and $n = 2r + 1$. The set $\{s_{i-r}(t), s_{i-r+1}(t), \dots, s_{i+r}(t)\}$ will be called a *neighbourhood* of $s_i(t)$.

Consider now a set of independent Boolean random variables $X_{i,\mathbf{u}}$, where $i = 0, 1, \dots L-1$ and $\mathbf{u} \in \mathcal{A}^n$. The probability that the random variable $X_{i,\mathbf{u}}$ takes the value $a \in \{0, 1\}$ will be denoted by $w(a|\mathbf{u})$,

$$
Pr(X_{i,\mathbf{u}} = a) = w(a|\mathbf{u}). \tag{9.2}
$$

Since $w(a|\mathbf{u})$ are probabilities, we must have

$$
\sum_{a \in \mathcal{A}} w(a|\mathbf{u}) = 1,
$$

for all $\mathbf{u} \in \mathcal{A}^n$. The update rule for PCA is defined by

$$
s_i(t+1) = X_{i,\{s_{i+l}(t), s_{i+l+1}(t), \dots, s_{i+r}(t)\}}. \tag{9.3}
$$

Note that values of random variables X used at time step t are independent of those used at previous time steps. Consequently, $\{\mathbf{s}(t) : t = 0, 1, 2, \dots\}$ is a Markov stochastic process, and its states are binary sequences $\mathbf{s}(t) = s_0(t)s_2(t)\dots s_{L-1}(t)$. The probability of transition from $\mathbf{x} \in \{0, 1\}^L$ to $\mathbf{y} \in \{0, 1\}^L$ in one time step is given by

$$
Pr(\mathbf{y}|\mathbf{x}) = \prod_{i=0}^{L-1} w(y_i|\{x_{i+l}, x_{i+l+1}, \dots, x_{i+r}\}). \tag{9.4}
$$

Example 9.2 For rule described in Example 9.1, $r = 1$ and $n = 3$, and each lattice site i is associated with eight random variables $X_{i,000}$, $X_{i,001}$, $X_{i,010}$, $X_{i,011}$, $X_{i,100}$, $X_{i,101}$, $X_{i,110}$, and $X_{i,111}$. Probability distributions of these r.v. are determined by the transition probabilities given in Eq. (9.1). For instance, $w(1|1, 0, 0) = \alpha$, and therefore we have $Pr\big(X_{i,100} = 1\big) = \alpha$, and $Pr\big(X_{i,100} = 0\big) = 1 - \alpha$. Similarly, $w(1|0, 0, 0) = 0$, hence $Pr\big(X_{i,000} = 1\big) = 0$ and $Pr\big(X_{i,000} = 0\big) = 1$, meaning that $X_{i,000}$ is in this case a deterministic variable.

9.2 Maps in the Space of Measures

Although the definition of the PCA as a Markov process is very useful in applications, for our purposes a different view will be more convenient. We will define PCA as maps in the space of probability measures. This definition does not require periodic lattice and will let us to define orbits of measures in a similar way as for deterministic rules (c.f. definition 8.1).

Definition 9.1 Let $w : \mathcal{A} \times \mathcal{A}^{2r+1} \to [0, 1]$, whose values are denoted by $w(a|\mathbf{b})$ for $a \in \mathcal{A}$, $\mathbf{b} \in \mathcal{A}^{2r+1}$, satisfying $\sum_{a \in \mathcal{A}} w(a|\mathbf{b}) = 1$, be called *local transition function* of *radius r*, and its values will be called *local transition probabilities. Probabilistic cellular automaton* with local transition function w is a map $F : \mathfrak{M}(\mathcal{A}^{\mathbb{Z}}) \to \mathfrak{M}(\mathcal{A}^{\mathbb{Z}})$ defined as

$$(F\mu)([\mathbf{a}]_i) = \sum_{\mathbf{b} \in \mathcal{A}^{|\mathbf{a}|+2r}} w(\mathbf{a}|\mathbf{b})\mu([\mathbf{b}]_{i-r}) \text{ for all } i \in \mathbb{Z}, \mathbf{a} \in \mathcal{A}^{\star}, \tag{9.5}$$

where we define

$$w(\mathbf{a}|\mathbf{b}) = \prod_{j=1}^{|\mathbf{a}|} w(a_j|b_j b_{j+1} \ldots b_{j+2r}). \tag{9.6}$$

Example 9.3 As an example, we will consider a PCA rule which generalizes some of the CA rules investigated in [2]. This PCA can be viewed as a simple model for diffusion of innovations, spread of rumors, or a similar process involving transport of information between neighbours. We consider an infinite one-dimensional lattice where each site is occupied by an individual who has already adopted the innovation (1) or who has not adopted it yet (0). Once the individual adopts the innovation, he remains in state 1 forever. Individuals in state 0 change can their states to 1 (adopt the innovation) with probabilities depending on the state of nearest neighbours. All changes of states take place simultaneously. This process can be formally described as a radius 1 binary PCA with the following transition probabilities,

$$\begin{aligned} &w(1|000) = 0,\ w(1|001) = \alpha,\ w(1|010) = 1,\ w(1|011) = 1, \\ &w(1|100) = \beta,\ w(1|101) = \gamma,\ w(1|110) = 1,\ w(1|111) = 1, \end{aligned} \tag{9.7}$$

where α, β, γ are fixed parameters of the model, $\alpha, \beta, \gamma \in [0, 1]$. As before, the remaining eight transition probabilities can be obtained using $w(0|abc) = 1 - w(1|abc)$ for $a, b, c \in \{0, 1\}$. Figure 9.1 shows a typical spatiotemporal pattern generated by this rule for parameter values $\alpha = 0.1$, $\beta = 0.2$, and $\gamma = 0.145$.

Given some measure $\mu \in \mathfrak{M}(\mathcal{A}^{\mathbb{Z}})$, let us compute the measure of the cylinder set $[1]_i$ after the application of F, that is, the probability of 1 in the measure $F\mu$. Equation (9.5) becomes

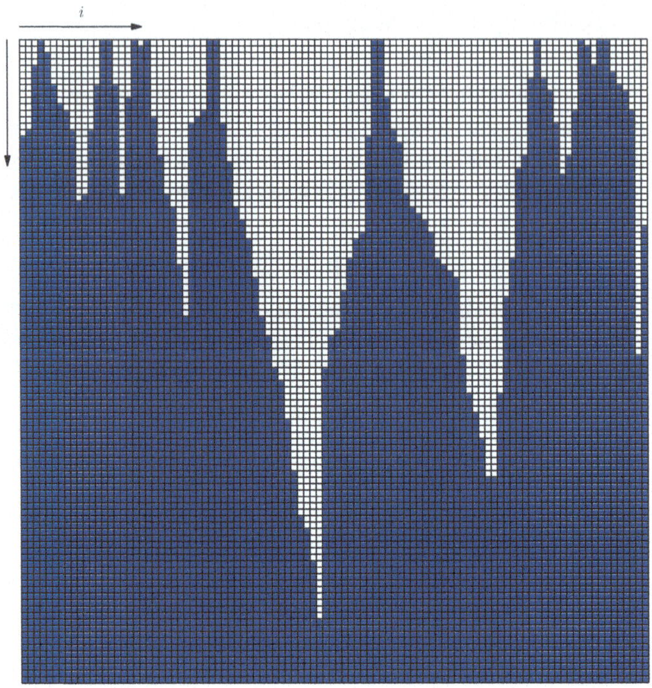

Fig. 9.1 Spatiotemporal pattern generated by PCA of example 9.3 for lattice of 100 sites with periodic boundary condition

$$
\begin{aligned}
(F\mu)([1]_i) = \sum_{\mathbf{b}\in\mathcal{A}^3} w(1|\mathbf{b})\mu([\mathbf{b}]_i) &= \alpha\mu([001]_i) + \mu([010]_i) + \mu([011]_i) \\
&+ \beta\mu([100]_i) + \gamma\mu([101]_i) + \mu([110]_i) + \mu([111]_i).
\end{aligned}
$$

If μ was a symmetric Bernoulli measure, then all 3-blocks would have the same measure $1/8$, thus we would have

$$
(F\mu)([1]_i) = \frac{1}{8}(\alpha + \beta + \gamma) + \frac{1}{2}.
$$

This is to be interpreted as follows: if initially all sites were randomly occupied by adopters with 50% probability to be an adopter, then after one iteration of the rule the probability that a given site is an adopter is $\frac{1}{8}(\alpha + \beta + \gamma) + \frac{1}{2}$.

In a similar fashion we could compute probabilities of longer blocks, except that the corresponding expressions would be longer. For example, for block 10 we would have

$$
(F\mu)([10]_i) = \sum_{\mathbf{b}\in\mathcal{A}^4} w(10|\mathbf{b})\mu([\mathbf{b}]_i).
$$

The sum has 16 terms, thus we will not write it here, only remarking that the transition probabilities $w(10|\mathbf{b})$ need to be computed using Eq. (9.6),

$$w(10|b_1b_2b_3b_4) = w(1|b_1b_2b_3)w(0|b_2b_3b_4).$$

In the interpretation of PCA as a dynamical system on a lattice, $w(10|b_1b_2b_3b_4)$ signifies the probability of transition from $b_1b_2b_3b_4$ to 10 in a single time step,

$$\begin{array}{ccccccc} \cdots & b_1 & b_2 & b_3 & b_4 & \cdots \\ \cdots & & 1 & 0 & & \cdots \end{array}$$

where the above shows a fragment of the lattice at time step t followed by the same fragment at time step $t+1$.

Example 9.4 For PCA rule of radius r, when the transition probabilities $w(a|\mathbf{b})$ are equal to 0 or 1 for all $a \in \mathcal{A}$ and $\mathbf{b} \in \mathcal{A}^{2r+1}$, the rule is completely deterministic. For example, consider the PCA of the previous example with $\alpha = \beta = 1$ and $\gamma = 0$. The transition probabilities to 1 are

$$\begin{aligned} &w(1|000) = 0,\ w(1|001) = 1,\ w(1|010) = 1,\ w(1|011) = 1,\\ &w(1|100) = 1,\ w(1|101) = 0,\ w(1|110) = 1,\ w(1|111) = 1, \end{aligned}$$

and the corresponding transitions to 0,

$$\begin{aligned} &w(0|000) = 1,\ w(0|001) = 0,\ w(0|010) = 0,\ w(0|011) = 0,\\ &w(0|100) = 0,\ w(0|101) = 1,\ w(0|110) = 0,\ w(0|111) = 0. \end{aligned}$$

It is clear that blocks 001, 010, 011, 100, 110 and 111 always produce 1 in one iteration of the rule, and blocks 000 and 101 always produce 0. This corresponds to the deterministic CA rule with the local function given by

$$\begin{aligned} &f(0,0,0) = 0,\ \ f(0,0,1) = 1,\ \ f(0,1,0) = 1,\ \ f(0,1,1) = 1,\\ &f(1,0,0) = 1,\ \ f(1,0,1) = 0,\ \ f(1,1,0) = 1,\ \ f(1,1,1) = 1. \end{aligned}$$

The reader can easily verify that this is elementary CA with Wolfram number 222.

In general, we can treat and ECA with the local function f as PCA with transition probabilities for $a, b, c \in \{0, 1\}$ given by

$$w(1|abc) = f(a, b, c), \tag{9.8}$$

$$w(0|abc) = 1 - f(a, b, c). \tag{9.9}$$

9.3 Orbits of Probability Measures

For any probabilistic measure $\mu \in \mathfrak{M}(\mathcal{A}^{\mathbb{Z}})$ and probabilistic CA F, we define the orbit of μ under F as

$$\{F^n \mu\}_{n=0}^{\infty}. \tag{9.10}$$

As in the case of deterministic CA, it is very difficult to compute $F^n \mu$ directly, and no general method for doing this is known. We can, however, construct Ruelle-Frobenius-Perron equation for PCA, in a similar way as we did for deterministic rules.

We will assume that the initial measure μ is shift invariant. If F is a PCA, then F^n is a PCA as well, thus Eq. (9.5) can be written for F^n,

$$(F^{n+1}\mu)([\mathbf{a}]_i) = \sum_{\mathbf{b}\in\mathcal{A}^{|\mathbf{a}|+2r}} w(\mathbf{a}|\mathbf{b}) F^n \mu([\mathbf{b}]_{i-r}). \tag{9.11}$$

Define

$$P_n(\mathbf{a}) = (F^n \mu)([\mathbf{a}]_i),$$

then

$$P_{n+1}(\mathbf{a}) = \sum_{\mathbf{b}\in\mathcal{A}^{|\mathbf{a}|+2r}} w(\mathbf{a}|\mathbf{b}) P_n(\mathbf{b}). \tag{9.12}$$

As before, $P_n(\mathbf{a})$ should be interpreted as the probability of occurrence of $\mathbf{a}$ after n iterations of F.

The probabilistic initial value problem for PCA can be posed in the same way as for deterministic CA.

Definition 9.2 Let $F : \mathfrak{M}(\mathcal{A}^{\mathbb{Z}}) \to \mathfrak{M}(\mathcal{A}^{\mathbb{Z}})$ be probabilistic cellular automaton. The *probabilistic initial value problem* for F is the problem of finding block probabilities $P_n(\mathbf{a})$ for all $\mathbf{a} \in \mathcal{A}^\star$ and all $n > 0$, assuming that all initial block probabilities $P_0(\mathbf{a})$ are given. The *k-th order probabilistic initial value problem* for F is the problem of finding block probabilities $P_n(\mathbf{a})$ for all $\mathbf{a} \in \mathcal{A}^k$ and all $n > 0$, assuming that all initial block probabilities $P_0(\mathbf{a})$ are given.

Using Eq. (9.12) we can obtain a form of the solution resembling Eq. (8.11) derived earlier for deterministic CA. Writing Eq. (9.12) for consecutive n values we obtain

$$P_1(\mathbf{a}) = \sum_{\mathbf{b}\in\mathcal{A}^{|\mathbf{a}|+2r}} w(\mathbf{a}|\mathbf{b}) P_0(\mathbf{b}),$$

$$P_2(\mathbf{a}) = \sum_{\mathbf{b}\in\mathcal{A}^{|\mathbf{a}|+2r}} w(\mathbf{a}|\mathbf{b}) P_1(\mathbf{b}),$$

$$\ldots.$$

For $n = 2$, we have

$$P_2(\mathbf{a}) = \sum_{\mathbf{b}\in\mathcal{A}^{|\mathbf{a}|+2r}} w(\mathbf{a}|\mathbf{b}) \sum_{\mathbf{c}\in\mathcal{A}^{|\mathbf{b}|+2r}} w(\mathbf{b}|\mathbf{c})P_0(\mathbf{c})$$

$$= \sum_{\mathbf{b}\in\mathcal{A}^{|\mathbf{a}|+2r}} \sum_{\mathbf{c}\in\mathcal{A}^{|\mathbf{a}|+4r}} w(\mathbf{a}|\mathbf{b})w(\mathbf{b}|\mathbf{c})P_0(\mathbf{c}) = \sum_{\substack{\mathbf{b}\in\mathcal{A}^{|\mathbf{a}|+2r}\\ \mathbf{c}\in\mathcal{A}^{|\mathbf{a}|+4r}}} w(\mathbf{a}|\mathbf{b})w(\mathbf{b}|\mathbf{c})P_0(\mathbf{c}).$$

It is, therefore, clear that for general n,

$$P_n(\mathbf{a}) = \sum_{\substack{\mathbf{b}_1\in\mathcal{A}^{|\mathbf{a}|+2r}\\ \mathbf{b}_2\in\mathcal{A}^{|\mathbf{a}|+4r}\\ \dots\\ \mathbf{b}_n\in\mathcal{A}^{|\mathbf{a}|+2nr}}} w(\mathbf{a}|\mathbf{b_1})w(\mathbf{b}_1|\mathbf{b_2})\dots w(\mathbf{b}_{n-1}|\mathbf{b_n})P_0(\mathbf{b}_n). \tag{9.13}$$

If we define

$$w^n(\mathbf{a}|\mathbf{b}) = \sum_{\substack{\mathbf{b}_1\in\mathcal{A}^{|\mathbf{a}|+2r}\\ \mathbf{b}_2\in\mathcal{A}^{|\mathbf{a}|+4r}\\ \dots\\ \mathbf{b}_{n-1}\in\mathcal{A}^{|\mathbf{a}|+2(n-1)r}}} w(\mathbf{a}|\mathbf{b_1})w(\mathbf{b}_1|\mathbf{b_2})\dots w(\mathbf{b}_{n-1}|\mathbf{b}),$$

then Eq. (9.13) becomes

$$P_n(\mathbf{a}) = \sum_{\mathbf{b}\in\mathcal{A}^{|\mathbf{a}|+2nr}} w^n(\mathbf{a}|\mathbf{b})P_0(\mathbf{b}). \tag{9.14}$$

We will call $w^n(\mathbf{a}|\mathbf{b})$ the the n-step transition probability. If $|\mathbf{a}| = m$, then $w^n(\mathbf{a}|\mathbf{b})$ can be intuitively understood as the conditional probability of seeing the block $\mathbf{a}$ on sites $[1, m]$ after n iterations of F, conditioned on the fact that the initial configuration contained the block $\mathbf{b}$ on sites $[1 - nr, m + nr]$.

Since some of the transition probabilities may be zero, we define, for any block $\mathbf{b} \in \mathcal{A}^\star$, the set

$$\operatorname{supp} w^n(\mathbf{a}|\cdot) = \{\mathbf{b} \in \mathcal{A}^{|\mathbf{a}|+2rn} : w^n(\mathbf{a}|\mathbf{b}) > 0\}. \tag{9.15}$$

Then we can write eq. (9.14) as

$$P_n(\mathbf{a}) = \sum_{\mathbf{b}\in\operatorname{supp} w^n(\mathbf{a}|\cdot)} w^n(\mathbf{a}|\mathbf{b})P_0(\mathbf{b}). \tag{9.16}$$

Similarly to Eq. (8.11), Eq. (9.14) is of limited use in practice, as it is usually impossible to find a "closed form" of the n-step transition probability. Nevertheless, in some simple cases, it is possible to obtain explicit formulae for $w^n(\mathbf{a}|\mathbf{b})$, as we will see in the next section.

9.4 Single Transition α-Asynchronous Rules

We will now introduce a very special type of a probabilistic CA, in which each cell is independently updated with some probability α. These rules were first studied experimentally in [3], and subsequently called *α-asynchronous rules* [4]. They are formally defined as follows. Let $f : \{0, 1\}^3 \to \{0, 1\}$ be a local function of an ECA function and let $\alpha \in [0, 1]$ be a given parameter (called the *synchrony rate*). The α-asynchronous PCA rule corresponding to f is defined by

$$
\begin{aligned}
w(1|b_1b_2b_3) &= \alpha f(b_1, b_2, b_3) + (1 - \alpha)b_2, \\
w(0|b_1b_2b_3) &= 1 - w(1|b_1b_2b_3),
\end{aligned}
$$

for all $(b_1, b_2, b_3) \in \{0, 1\}^3$. Note that when $\alpha = 1$, the α-asynchronous PCA is equivalent to the deterministic rule with local function f (*cf.* Eq. 9.8). When $\alpha = 0$, it becomes the deterministic identity rule (ECA 204). We will denote α-asynchronous rules corresponding to ECA with Wolfram number nnn by nnnA.

The simplest α-asynchronous rules are those corresponding to ECA with a single active transition (*cf.* Sect. 1.2 for the relevant definitions). There are 8 such rules, shown in Table 9.1, but only three are minimal, namely rules 200, 140 and 76.

9.4.1 Rule 200A

We will first consider an α-asynchronous PCA corresponding to ECA 200. It is defined by

$$
\begin{aligned}
w(1|b_1b_2b_3) &= \alpha f_{200}(b_1, b_2, b_3) + (1 - \alpha)b_2, \\
w(0|b_1b_2b_3) &= 1 - w(1|b_1b_2b_3),
\end{aligned}
$$

Table 9.1 Single transition ECA rules. Minimal rules are shown in bold

Wolfram number	Fatès transition code
205	A
206	B
220	C
236	D
200	E
196	F
140	G
76	H

which, using definition of f_{200}, can be explicitly written as

$$w(1|b_1b_2b_3) = \begin{cases} 0 & \text{if } b_1b_2b_3 \in \{000, 001, 100, 101\}, \\ 1 & \text{if } b_1b_2b_3 \in \{011, 110, 111\}, \\ 1-\alpha & \text{if } b_1b_2b_3 = 010, \end{cases} \tag{9.17}$$

and $w(0|b_1b_2b_3) = 1 - w(1|b_1b_2b_3)$ for all $b_1b_2b_3 \in \{0, 1\}^3$. As we already remarked, for $\alpha = 1$ this rule is equivalent to the deterministic rule 200, which has transition code E.

Proposition 9.1 *The α-asynchronous rule 200A has the following properties:*

(i) the set $\operatorname{supp} w^n(1|\cdot)$ *consists of all blocks of the form*

$$\{\underbrace{\star \cdots \star}_{n} \, 1 \, \underbrace{\star \cdots \star}_{n}\}.$$

(ii) Let C_n be the set of blocks of the form

$$\{\underbrace{\star \cdots \star}_{n-1} \, 010 \, \underbrace{\star \cdots \star}_{n-1}\}.$$

For any block $\mathbf{b} \in C_n$, *we have* $w^n(1|\mathbf{b}) = (1-\alpha)^n$.

(iii) For any block $\mathbf{b} \in \operatorname{supp} w^n(1|\cdot)$ *which does not belong to* C_n, *we have* $w^n(1|\mathbf{b}) = 1$.

Proof For (i), from the definition of the rule, we can see that a site in state 0 will always remain 0, so any block which does not have 1 in the center will never be transformed to a single 1 under n iterations of the rule. Moreover, any block with 1 in the center could produce a single 1 after n iterations, with some non-zero probability.

For (ii), note that the two zeros surrounding the center 1 will be always preserved, so we only need to make sure that the transition $010 \to 1$ occurs in each time step. This will happen with probability $(1-\alpha)^n$.

For (iii), we note that for every block $\mathbf{b} \in \operatorname{supp} w^n(1|\cdot)$ which does not belong to C_n must have 011, 110 or 111 in the center. All of these central 3-blocks will be preserved when the rule 200A is iterated, thus $w^n(1|\mathbf{b}) = 1$. □

We can now use Eq. (9.16) to compute $P_n(1)$ assuming Bernoulli initial measure,

$$\begin{aligned}
P_n(1) &= \sum_{\mathbf{b}\in \operatorname{supp} w^n(1|\cdot)} w^n(1|\mathbf{b}) P_0(\mathbf{b}) \\
&= \sum_{\mathbf{b}\in C_n} w^n(1|\mathbf{b}) P_0(\mathbf{b}) + \sum_{\mathbf{b}\in \operatorname{supp} w^n(1|\cdot)\setminus C_n} w^n(1|\mathbf{b}) P_0(\mathbf{b}) \\
&= \sum_{\mathbf{b}\in C_n} (1-\alpha)^n P_0(\mathbf{b}) + \sum_{\mathbf{b}\in \operatorname{supp} w^n(1|\cdot)\setminus C_n} P_0(\mathbf{b}) \\
&= p(1-p)(1-\alpha)^n + 2(1-p)p^2 + p^3 = p^2(2-p) + (1-\alpha)^n p(1-p)^2.
\end{aligned}$$

This give us the solution of the first order probabilistic initial value problem for the fundamental block 0,

$$P_n(0) = 1 - p^2(2-p) - (1-\alpha)^n p(1-p)^2.$$

As we already remarked, blocks 011, 110 and 111 are preserved, thus $P_n(011) = P_n(110) = p^2(1-p)$ and $P_n(111) = p^3$, and by consistency conditions

$$P_n(11) = P_n(111) + P_n(110) = p^3 + p^2(1-p) = p^2.$$

From Eq. (7.41) we obtain

$$P_n(00) = P_n(11) - 1 + 2P_n(0) = 2\,p^3 - 3\,p^2 + 1 - 2\,(1-\alpha)^n\,p\,(1-p)^2,$$

and this is the only fundamental block probability we need for the second order initial value problem. For 3-blocks, we already know that $P_n(010) = (1-\alpha)^n$, and $P_n(000)$ can be computed similarly as $P_n(0)$, by examining $\operatorname{supp} w^n(000|\cdot)$. Details can be found in [5], here we will only summarize the results by listing probabilities of all fundamental blocks of length of up to 3 for rule 200A,

$$\begin{aligned}
P_n(0) &= 1 - p^2(2-p) - (1-\alpha)^n p(1-p)^2. \\
P_n(00) &= 2\,p^3 - 3\,p^2 + 1 - 2\,(1-\alpha)^n\,p\,(1-p)^2, \\
P_n(000) &= (1-p)^2 \\
&\quad \times \left((1-\alpha)^n p\left(2p^2-3\right) + (1-\alpha)^{2n} p^2\,(1-p) - p^3 - p^2 + 2p + 1\right), \\
P_n(010) &= (1-\alpha)^n.
\end{aligned}$$

Interested readers should consult [5] for solutions of the probabilistic initial value problem for other single-transition rules, 140 and 76 A. For rule 140A full solutions can be obtained, while for rule 76 A the solution of the first order problem can be given in a recursive form.

9.5 Cluster Expansion

We will now demonstrate a method for computing block probabilities [6] using the cluster expansion method originating from statistical physics. The heart of the method is the cluster expansion formula,

$$P_n(0) = \sum_{k=1}^{\infty} k P_n(10^k 1), \tag{9.18}$$

which expresses $P_n(0)$ in terms of probabilities of blocks of zeros bounded by 1's., to be referred to as "clusters". Various proofs of this formula can be given (see, for example, [7]), but we will show here that it is a direct consequence of additivity of measure.

Let F be a binary PCA rule and ν be the initial measure. Consider a cylinder set of a single zero anchored at i, $[0]_i$. A single zero must belong to a cluster of zeros of size k, where $k \geq 1$. For a given k, if 0 belongs to a cluster of k zeros, than it must be the j-th zero of the cluster, with $1 \leq j \leq k$. This implies

$$[0]_i = \bigcup_{k=1}^{\infty} \bigcup_{j=1}^{k} [10^k 1]_{i-j}. \tag{9.19}$$

Since all the cylinder sets on the right hand side are mutually disjoint, the measure of the union is the sum of measures if individual cylinder sets,

$$(F^n \nu)([0]_i) = \sum_{k=1}^{\infty} \sum_{j=1}^{k} (F^n \nu)([10^k 1]_{i-j}). \tag{9.20}$$

The measure is shift-invariant, therefore $(F^n \nu)([10^k 1]_{i-j}) = P_n(10^k 1)$, and we obtain

$$P_n(0) = \sum_{k=1}^{\infty} \sum_{j=1}^{k} P_n(10^k 1). \tag{9.21}$$

Since $\sum_{j=1}^{k} 1 = k$, this yields eq. (9.18).

Consider again the PCA rule described in Example 9.3, with transition probabilities

$$\begin{aligned} &w(1|000) = 0,\ w(1|001) = \alpha,\ w(1|010) = 1,\ w(1|011) = 1, \\ &w(1|100) = \beta,\ w(1|101) = \gamma,\ w(1|110) = 1,\ w(1|111) = 1, \end{aligned} \tag{9.22}$$

For this rule, one can show that every $P_{n+1}(10^k 1)$ is expressible as a linear combination of probabilities of the type $P_n(10^k 1)$. Let us write Eq. (9.12) for $\mathbf{a} = 10^k 1$.

Two cases must be distinguished, $k = 1$ and $k > 1$. For $k = 1$, we have

$$\begin{aligned} P_{n+1}(101) &= \alpha(1-\beta)P_n(01001) + (1-\gamma)P_n(01010) + (1-\gamma)P_n(01011) \\ &\quad + \alpha\beta P_n(10001) + (1-\alpha)\beta P_n(10010) + (1-\alpha)\beta P_n(10011) \\ &\quad + \alpha(1-\beta)P_n(11001) + (1-\gamma)P_n(11010) + (1-\gamma)P_n(11011). \end{aligned}$$

By consistency conditions, $P_n(10010) + P_n(10011) = P_n(1001)$ and $P_n(11010) + P_n(11011) = P_n(1101)$, as well as $P_n(01001) + P_n(11001) = P_n(1001)$. This yields

$$\begin{aligned} P_{n+1}(101) &= (1-\gamma)P_n(01010) + (1-\gamma)P_n(01011) + \alpha\beta P_n(10001) \\ &\quad + (1-\alpha)\beta P_n(1001) + \alpha(1-\beta)P_n(1001) + (1-\gamma)P_n(1101), \end{aligned}$$

and further reduction is possible using $P_n(01010) + P_n(01011) = P_n(0101)$ and $P_n(0101) + P_n(1101) = P_n(101)$. The final result is

$$P_{n+1}(101) = (1-\gamma)P_n(101) + (\alpha - 2\alpha\beta + \beta)P_n(1001) + \alpha\beta P_n(10001). \tag{9.23}$$

For $k > 1$, using a similar reasoning, we obtain

$$P_{n+1}(10^k1) = (1-\alpha)(1-\beta)P_n(10^k1) + (\alpha - 2\alpha\beta + \beta)P_n(10^{n+1}1) + \alpha\beta P_n(10^{n+2}1). \tag{9.24}$$

Equations (9.23) and (9.24) clearly show $P_{n+1}(10^k1)$ is expressible as a linear combination of probabilities of the type $P_n(10^k1)$.

We can now write Eqs. (9.23) and (9.24) in matrix form,

$$\begin{bmatrix} P_{n+1}(101) \\ P_{n+1}(1001) \\ \vdots \\ P_{n+1}(10^n1) \\ \vdots \end{bmatrix} = \begin{bmatrix} \tilde{a} & b & c & 0 & 0 & 0 & \cdots \\ 0 & a & b & c & 0 & 0 & \\ 0 & 0 & a & b & c & 0 & \\ 0 & 0 & 0 & a & b & c & \\ \vdots & & & & & & \ddots \end{bmatrix} \begin{bmatrix} P_n(101) \\ P_n(1001) \\ \vdots \\ P_n(10^k1) \\ \vdots \end{bmatrix}, \tag{9.25}$$

where $a = (1-\alpha)(1-\beta)$, $\tilde{a} = 1-\gamma$, $b = \alpha - 2\alpha\beta + \beta$, and $c = \alpha\beta$.

Let us define infinite matrices

$$\mathbf{M} = \begin{bmatrix} \tilde{a} & b & c & 0 & 0 & 0 & \cdots \\ 0 & a & b & c & 0 & 0 & \\ 0 & 0 & a & b & c & 0 & \\ 0 & 0 & 0 & a & b & c & \\ \vdots & & & & & & \ddots \end{bmatrix}, \quad \mathbf{P}_n = \begin{bmatrix} P_n(101) \\ P_n(1001) \\ \vdots \\ P_n(10^k1) \\ \vdots \end{bmatrix}. \tag{9.26}$$

With this notation,

$$\mathbf{P}_{n+1} = \mathbf{M}^n\mathbf{P}_n, \tag{9.27}$$

therefore

$$\mathbf{P}_n = \mathbf{M}^n \mathbf{P}_0. \tag{9.28}$$

We need to compute $\mathbf{M}^n$, and in order to do this, we will write $\mathbf{M}$ as as sum of three matrices,

$$\mathbf{M} = \mathbf{A} + \mathbf{B} + \mathbf{C}, \tag{9.29}$$

where

$$\begin{aligned}
\mathbf{A} &= \mathrm{diag}(\tilde{a}, a, a, \ldots),\\
\mathbf{B} &= {}^{1}\mathrm{sdiag}(b, b, b, \ldots),\\
\mathbf{C} &= {}^{2}\mathrm{sdiag}(c, c, c, \ldots).
\end{aligned}$$

In the above, $\mathrm{diag}(x_1, x_2, x_3, \ldots)$ denotes an infinite matrix with $x_1, x_2, x_3, \ldots$ on the diagonal and zeros elsewhere. Similarly, ${}^{k}\mathrm{sdiag}(x_1, x_2, x_3, \ldots)$ will denote k-shifted diagonal matrix, with $x_1, x_2, x_3, \ldots$ on the k-th line above the diagonal and zeros elsewhere.

9.5.1 Special Case: $\tilde{a} = a$

It will be easier to start with a special case $\tilde{a} = a$, when all diagonal elements of $\mathbf{A}$ are the same. If this holds, matrices $\mathbf{A}$, $\mathbf{B}$, and $\mathbf{C}$ pairwise commute, thus we can use the trinomial expansion formula,

$$\mathbf{M}^n = (\mathbf{A} + \mathbf{B} + \mathbf{C})^n = \sum_{i+j+k=n} \frac{n!}{i!j!k!} \mathbf{A}^i \mathbf{B}^j \mathbf{C}^k, \tag{9.30}$$

Powers of matrices $\mathbf{A}$, $\mathbf{B}$ and $\mathbf{C}$ are easy to compute,

$$\mathbf{A}^i = \mathrm{diag}(a^i, a^i, a^i, \ldots), \tag{9.31}$$

$$\mathbf{B}^j = {}^{j}\mathrm{sdiag}(b^j, b^j, b^j, \ldots), \tag{9.32}$$

$$\mathbf{C}^k = {}^{2k}\mathrm{sdiag}(c^k, c^k, c^k, \ldots), \tag{9.33}$$

and, therefore,

$$\mathbf{A}^i \mathbf{B}^j \mathbf{C}^k =^{j+2k} \mathrm{sdiag}(a^i b^j c^k, a^i b^j c^k, a^i b^j c^k, \ldots). \tag{9.34}$$

In the first row of the above matrix, the only non-zero element $(a^i b^j c^k)$ is in the column $1 + j + 2k$. In the second row, the only non-zero element $(a^i b^j c^k)$ is in the column $2 + j + 2k$, and so on. This means that

$$\mathbf{A}^i\mathbf{B}^j\mathbf{C}^k\mathbf{P}_0 = \begin{bmatrix} a^i b^j c^k P_0(10^{1+j+2k}1) \\ a^i b^j c^k P_0(10^{2+j+2k}1) \\ a^i b^j c^k P_0(10^{3+j+2k}1) \\ \vdots \end{bmatrix}. \tag{9.35}$$

If the initial measure is Bernoulli with $P_0(1) = p$, then $P_0(10^n 1) = p^2(1-p)^n$, thus we can now write

$$\mathbf{P}_n = \mathbf{M}^n\mathbf{P}_0 = \sum_{i+j+k=n} \frac{n!}{i!j!k!} \begin{bmatrix} \tilde{a}^i b^j c^k p^2(1-p)^{1+j+2k} \\ a^i b^j c^k p^2(1-p)^{2+j+2k} \\ a^i b^j c^k p^2(1-p)^{3+j+2k} \\ \vdots \end{bmatrix}. \tag{9.36}$$

We finally obtain

$$\begin{aligned} P_n(10^l 1) &= \sum_{i+j+k=n} \frac{n!}{i!j!k!} a^i b^j c^k p^2(1-p)^{l+j+2k} \\ &= p^2(1-p)^l \sum_{i+j+k=n} \frac{n!}{i!j!k!} a^i [b(1-p)]^j [c(1-p)^2]^k \\ &= p^2(1-p)^l \left(a + b(1-p) + c(1-p)^2\right)^n . \end{aligned} \tag{9.37}$$

Defining $\theta = a + b(1-p) + c(1-p)^2$, this can be written in a more compact form,

$$P_n(10^l 1) = p^2(1-p)^l \theta^n. \tag{9.38}$$

9.5.2 General Case

In order to deal with the case of $\tilde{a} \neq a$, let us first note that n-th powers of matrices

$$\begin{bmatrix} \tilde{a} & b & c & 0 & 0 & 0 & \cdots \\ 0 & a & b & c & 0 & 0 & \\ 0 & 0 & a & b & c & 0 & \\ 0 & 0 & 0 & a & b & c & \\ \vdots & & & & & & \ddots \end{bmatrix}^n, \quad \begin{bmatrix} a & b & c & 0 & 0 & 0 & \cdots \\ 0 & a & b & c & 0 & 0 & \\ 0 & 0 & a & b & c & 0 & \\ 0 & 0 & 0 & a & b & c & \\ \vdots & & & & & & \ddots \end{bmatrix}^n$$

differ only in their first row. This means that the expression for $P_n(10^l 1)$ given in Eq. (9.38) remains valid for $l > 1$. We only need, therefore, to consider the $l = 1$ case, that is, to compute $P_n(101)$. This can be done by using Eq. (9.23) and substituting for $P_n(1001)$ and $P_n(10001)$ by appropriate expressions obtained from Eq. (9.38),

$$\begin{aligned} P_{n+1}(101) = \tilde{a}P_n(101) + bp^2(1-p)^2\left(a + b(1-p) + c(1-p)^2\right)^n \\ + cp^2(1-p)^3\left(a + b(1-p) + c(1-p)^2\right)^n . \end{aligned} \tag{9.39}$$

This is a first-order non-homogeneous difference equation for $P_n(101)$,

$$P_{n+1}(101) = \tilde{a}P_n(101) + K\theta^n, \tag{9.40}$$

where

$$K = bp^2(1-p)^2 + cp^2(1-p)^3, \tag{9.41}$$

$$\theta = a + b(1-p) + c(1-p)^2. \tag{9.42}$$

Non-homogeneous equations like this can be easily solved by standard methods [8]. In our case the solution is

$$P_n(101) = P_0(101)\tilde{a}^n + K\sum_{i=1}^{n} \tilde{a}^{n-i}\theta^{i-1}. \tag{9.43}$$

The sum on the right hand side is a partial sum of geometric series if $\tilde{a} \neq \theta$, or of an arithmetic series when $\tilde{a} = \theta$. Using $P_n(101) = p^2(1-p)$ as well as appropriate formulae for partial sums of geometric and arithmetic series one obtains

$$P_n(101) = \begin{cases} p^2(1-p)\tilde{a}^n + K(\tilde{a}^n - \theta^n)/(\tilde{a}-\theta) & \text{if } \tilde{a} \neq \theta, \\ p^2(1-p)\tilde{a}^n + K\tilde{a}^{n-1}n & \text{if } \tilde{a} = \theta. \end{cases} \tag{9.44}$$

We now have expressions for $P_n(10^l1)$ for $l = 1$ (Eq. 9.44) and for $l > 1$ (Eq. 9.38). We are thus ready to use the cluster expansion formula. Equation (9.18) becomes

$$\begin{aligned} P_n(0) &= \sum_{k=1}^{\infty} kP_n(10^k1) = P_n(101) + \sum_{k=2}^{\infty} kP_n(10^k1) \\ &= P_n(101) + \sum_{k=2}^{\infty} kp^2(1-p)^k\theta^n = P_n(101) + \theta^n p^2 \sum_{k=2}^{\infty} k(1-p)^k \\ &= P_n(101) + \theta^n(1+p)(1-p)^2. \end{aligned}$$

The final expression for $P_n(0)$ is, therefore,

$$P_n(0) = \begin{cases} \theta^n(1+p)(1-p)^2 + p^2(1-p)\tilde{a}^n + \dfrac{K(\tilde{a}^n - \theta^n)}{\tilde{a}-\theta} & \text{if } \tilde{a} \neq \theta, \\ \theta^n(1+p)(1-p)^2 + p^2(1-p)\tilde{a}^n + K\tilde{a}^{n-1}n & \text{if } \tilde{a} = \theta, \end{cases}$$

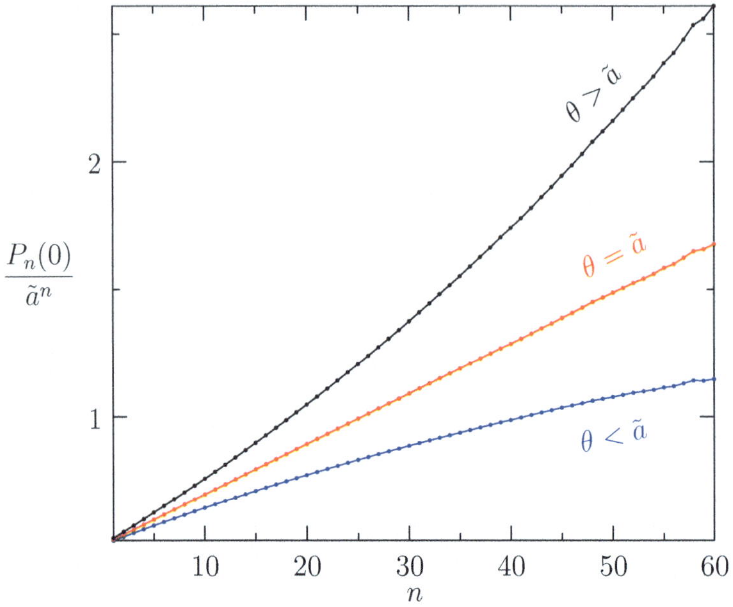

Fig. 9.2 Graph of $P_n(0)/\tilde{a}^n$ obtained numerically for lattice of 10^6 sites for $\alpha = 0.1$, $\beta = 0.2$, $\gamma = 0.135$ (bottom), $\gamma = 0.145$ (middle), and $\gamma = 0.155$ (top)

which further simplifies to

$$P_n(0) = \begin{cases} \theta^n(1+p)(1-p)^2 + p^2(1-p)\tilde{a}^n + \dfrac{K(\tilde{a}^n - \theta^n)}{\tilde{a} - \theta} & \text{if } \tilde{a} \neq \theta, \\ (1-p)\tilde{a}^n + K\tilde{a}^{n-1}n & \text{if } \tilde{a} = \theta, \end{cases} \tag{9.45}$$

where

$$\begin{aligned}
K &= bp^2(1-p)^2 + cp^2(1-p)^3, \\
\theta &= a + b(1-p) + c(1-p)^2, \\
a &= (1-\alpha)(1-\beta), \\
\tilde{a} &= 1 - \gamma, \\
b &= \alpha - 2\alpha\beta + \beta, \\
c &= \alpha\beta.
\end{aligned}$$

The above result exhibits a feature which could be described as "degeneracy". When $\tilde{a} = \theta$, the ratio $P_n(0)/\tilde{a}^n$ becomes a linear function of n, while for $\tilde{a} \neq \theta$ it is never linear. This is illustrated in Fig. 9.2. It would be nearly impossible to discover this subtle phenomenon without having the explicit formula for $P_n(0)$.

Table 9.2 Parameters corresponding to special cases of the PCA rule of Eq. (9.22) for which the rule becomes deterministic

F	α	β	γ	a	b	c	$\tilde{a}$	K	θ
204	0	0	0	1	0	0	1	0	1
236	0	0	1	1	0	0	0	0	1
220	0	1	0	0	1	0	1	$p^2(1-p)^2$	$1-p$
252	0	1	1	0	1	0	0	$p^2(1-p)^2$	$1-p$
206	1	0	0	0	1	0	1	$p^2(1-p)^2$	$1-p$
238	1	0	1	0	1	0	0	$p^2(1-p)^2$	$1-p$
222	1	1	0	0	0	1	1	$p^2(1-p)^3$	$(1-p)^2$
254	1	1	1	0	0	1	0	$p^2(1-p)^3$	$(1-p)^2$

Since deterministic CA are nothing else but special cases of PCA, we can choose integer values of α, β and γ to obtain eight different deterministic rules for which the rule defined by Eq. (9.22) becomes deterministic. Table 9.2 shows all needed parameters for these eight cases. Expressions for $P_n(0)$ given by Eq. (9.45) simplify to those shown below.

- Rule 206 ($\alpha = 1, \beta = 0, \gamma = 0$) or rule 220 ($\alpha = 0, \beta = 1, \gamma = 0$)

$$P_n(0) = p(1-p) + (1-p)^{n+2}, \tag{9.46}$$

- Rule 222 ($\alpha = \beta = 1, \gamma = 0$)

$$P_n(0) = \frac{2(1-p)^{2n+2} + p(1-p)}{2-p},$$

- Rule 236 ($\alpha = 0, \beta = 0, \gamma = 1$)

$$P_n(0) = (p+1)(1-p)^2,$$

- Rule 238 ($\alpha = 1, \beta = 0, \gamma = 1$) or rule 252 ($\alpha = 0, \beta = 1, \gamma = 1$),

$$P_n(0) = (1-p)^{n+1},$$

- Rule 254 ($\alpha = \beta = \gamma = 1$)

$$P_n(0) = (1-p)^{2n+1}.$$

These formulae agree with expressions given in the Appendix C if one takes into the account the fact that for a rule which is a Boolean conjugate of a given rule, one needs to replace $P_n(1)$ by $1 - P_n(0)$ as well as replace p by $1 - p$. For example, rule 206 is not listed in the Appendix C because it is not minimal. However, its Boolean conjugate, rule 140, is listed, and one finds the probabilistic solution for it given by

$$P_n(0) = p^2 - p + 1 - p^{2+n}.$$

If we now take $1 - P_n(0)$ and replace p by $1 - p$, we obtain

$$P_n(0) = -(1-p)^2 + 1 - p + (1-p)^{2+n} = p(1-p) + (1-p)^{n+2},$$

which agrees with the expression obtained for rule 206 in Eq. (9.46).

References

1. Fukś, H.: Non-deterministic density classification with diffusive probabilistic cellular automata. Phys. Rev. E **66**, 066106 (2002). https://doi.org/10.1103/PhysRevE.66.066106
2. Boccara, N., Fukś, H.: Modeling diffusion of innovations with probabilistic cellular automata. In: M. Delorme, J. Mazoyer (eds.) Cellular Automata: A Parallel Model. Kluwer Academic Publishers, Dordrecht (1998)
3. Fatès, N., Morvan, M.: An experimental study of robustness to asynchronism for elementary cellular automata. Complex Systems **16**, 1–27 (2005)
4. Fatès, N., Regnault, D., Schabanel, N., Thierry, É.: Asynchronous behavior of double-quiescent elementary cellular automata. In: J. Correa, A. A. Hevia, M. Kiwi (eds.) LATIN 2006: Theoretical Informatics, *LNCS*, vol. 3887, pp. 455–466 (2006)
5. Fukś, H., Skelton, A.: Orbits of Bernoulli measure in single-transition asynchronous cellular automata. Dis. Math. Theor. Comp. Science **AP**, 95–112 (2012)
6. Fukś, H.: Computing the density of ones in probabilistic cellular automata by direct recursion. In: P.Y. Louis, F.R. Nardi (eds.) Probabilistic Cellular Automata - Theory, Applications and Future Perspectives, Lecture Notes in Computer Science, pp. 131–144. Springer (2018). https://doi.org/10.1007/978-3-319-65558-1
7. Stauffer, D., Aharony, A.: Introduction to percolation theory. Taylor and Francis (1994)
8. Cull, P., Flahive, M., Robson, R.: Difference equations. Springer (2004)

Chapter 10
Applications

10.1 Asymptotic Emulation

In Sect. 2.3, we introduced the concept of emulation. Recall that that a CA rule F emulates rule G in k iterations ($k \geq 0$) or F is *a* kth level emulator of G if

$$F^{k+1} = GF^k. \tag{10.1}$$

Emulation simply means that after k time steps we can replace the rule F by G and we will obtain the same result as if we had kept rule F. If rule F emulates identity, then the spatiotemporal patterns generated by it after a few time steps become identical with the pattern generated by the identity rule (vertical strips), as shown, for example, in Fig. 2.1 for rules 4, 36, 72 or 200—all of which are emulators of identity.

Visual examination of patterns generated by elementary cellular automata reveals that not only emulators of identity produce patterns resembling rule 204 (identity rule). There exist rules, like rule 172 shown in Fig. 4.1, for which the spatiotemporal pattern eventually becomes vertical strips, but time required to achieve such a state may be quite long. Rule 172 does not emulate identity in the sense of Eq. 10.1, but we could say, that it emulates identity "approximately", with the approximation getting better and better with increasing number of time steps.

In order to describe this this phenomenon more formally, we will introduce a distance between rules.

Definition 10.1 Let F, G be binary CA rules of radius r, with the corresponding block mappings, respectively, $\mathbf{f}$ and $\mathbf{g}$. Distance between F and G is defined as

$$d(F, G) = 2^{-2r-1} \sum_{\mathbf{b} \in \{0,1\}^{2r+1}} |\mathbf{f}(\mathbf{b}) - \mathbf{g}(\mathbf{b})| . \tag{10.2}$$

The original version of the chapter has been revised. A correction to this chapter can be found at https://doi.org/10.1007/978-3-031-38700-5_13

H. Fukś, *Solvable Cellular Automata*, Understanding Complex Systems, https://doi.org/10.1007/978-3-031-38700-5_10

It is straightforward to show that this function is a metric in the space of cellular automata rules. Obviously, $d(F, G) \geq 0$ and $d(F, G) = 0 \Leftrightarrow F = G$. Triangle inequality holds too since $|x + y| \leq |x| + |y|$ for all $x, y \in \{0, 1\}$.

Definition 10.2 A cellular automaton rule F *asymptotically emulates rule* G if

$$\lim_{n\to\infty} d(F^{n+1}, GF^n) = 0. \tag{10.3}$$

Clearly, if F is a kth level emulator of G in the sense of Eq. 10.1, then F emulates G asymptotically. We may think about asymptotic emulation as ∞-th level emulation.

Proposition 10.1 *Let F, G be two elementary CA rules with associated block mappings* $\mathbf{f}$ *and* $\mathbf{g}$, *and let* $\mathbf{h}$ *be defined as* $\mathbf{h}(\mathbf{b}) = |\mathbf{f}(\mathbf{b}) - \mathbf{g}(\mathbf{b})|$ *for any* $\mathbf{b} \in \{0, 1\}^3$. *Let* $A_0 = \mathbf{h}^{-1}(1)$ *and* $A_n = \mathbf{f}^{-n}(A_0)$. *Then*

$$d(F^{n+1}, GF^n) = \frac{\text{card } A_n}{2^{2n+3}}. \tag{10.4}$$

Proof The mapping F^{n+1} is a CA rule of radius $n + 2$, therefore using the definition of the distance (10.2) and properties of block mappings we obtain

$$d(F^{n+1}, GF^n) = 2^{-2n-3} \sum_{\mathbf{b}\in\{0,1\}^{2n+3}} \left|\mathbf{f}^{n+1}(\mathbf{b}) - \mathbf{g}(\mathbf{f}^n(\mathbf{b}))\right|, \tag{10.5}$$

or $d(F^{n+1}, GF^n) = 2^{-2n-3}c_n$, where c_n is a number of blocks $\mathbf{b} \in \{0, 1\}^{2n+3}$ such that $\mathbf{f}^{n+1}(\mathbf{b}) \neq \mathbf{g}(\mathbf{f}^n(\mathbf{b}))$. Similarly, the set A_0 is a set of all blocks $\mathbf{b} \in \{0, 1\}^3$ such that $\mathbf{f}(\mathbf{b}) \neq \mathbf{g}(\mathbf{b})$. Let us now consider a block $\mathbf{a} \in \{0, 1\}^{2n+3}$ such that $\mathbf{f}^{n+1}(\mathbf{a}) \neq \mathbf{g}(\mathbf{f}^n(\mathbf{a}))$. The last relation can be written as $\mathbf{f}(\mathbf{f}^n(\mathbf{a})) \neq \mathbf{g}(\mathbf{f}^n(\mathbf{a}))$, and this is possible if and only if $\mathbf{f}^n(\mathbf{a}) \in A_0$, which is equivalent to $\mathbf{a} \in \mathbf{f}^{-n}(A_0)$. This proves that block $\mathbf{a} \in \{0, 1\}^{2n+3}$ satisfies $\mathbf{f}^{n+1}(\mathbf{a}) \neq \mathbf{g}(\mathbf{f}^n(\mathbf{a}))$ if and only if $\mathbf{a} \in A_n$, so finally $c_n =$ card A_n. □

Equation (10.4) can be written as

$$d(F^{n+1}, GF^n) = \sum_{\mathbf{a}\in\mathbf{h}^{-1}(1)} \frac{\text{card } \mathbf{f}^{-n}(\mathbf{a})}{2^{2n+3}}. \tag{10.6}$$

Recall that for elementary rules and symmetric Bernoulli initial measure we derived probabilistic solution formula in Eq. (8.57),

$$P_n(\mathbf{a}) = \frac{\text{card } \mathbf{f}^{-n}(\mathbf{a})}{2^{2n+|\mathbf{a}|}}. \tag{10.7}$$

If $\mathbf{a} \in \mathbf{h}^{-1}(1)$ then $|\mathbf{a}| = 3$, therefore

$$d(F^{n+1}, GF^n) = \sum_{\mathbf{a}\in\mathbf{h}^{-1}(1)} P_n(\mathbf{a}). \tag{10.8}$$

This means that we can use probabilistic solution formulae for computing $d(F^{n+1}, GF^n)$ and proving asymptotic emulation.

Example 10.1 For rule 140, in the Appendix C we find the solution formulae for fundamental probabilities given by

$$\begin{aligned}
P_n(0) &= p^2 - p + 1 - p^{2+n}, \\
P_n(00) &= 2\,p^2 - 2\,p + 1 - p^{2+n}, \\
P_n(000) &= p^4 - 2\,p^3 + 4\,p^2 - 3\,p + 1 - p^{2+n}, \\
P_n(010) &= p^{2+n}\,(p-1) - p\,(p-1)\,.
\end{aligned}$$

For $p = 1/2$, this becomes

$$\begin{aligned}
P_n(0) &= \frac{3}{4} - \left(\frac{1}{2}\right)^{2+n}, \\
P_n(00) &= \frac{1}{2} - \left(\frac{1}{2}\right)^{2+n}, \\
P_n(000) &= \frac{5}{16} - \left(\frac{1}{2}\right)^{2+n}, \\
P_n(010) &= -\frac{1}{2}\left(\frac{1}{2}\right)^{2+n} + \frac{1}{4}.
\end{aligned}$$

We will show that rule 140 asymptotically emulates identity. For this we will need to compute $d(F_{140}^{n+1}, F_{204}F_{140}^n)$. The local function of rule 140 differs from the local function of rule 204 only on one block 110, therefore

$$\mathbf{h}^{-1}(1) = \{110\}.$$

Equation (10.8) becomes

$$d(F_{140}^{n+1}, F_{204}F_{140}^n) = P_n(110),$$

and the only thing left is to express $P_n(110)$ in terms of fundamental block probabilities given in Eq. (7.41),

$$P_n(110) = P_n(0) - P_n(00) - P_n(010).$$

We obtain

$$d(F_{140}^{n+1}, F_{204}F_{140}^n) = \frac{3}{4} - \left(\frac{1}{2}\right)^{2+n} - \frac{1}{2} + \left(\frac{1}{2}\right)^{2+n} + \frac{1}{2}\left(\frac{1}{2}\right)^{2+n} - \frac{1}{4} = \left(\frac{1}{2}\right)^{3+n},$$

Table 10.1 Distance $d(F^{n+1}, F_{204}F^n)$ for rules asymptotically emulating identity

F	$d(F^{n+1}, F_{204}F^n)$	F	$d(F^{n+1}, F_{204}F^n)$
13	$7 \cdot 2^{-n-4}$	132	2^{-2n-2}
32	$5 \cdot 2^{-2n-3}$	136	2^{-n-2}
40	2^{-n-1}	140	2^{-n-3}
44	$7 \cdot 2^{-2n-3}$	160	$3 \cdot 2^{-n-2} - 4^{-n-1}$
77	2^{-2n-2}	164	$5 \cdot 2^{-n-3} - 4^{-n-1}$
78	4^{-1} if $n = 1$ $15 \cdot 2^{-n-6}$ if $n > 1$	168	$3^{n+1} \cdot 2^{-2n-3}$
128	$3 \cdot 2^{-2n-3}$	172	$\dfrac{-(1-\sqrt{5})^{n+3} + (1+\sqrt{5})^{n+3}}{2^{2n+6}\sqrt{5}}$
		232	2^{-2n-2}

which immediately yields

$$\lim_{n\to\infty} d(F_{140}^{n+1}, F_{204}F_{140}^n) = 0,$$

meaning that rule 140 asymptotically emulates identity.

For many other rules asymptotic emulation can be proved is a similar fashion. Table 10.1 shows elementary rules which are asymptotic emulators of identity, with the expressions for the distance $d(F^{n+1}, F_{204}F^n)$ computed using Eq. (10.8), just like in the preceding example. Only rules not included in Table 2.1 are listed. The reader can verify that in all cases the distance $d(F^{n+1}, F_{204}F^n)$ tends to zero as $n \to \infty$. Spatiotemporal patters for these rules are shown in Fig. 10.1.

10.2 Jamming Transitions

In Sect. 6.2, we remarked that the dynamics of rule 184 can be interpreted as as an assembly of interacting particles, following the rule

$$\overset{\curvearrowright}{1}0, \quad \overset{\circlearrowleft}{1}1. \tag{10.9}$$

This means that 1 followed by 0 moves to the right and 1 followed by 1 stays in the same place. Rule 184 is also sometimes called a "traffic rule" because we can interpret 1's as cars instead of particles, obtaining a simple model of road traffic, as shown in Fig. 10.2. In spite of its simplicity, this model exhibits some features of real traffic flow, including jamming transition. It can also serve as an illustrative model of a second order phase transition, featuring critical slowing down and power law convergence to the steady state a the critical point. We will discuss all these phenomena below, using the probabilistic solution of the initial value problem for rule 184.

Fig. 10.1 Spatiotemporal patterns for rules asymptotically emulating identity shown in Table 10.1

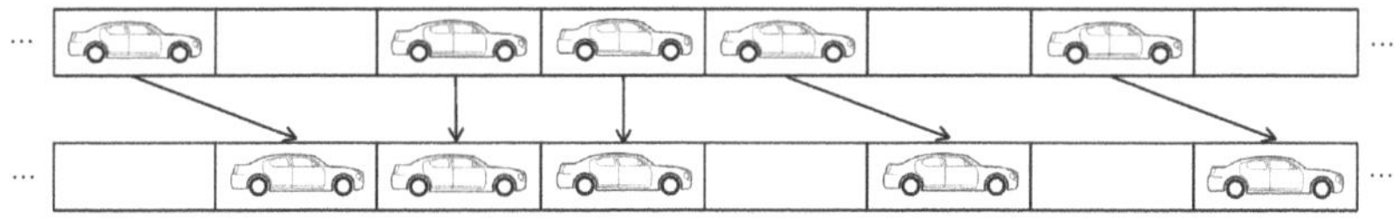

Fig. 10.2 Simple traffic model using rule 184. Two time steps are shown with arrows indicating where the given car moves in one time step

Recall that for rule 184, we obtained

$$P_n(00) = \sum_{j=1}^{n+1} \frac{j}{n+1} \binom{2n+2}{n+1-j} p^{n+1-j} (1-p)^{n+1+j}. \tag{10.10}$$

This expression exhibits interesting behaviour for large n. In order to find the limit $\lim_{n\to\infty} P_n(00)$, we can write the above equation in the form

$$P_n(00) = \sum_{j=1}^{n+1} \frac{j}{n+1} g(n+1-j, 2(n+1), p), \tag{10.11}$$

where

$$g(k, n, p) = \binom{n}{k} p^k (1-p)^{n-k} \tag{10.12}$$

is the distribution function of the binomial distribution. Using de Moivre-Laplace limit theorem, binomial distribution for large n and for k in the neighbourhood of np can be approximated by the normal distribution

$$g(k, n, p) \simeq \frac{1}{\sqrt{2\pi np(1-p)}} \exp \frac{-(k-np)^2}{2np(1-p)}, \tag{10.13}$$

where $\simeq$ means that the ratio of the right hand and the left hand side tends to 1 as $n \to \infty$. To simplify notation, let us define $N = n + 1$. Now, using (10.13) to approximate $g(N - j, 2N, p)$, and approximating sum by an integral, we obtain

$$P_n(00) \simeq \int_1^N \frac{x}{N} \frac{1}{\sqrt{4\pi Np(1-p)}} \exp \frac{-(N-x-2Np)^2}{4Np(1-p)} dx. \tag{10.14}$$

Integration yields

$$\begin{aligned} P_n(00) \simeq & \sqrt{\frac{2p(1-p)}{2\pi N}} \left\{ \exp\left(\frac{-(1-N+2pN)^2}{4Np(1-p)} \right) - \exp\left(\frac{-2pN}{2(1-p)} \right) \right\} \\ & + \frac{1}{2}(1-2p) \left\{ \operatorname{erf}\left(\frac{2pN}{\sqrt{4p(1-p)N}} \right) - \operatorname{erf}\left(\frac{1-N+2pN}{\sqrt{4p(1-p)N}} \right) \right\}, \end{aligned}$$

where $\operatorname{erf}(x)$ denotes the error function

$$\operatorname{erf}(x) = \frac{2}{\sqrt{\pi}} \int_0^x e^{-t^2} dt.$$

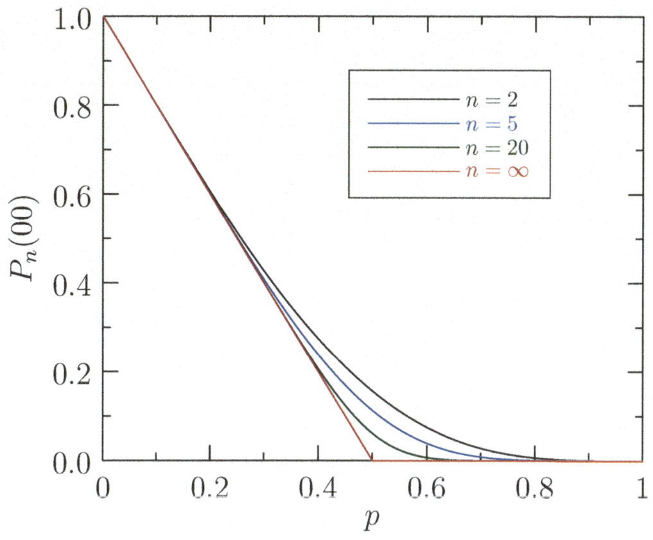

Fig. 10.3 Graphs of $P_n(00)$ versus p for rule 184

The first line in the above equation, involving two exponential functions, tends to 0 with $N \to \infty$. Furthermore, since $\lim_{x\to\infty} \operatorname{erf}(x) = 1$, we obtain

$$\lim_{n\to\infty} P_n(00) = \frac{1}{2}(1-2p)\left\{1 - \lim_{N\to\infty} \operatorname{erf}\left(\frac{1-N+2pN}{\sqrt{4p(1-p)N}}\right)\right\}.$$

Now, noting that

$$\lim_{N\to\infty} \operatorname{erf}\left(\frac{1-N+2pN}{\sqrt{4p(1-p)N}}\right) = \begin{cases} 1, & \text{if } 2p \geq 1, \\ -1, & \text{otherwise,} \end{cases}$$

we obtain

$$P_\infty(00) = \lim_{n\to\infty} P_n(00) = \begin{cases} 1-2p & \text{if } p < 1/2, \\ 0 & \text{otherwise.} \end{cases} \tag{10.15}$$

The behaviour of $P_\infty(00)$ resembles second order phase transition often encountered in statistical physics. When p increases, $P_\infty(00)$ decreases, and at $p = 1/2$ it becomes zero and stays zero for $p > 1/2$. The value of $p_c = 1/2$ will be called *critical*. Using again the terminology of statistical physics, we can call p the *control parameter* and $P_\infty(00)$ the *order parameter*. The order parameter is non-zero when the control parameter is below the critical point, and becomes zero above the critical point. A good example of this phenomenon in the real world is the ferromagnetic phase transition, with the temperature T as the control parameter and magnetization M as the order parameter (Fig. 10.4). When the ferromagnetic material is heated, its net magnetization decreases with temperature and at the critical point, called the

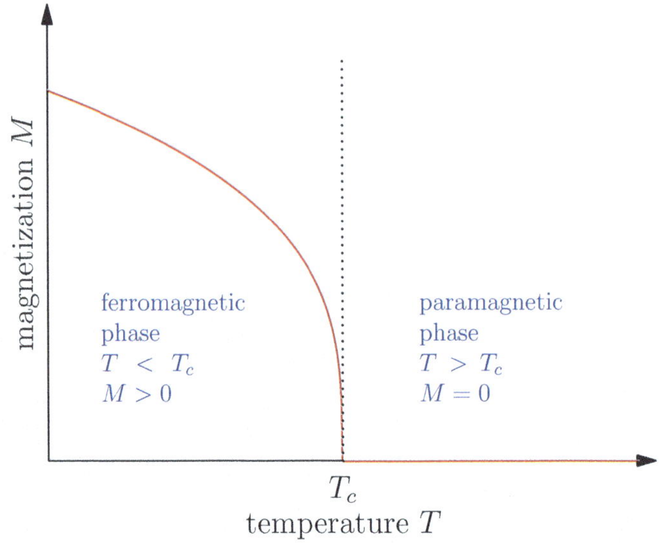

Fig. 10.4 Ferromagnetic phase transition

Curie point T_c, it looses its magnetization completely so that $M = 0$ for $T > T_c$. This transition is called *second order*, because while $M(T)$ is continuous at $T = T_c$, its derivative $\partial M/\partial T$ is not. We can see it in our model as well: $P_\infty(00)$ is a continuous function of p, but at $p = p_c$ its derivative with respect to p is not continuous. On should note, however, that the slope of $P_\infty(00)$ to the left of p_c is finite, while the magnetization has infinite slope to the left of T_c. This indicates that the second order phase transition in rule 184 is of somewhat "simpler" type - it is called the *mean field type* phase transition, but detailed discussion of this subject is beyond the scope of this book.

Figure 10.3 shows several graphs of $P_n(00)$ as a function of p for different n values. We can see that $P_n(00)$ converges toward $P_\infty(00)$ rather quickly, but the speed of convergence depends on the value of p. The slowest convergence happens at the critical point $p = 1/2$. This phenomenon is called *critical slowing down*: the convergence rate decreases as we are getting closer to the critical point p_c.

The behaviour of $P_n(00)$ at the critical point is also worth investigating. When $p = 1/2$, the sum in Eq. (10.10) can be computed in a closed form, and the whole expression simplifies to

$$P_n(00) = 2^{-2n-3}\binom{2n+2}{n+1}.$$

When n is sufficiently large, Stirling's formula can be used to approximate the binomial coefficient, yielding

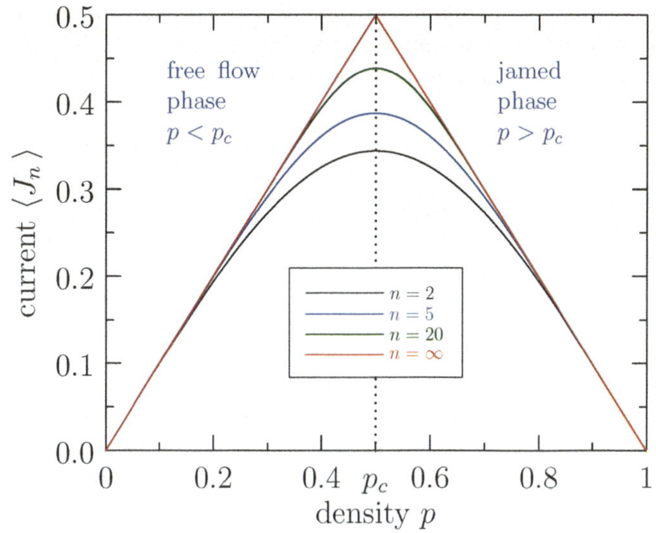

Fig. 10.5 Fundamental diagram for rule 184

$$\binom{2n}{n} \simeq \frac{2^{2n}}{\sqrt{n\pi}}.$$

Applying this to $P_n(00)$ for $p = 1/2$, we obtain

$$P_n(00) \simeq \frac{1}{2\sqrt{\pi n}},$$

meaning that at the critical point, when $n \to \infty$, $P_n(00)$ converges toward $P_\infty(00)$ as $n^{-1/2}$. This type of power law convergence with exponent $-1/2$ has been numerically observed in other cellular automata rules with additive invariants [1] and is very likely universal. Using probabilistic solution formulae given in the Appendix C and Stirling's approximation, the reader can verify that rules 14, 43 and 142 all exhibit $n^{-1/2}$ convergence for some (or all) of their fundamental probabilities at $p = 1/2$.

In Sect. 6.5 we derived the expression for the current for number-conserving rules,

$$J(x_0, x_1) = -f(0, 0, x_0) - f(0, x_0, x_1) - x_0. \tag{10.16}$$

For rule 184, this yields

$$J(x_0, x_1) = x_0(1 - x_1). \tag{10.17}$$

The current $J(y_0, y_1)$ can be interpreted as the number of particles or cars passing from i to $i+1$ per unit time, where $y_i = [F^n_{184}(x)]_i$, $y_{i+1} = [F^n_{184}(x)]_{i+1}$. We can see that movement of a car is only possible when an occupied site has an empty site immediately to its right. Let us compute the expected value of the current after n iterations of rule 184, assuming that the the initial measure is Bernoulli with parameter p. As usual, it suffices to do this for $i = 0$,

$$\langle J_n \rangle := \langle J(y_0, y_1) \rangle = \langle y_0(1 - y_1) \rangle = P_n(10) \cdot 1 + (1 - P_n(10)) \cdot 0 = P_n(10).$$

Since $P_n(10) = P_n(0) - P_n(00)$, this yields

$$\langle J_n \rangle = 1 - p - \sum_{j=1}^{n+1} \frac{j}{n+1} \binom{2n+2}{n+1-j} p^{n+1-j} (1-p)^{n+1+j}.$$

The quantity $\langle J_n \rangle$ represents the average flow of cars, or in other words, average number of cars passing a given point per unit time. Recall that for rule 184 $\langle [F_{184}^n(x)]_j \rangle = p$, therefore p could be interpreted as *the density of cars*. Plot of the expected value of the current as a function of the density of cars is known in traffic engineering as the *fundamental diagram*. The fundamental diagram of rule 184 is shown in Fig. 10.5. In the steady state ($n \to \infty$) it assumes the characteristic tent shape (inverted V), often observed in real traffic data. We can see that there are two phases of the flow, depending on the density. Below the critical density p_c, the flow increases linearly with density. This is called the *free flow phase*. Above the critical density, when the density increases, the flow decreases. This is called the *jammed phase*.

The jamming transition occurring in rule 184 does appear in the real traffic flow, although in order to model the traffic flow more realistically, significantly more complex CA rules are needed. Most of the rules proposed for this purpose are not solvable, although for some generalizations of rule 184 the probabilistic initial value problem can be solved [2]. Interested reader should consult [3] for an extensive review of cellular automata traffic models and their properties.

10.3 Critical Slowing down

In Sect. 10.2 we gave an example of a second order phase transition occurring in rule 184. For the purpose of further considerations, we will define a more general concept of density-parameterized phase transitions in cellular automata. Let $P_n(\mathbf{a})$ be the probability of block $\mathbf{a}$ after n iterations of a cellular automaton F starting from a Bernoulli measure μ_p. Moreover, suppose that the limit $P_\infty(\mathbf{a}) = \lim_{n\to\infty} P_n(\mathbf{a})$ exists.

Definition 10.3 If the probability $P_\infty(\mathbf{a})$ treated as a function of p is discontinuous at some point $p = p_c$, we will say that it exhibits the *first order phase transition* at $p = p_c$. If $P_\infty(\mathbf{a})$ it is continuous but $dP_\infty(\mathbf{a})/dp$ is discontinuous at $p = p_c$, we will say that $P_\infty(\mathbf{a})$ exhibits the *second order phase transition* at $p = p_c$. The point $p = p_c$ will be called the *critical point*.

In statistical physics it is a well know phenomenon that systems exhibiting phase transitions slow down in the vicinity of the critical point. Usually one demonstrates this by computer simulations. Since there are many examples of CA with phase transition for which explicit formulae for block probabilities have been obtained in this book, we will be able to investigate the phenomenon of critical slowing down without resorting to computer simulations.

10.3.1 First Order Transitions

Among elementary CA, there is no rule with the first order phase transition for which the critical point would be inside the interval [0, 1]. There are, however, many cases when the transition happens at the end of the interval [0, 1], with $p_c = 0$ or $p_c = 1$. Such transitions have been studied in detail by J. Zabolitzky in [4] using ECA rule 22 as an example.

Before we give our own example, we need to introduce a way to quantify the rate of convergence of $P_n(\mathbf{a})$ toward $P_\infty(\mathbf{a})$. Our motivation will be continuous-time dynamics. Suppose that a time-dependent quantity $m(t)$ converges toward its limiting value $m(\infty)$ exponentially,

$$m(t) = m(\infty) + e^{-t/T},$$

where $T > 0$ is a constant characterizing the rate of convergence. Obviously for large T the of convergence is slow, and for smaller T it is more rapid. By direct integration is easy to verify that

$$\int_0^\infty |m(t) - m(\infty)| dt = \int_0^\infty e^{-t/T} dt = \left[-Te^{-t/T}\right]_0^\infty = T.$$

The integral $\int_0^\infty |m(t) - m(\infty)| dt$ might, therefore, serve as a measure of the rate of convergence. If the convergence toward the limiting value is not exponential but follows a power law, for example

$$m(t) = m(\infty) + t^{-a} \quad 0 < a < 1,$$

the integral $\int_{t_0}^\infty |m(t) - m(\infty)| dt$ diverges for any $t_0 \geq 0$. This feature can be useful for distinguishing between the exponential and power law convergence toward the limiting value.

For a system with discrete time, we can replace the integral by the sum. Define, therefore, in analogy to the continuous case discussed above,

$$\tau_n(p) = \sum_{k=1}^{n} |P_k(\mathbf{a}) - P_\infty(\mathbf{a})|,$$

and

$$\tau(p) = \lim_{n\to\infty} \tau_n(p).$$

We will now show that for ECA exhibiting the first order phase transition, the rate of convergence of $P_n(\mathbf{a})$ toward $P_\infty(\mathbf{a})$ increases as p approaches p_c, and becomes infinite at $p = p_c$, meaning that $\lim_{p\to p_c} \tau(p) = \infty$.

Example 10.2 Consider rule 160 withe the local function

$$f_{160}(x_1, x_2, x_3) = x_1 x_3.$$

The probabilistic solution for this rule listed in the Appendix C is

$$P_n(0) = 1 - p^{n+1},$$

hence

$$P_\infty(0) = \begin{cases} 1 & p \in [0, 1), \\ 0 & p = 1. \end{cases}$$

$P_\infty(0)$ is discontinuous at $p = 1$, thus we have a first order phase transition with $p_c = 1$. Let us compute the convergence rate. For $p < 1$ we have

$$\tau_n(p) = \sum_{k=1}^{n} |P_k(0) - P_\infty(0)| = \sum_{k=1}^{n} p^{k+1} = \frac{p^{2+n} - p^2}{p - 1},$$

$$\tau(p) = \lim_{n\to\infty} \frac{p^{2+n} - p^2}{p - 1} = \frac{p^2}{1 - p}$$

We can clearly see that $\tau(p)$ increases as $p \to p_c$ and that $\lim_{p\to p_c} \tau(p) = \infty$. Critical slowing down near p_c is apparent.

Example 10.3 A more complex example is the ECA rule 60 with the local function

$$f_{60}(x_1, x_2, x_3) = x_2 + x_1 - 2\, x_1 x_2 = x_1 + x_2 \mod 2.$$

For this rule we found

$$P_n(0) = \frac{1}{2} + \frac{1}{2}\, (1 - 2\, p)^{G(n)},$$

where $G(n)$ is the Gould's sequence,

$$G(n) = \sum_{k=0}^{n} \left(\binom{n}{k} \mod 2 \right).$$

Since for $n_m = 2^m - 1$ we have $G(n_m) \to \infty$ as $m \to \infty$, we will define $\tilde{p}_m(0) = p_{n_m}(0)$. The limiting value of $\tilde{P}_m(0)$ is

$$\tilde{P}_\infty(0) = \begin{cases} \frac{1}{2} & p \in (0, 1), \\ 1 & p \in \{0, 1\}, \end{cases}$$

meaning that we have two discontinuities and two phase transitions of the first order, at $p = 0$ and $p = 1$. The rate of convergence of $\tilde{p}_m(0)$ for $p \in (0, 1)$ becomes

$$\tau_m(p) = \sum_{k=1}^{m} |\tilde{P}_k(0) - \tilde{P}_\infty(0)| = \sum_{k=1}^{m} (1 - 2p)^{G(n_k)},$$

$$\tau(p) = \lim_{m \to \infty} \tau_m(p) = \sum_{k=1}^{\infty} (1 - 2p)^{G(n_k)}.$$

Although it is unfortunately not possible to obtain the closed form expression for $\tau(p)$, we can still compute $\tau_m(p)$ for a fixed m and observe how it changes as m increases. For $m = 3, 4, 8$, the expressions are

$$\begin{aligned} \tau_3(p) &= (1 - 2\,p)^2 + (1 - 2\,p)^4 + (1 - 2\,p)^8\,, \\ \tau_4(p) &= (1 - 2\,p)^2 + (1 - 2\,p)^4 + (1 - 2\,p)^8 + (1 - 2\,p)^{16}\,, \\ \tau_8(p) &= (1 - 2\,p)^2 + (1 - 2\,p)^4 + (1 - 2\,p)^8 + (1 - 2\,p)^{16} + (1 - 2\,p)^{32} \\ &\quad + (1 - 2\,p)^{64} + (1 - 2\,p)^{128} + (1 - 2\,p)^{256}\,. \end{aligned}$$

Graphs of the above are shown in Fig. 10.6. Again, it is apparent that near the critical points the convergence rate develops a singularity. Keeping in mind that $G(n)$ is always even, we obtain for $p = 0$

$$\tau(0) = \sum_{k=1}^{\infty} 1^{G(n_k)} = \infty,$$

and for $p = 1$

$$\tau(1) = \sum_{k=1}^{\infty} (-1)^{G(n_k)} = \infty,$$

confirming our observations.

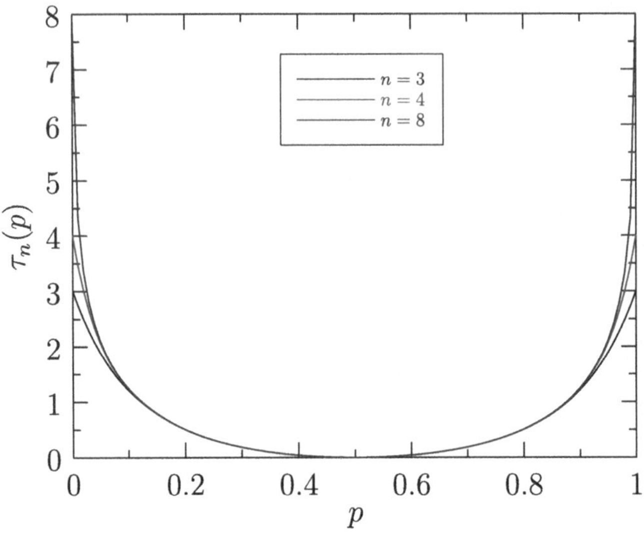

Fig. 10.6 Critical slowing down near $p = 0$ and $p = 1$ for ECA rule 60

10.3.2 Second Order Transitions

Second order transition appears in at least four elementary cellular automata [5], namely in rules 14, 35, 43, 142 and 184. Only for rule 184 we have explicit expressions for fundamental block probabilities to our disposal, thus we will use this rule as an example. The transition has already been described in Sect. 10.2, recall that

$$P_n(00) = \sum_{j=1}^{n+1} \frac{j}{n+1} \binom{2n+2}{n+1-j} p^{n+1-j} (1-p)^{n+1+j},$$

and

$$P_\infty(00) = \lim_{n\to\infty} P_n(00) = \begin{cases} 1-2p & \text{if } p < 1/2, \\ 0 & \text{otherwise.} \end{cases}$$

The probability $P_\infty(0)$ is continuous at $p_c = 1/2$, but its derivative with respect to p is not,

$$\frac{dP_\infty(00)}{dp} = \begin{cases} -2 & \text{if } p < 1/2, \\ \text{undefined} & \text{if } p = 1/2, \\ 0 & \text{f } p > 1/2, \end{cases}$$

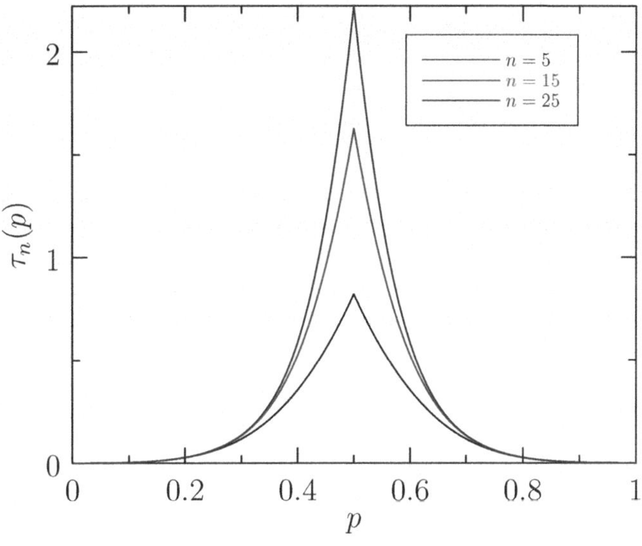

Fig. 10.7 Critical slowing down near $p = 1/2$ for ECA rule 184

confirming that the transition is of second order. Let us compute τ,

$$\tau_n(p) = \begin{cases} \sum_{k=1}^{n} \left(\sum_{j=1}^{k+1} \frac{j}{k+1} \binom{2k+2}{k+1-j} p^{k+1-j} (1-p)^{k+1+j} - 1 + 2p \right) & p < 1/2, \\ \sum_{k=1}^{n} \sum_{j=1}^{k+1} \frac{j}{k+1} \binom{2k+2}{k+1-j} p^{k+1-j} (1-p)^{k+1+j} & p \geq 1/2. \end{cases}$$

As in the case of rule 60, it is not possible to obtain a closed form expression for the above sums, but we can still plot $\tau_n(p)$ for fixed values of n. Figure 10.7 shows such graphs for $n = 5$, 15 and 25. Development of the singularity at $p = 1/2$ is again clearly visible.

We can actually compute exact value of $\tau_n(p)$ for $p = p_c = 1/2$,

$$\begin{aligned} \tau_n(1/2) &= \sum_{k=1}^{n} \sum_{j=1}^{k+1} \frac{j}{k+1} \binom{2k+2}{k+1-j} 2^{-2-2k} \\ &= \sum_{k=1}^{n} \frac{k+2}{k+1} \binom{2k+2}{k} 2^{-3-2k} = \frac{(n+3)}{4^{n+2}} \binom{2n+4}{n+1} - \frac{3}{4} \\ &= \frac{(n+3)}{4^{n+2}} \frac{(2n+4)!}{(n+1)!\,(n+3)!} - \frac{3}{4}. \end{aligned}$$

Using Stirling's approximation for factorials in the above, it is straightforward to demonstrate that

$$\tau(1/2) = \lim_{n\to\infty} \tau_n(1/2) = \infty,$$

which we leave as an exercise to the reader. This confirms that the convergence slows down as p approaches $p_c = 1/2$, with τ becoming infinite at the critical point.

In cellular automata with radius larger than 1, second order phase transitions like the above sometimes occur at more than one value of p. In particular, rules with additive invariants often exhibit multiple phase transitions. Detailed numerical studies exploring the critical behaviour of such rules are described in Refs. [1, 6].

10.4 Density Classification

Finding cellular automata capable of performing specific computational tasks has been a subject of active research since the inception of the field of cellular automata. One such a task is the so-called *density classification task*. The CA performing this task should converge to a configuration of all 1's if the initial finite configuration with periodic boundary conditions contains more 1's than 0's, and to all 0's if the initial configuration contains more 0's than 1's. This should happen within k time steps, where, in general, k can depend on the lattice size L.

In 1995, M. Land and R. K. Belew demonstrated that no binary CA exist which would solve the density classification problem. For this reason, many modified versions of the problem have been proposed. One of such modifications is to allow more than one rule, and indeed in this formulation the problem is solvable, as demonstrated in [7]. The solution uses ECA rules 184 and 232, and we will now show how to prove the validity of this method using solutions of the initial value problem for rules 184 and 232 obtained in this book.

Let $n > 0$ and let $\mathbf{a} \in \{0, 1\}^{2n+1}$ be a binary string we want to classify. We will identify periodic configuration $x \in \{0, 1\}^{\mathbb{Z}}$ of period $L = 2n + 1$ with the finite string $\mathbf{a}$ by setting $x_0 = a_1, x_1 = a_2, \ldots, x_{2n} = a_{2n+1}$ and $x_i = x_{i \bmod L}$ for all remaining indices i. If the string $\mathbf{a}$ contains more 1's than 0's, we will call the string and the corresponding configuration *1-dense*, otherwise we will call them *0-dense*.

Proposition 10.2 *If $x \in \{0, 1\}^{\mathbb{Z}}$ is a periodic configuration with period $2n + 1$ and is 1-dense, then $[F_{232}^n F_{184}^n(x)]_j = 1$ for all $j \in \mathbb{Z}$. If x is 0-dense, then $[F_{232}^n F_{184}^n(x)]_j = 0$ for all $j \in \mathbb{Z}$.*

Two results will be needed for the proof. The first is Proposition 8.3 which says that the string $\mathbf{b} \in \{0, 1\}^{2n+2}$ belongs to $\mathbf{f}_{184}^{-n}(00)$ if and only if

$$\sum_{i=1}^{k} b_i < k/2, \tag{10.18}$$

for all $k \in \{1, \ldots, 2n+2\}$. Suppose that **a** identified with x is 1-dense and let $y = F_{184}^n(x)$. If, for some j, $y_j y_{j+1} = 00$, then the string $x_{j-n}x_{j-n+1}\ldots x_{j+n+1}$ must belong to n-step preimages of 00 under the rule 184. As such, by the virtue of Proposition 8.3, it must have the property

$$\sum_{i=1}^{2n+1} x_{j-n+i-1} < (2n+1)/2, \tag{10.19}$$

where we wrote Eq. (10.18) for $k = 2n+1$. The above inequality is impossible to be satisfied by the 1-dense string, thus we conclude that y contains no 00 pairs.

The second result is the solution of the initial value problem for rule 232,

$$\begin{aligned}[F_{232}^n(x)]_j &= x_{j-n}\prod_{i=1}^{n}\overline{x}_{j+2i-n-1}x_{j+2i-n}\\ &+\sum_{k=1}^{n} x_{j-k}x_{j-k+1}\left(\prod_{i=1}^{k-1}\overline{x}_{j-k+2i}x_{j-k+2i+1}\right)\overline{x}_{j+k}\end{aligned}$$

We use the convention that $\prod_{i=a}^{b} f(i) = 1$ if $b < a$.

For $j = 0$, the above can be written as

$$\begin{aligned}[F_{232}^n(x)]_0 &= \underbrace{x_{-n}\overline{x}_{-n+1}x_{-n+2}\ldots\overline{x}_{n-1}x_n}_{2n+1 \text{ alternating } x \text{ and } \overline{x}}\\ &+\sum_{k=1}^{n} x_{-k}x_{-k+1}\underbrace{\overline{x}_{-k+2}x_{-k+3}\ldots\overline{x}_{k-2}x_{k-1}\overline{x}_k}_{2k-1 \text{ alternating } \overline{x} \text{ and } x}\\ &+\sum_{k=1}^{n}\underbrace{x_{-k+1}\overline{x}_{-k+2}\ldots x_{k-3}\overline{x}_{k-2}}_{2k-2 \text{ alternating } x \text{ and } \overline{x}}x_{k-1}x_k.\end{aligned} \tag{10.20}$$

We will prove the following lemma.

Lemma 10.1 *The set of all blocks* $x_{-n}x_{-n+1}\ldots x_n \in \{0,1\}^{2n+1}$ *containing no 00 pairs is represented by the density polynomial given by the right hand side of eq. (10.20).*

Proof If the string $x_{-n}x_{-n+1}\ldots x_n$ has no pair of 00, then it must have at least one pair 11 or be of the form 1010…01. Let us consider the first possibility. If the string has at least one pair of 11, we can always choose the pair which is closest to the origin in the sense that among pairs $x_i x_{i+1} = 11$ we choose the one that has the smallest absolute value of i. That smallest index value will be denoted m, so that the closest pair to the origin is $x_m x_{m+1} = 11$.

If $m \geq 0$, then there must be no other 11 pair on the left within the same distance, thus we must have

$$x_{-m}x_{-m+1}\dots x_m x_{m+1} = \underbrace{1010\dots 10}_{2m}11.$$

Putting $k = m + 1$, the density polynomial for the above string is

$$\underbrace{x_{-k+1}\overline{x}_{-k+2}\dots x_{k-3}\overline{x}_{k-2}}_{2k-2 \text{ alternating } x \text{ and } \overline{x}} x_{k-1}x_k.$$

Since k can vary from 1 to n, the density polynomial of all such strings is

$$\sum_{k=1}^{n} \underbrace{x_{-k+1}\overline{x}_{-k+2}\dots x_{k-3}\overline{x}_{k-2}}_{2k-2 \text{ alternating } x \text{ and } \overline{x}} x_{k-1}x_k. \tag{10.21}$$

If $m < 0$, let $k = -m$. There must be no other 11 pair on the right within the same distance, therefore we must have

$$x_{-k}x_{-k+1}\dots x_{k-1}x_k = 11\underbrace{1010\dots 10}_{2k-1}.$$

Again, the density polynomial for this string is

$$x_{-k}x_{-k+1}\underbrace{\overline{x}_{-k+2}x_{-k+3}\dots \overline{x}_{k-2}x_{k-1}\overline{x}_k}_{2k-1 \text{ alternating } \overline{x} \text{ and } x},$$

and the sum of all of them

$$\sum_{k=1}^{n} x_{-k}x_{-k+1}\underbrace{\overline{x}_{-k+2}x_{-k+3}\dots \overline{x}_{k-2}x_{k-1}\overline{x}_k}_{2k-1 \text{ alternating } \overline{x} \text{ and } x}. \tag{10.22}$$

We are left with the case when there is no 11 pair at all. The only possibility for this is to have alternating sequence of $2n + 1$ 1's and 0's starting and ending with 1, for which the density polynomial is

$$\underbrace{x_{-n}\overline{x}_{-n+1}x_{-n+2}\dots \overline{x}_{n-1}x_n}_{2n+1 \text{ alternating } x \text{ and } \overline{x}} \tag{10.23}$$

Combining expressions (10.21),(10.22), and (10.23), we obtain the right hand side of eq. (10.20). □

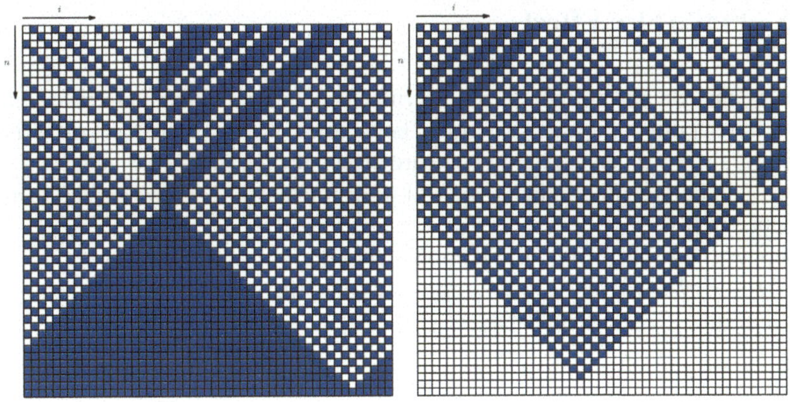

Fig. 10.8 Density classification with rules 184 and 232, for 1-dense string (left) and 0-dense string (right)

We already know that for 1-dense x the configuration $y = F^n_{184}(x)$ contains no 00 pairs. Let $z = F^n_{232}(y)$. The lemma says that n-step preimages of 1 under the rule 232 and stings without 00 pairs are one and the same thing, yet every z_i is an n-step image of a string with no 00 pairs, thus $z_i = 1$ must hold for every i.

If x is 0-dense, an almost identical reasoning can be constructed to proof that $[F^n_{232}F^n_{184}(x)]_j = 0$ for all $j \in \mathbb{Z}$, except that one needs to use Proposition 8.2 instead of 8.3.

Let us illustrate the process of density classification with two examples of strings of length 51, one 1-dense and the other 0-dense. Their classification is shown in Fig. 10.8. Rules are switched from 184 to 232 after 25 iterations, and we can see that the bottom row is all 1's in the first case and all 0's in the second case. The first example also shows that the time required to classify strings cannot be improved—the configuration of all 1's is reached at the very last row.

As a final remark, let us remark that an immediate consequence of Proposition 10.2 is that rule 232, when restricted to a particular subset of strings, can perform the density classification task alone.

Corollary 10.1 *Let* $\mathbf{a} = a_1a_2\ldots a_{2n+1}$ *be a binary string which contains only one type of matching pairs, that is, only some 00 pairs but no 11, or conversely, only some 11 pairs but no 00. Then*

$$\mathbf{f}^n_{232}(\mathbf{a}) = \text{majority}\,(a_1, a_2, \ldots, a_{2n+1}).$$

10.5 Finite Size Effects

In many models and in numerical experiments cellular automata are investigated using finite configurations with periodic boundaries. Knowing the solution formula for a given CA we can explicitly compare orbits of such finite configurations with orbits of infinite configurations. We will do it for rule 172, following the method described in [8].

Suppose now that the initial condition is periodic with period L, so that $x_i = x_{i+L}$ for all $i \in \mathbb{Z}$. We will take $i = 0, \dots L-1$ as the principal period of this configuration. For rule 172, we found that

$$[F^n_{172}(x)]_j = \overline{x}_{j-2}\overline{x}_{j-1}x_j + \left(\overline{x}_{j+n-2}x_{j+n-1} + x_{j+n-2}x_{j+n}\right) \prod_{i=j-2}^{j+n-3} (1 - \overline{x}_i\overline{x}_{i+1}).$$

We can of course apply this solution formula to our periodic configuration, except that all indices are to be taken modulo L. In order to obtain probabilistic solution for periodic configuration, let us further assume that $x_j = X_j$ for $j \in \{0, 1, \dots, L-1\}$, where X_j are independent and identically distributed random variables such that $Pr(X_j = 1) = p$, $Pr(X_j = 0) = 1 - p$, where $p \in [0, 1]$.

The expected value of a site with index j after n iterations of rule 172 can be obtained by taking the expected value of both sides of Eq. (10.24), yielding

$$\begin{aligned}\langle [F^n(x)]_j \rangle &= (1-p)^2 p \\ &+ \left\langle \left(\prod_{i=j-2}^{j+n-3} (1 - \bar{x}_i\bar{x}_{i+1}) \right) (\bar{x}_{j+n-2}x_{j+n-1} + x_{j+n-2}x_{j+n}) \right\rangle.\end{aligned} \tag{10.24}$$

The expected value must be j-independent, so the choice of j does not matter. We can choose $j = 2$ without the loss of generality and for this choice of j the only indices of x occurring on the right hand side of the above will be in the range from 0 to $n+2$. This means that for for $n \le L-3$, we will not need to use modulo L operation to bring the index to the principal period range. For this reason, the probabilistic solution for rule 172 obtained in Sect. 8.5 remains valid in the periodic case as long as $n \le L-3$.

Let us now consider the case of $n \ge L-3$. We start from a slightly stronger condition, $n \ge L$, leaving the additional cases of $n = L-1$ and $n = L-2$ for later. For $n \ge L$ we have

$$\prod_{i=j-2}^{j+n-3} (1 - \bar{x}_i\bar{x}_{i+1}) = \prod_{i=0}^{L-1}(1 - \bar{x}_i\bar{x}_{i+1}), \tag{10.25}$$

because in the product on the left hand side there are only L different factors, and $(1 - \bar{x}_i\bar{x}_{i+1})^m = (1 - \bar{x}_i\bar{x}_{i+1})$ for any positive integer m. Since $\langle [F^n(x)]_j \rangle$ should be the same for all j, we can take $j = L - n$,

$$\begin{aligned}\langle [F^n(x)]_j \rangle = \langle [F^n(x)]_{L-n} \rangle &= (1-p)^2 p \\ &+ \left\langle \left(\prod_{i=0}^{L-1} (1 - \bar{x}_i\bar{x}_{i+1}) \right) (\bar{x}_{L-2}x_{L-1} + x_{L-2}x_0) \right\rangle . \end{aligned} \tag{10.26}$$

Before we proceed any further, let us note that

$$\begin{aligned}\bar{x}_{L-2}x_{L-1} + x_{L-2}x_0 &= \bar{x}_{L-2}(1 - \bar{x}_{L-1}) + (1 - \bar{x}_{L-2})(1 - \bar{x}_0) \\ &= 1 - \bar{x}_{L-2}\bar{x}_{L-1} - \bar{x}_0 + \bar{x}_0\bar{x}_{L-2}.\end{aligned}$$

Since $(1 - \bar{x}_{L-2}\bar{x}_{L-1}) \prod_{i=0}^{L-1} (1 - \bar{x}_i\bar{x}_{i+1}) = \prod_{i=0}^{L-1} (1 - \bar{x}_i\bar{x}_{i+1})$, we obtain

$$\begin{aligned}\langle [F^n(x)]_j \rangle = (1-p)^2 p &+ \left\langle \prod_{i=0}^{L-1} (1 - \bar{x}_i\bar{x}_{i+1}) \right\rangle \\ &+ \left\langle \bar{x}_0(\bar{x}_{L-2} - 1) \prod_{i=0}^{L-1} (1 - \bar{x}_i\bar{x}_{i+1}) \right\rangle .\end{aligned} \tag{10.27}$$

We will consider the the two expected values on the right hand side separately. Let us start from the first one.

$$\begin{aligned}\left\langle \prod_{i=0}^{L-1} (1 - \bar{x}_i\bar{x}_{i+1}) \right\rangle &= \left\langle (1 - \bar{x}_0\bar{x}_1) \prod_{i=1}^{L-2} (1 - \bar{x}_i\bar{x}_{i+1})(1 - \bar{x}_{L-1}\bar{x}_0) \right\rangle \\ &= \left\langle (1 - \bar{x}_0\bar{x}_1)(1 - \bar{x}_{L-1}\bar{x}_0) \prod_{i=1}^{L-2} (1 - \bar{x}_i\bar{x}_{i+1}) \right\rangle \\ &= \left\langle (1 - \bar{x}_0\bar{x}_{L-1} - \bar{x}_0\bar{x}_1 + \bar{x}_0\bar{x}_1\bar{x}_{L-1}) \prod_{i=1}^{L-2} (1 - \bar{x}_i\bar{x}_{i+1}) \right\rangle \\ &= \left\langle \prod_{i=1}^{L-2} (1 - \bar{x}_i\bar{x}_{i+1}) \right\rangle - \langle \bar{x}_0 \rangle \left\langle \bar{x}_{L-1} \prod_{i=1}^{L-2} (1 - \bar{x}_i\bar{x}_{i+1}) \right\rangle \\ &\quad - \langle \bar{x}_0 \rangle \left\langle \bar{x}_1 \prod_{i=1}^{L-2} (1 - \bar{x}_i\bar{x}_{i+1}) \right\rangle + \langle \bar{x}_0 \rangle \left\langle \bar{x}_1\bar{x}_{L-1} \prod_{i=1}^{L-2} (1 - \bar{x}_i\bar{x}_{i+1}) \right\rangle .\end{aligned}$$

Similarly as in Sect. 8.5 we define

$$U_n = \prod_{i=1}^{n-1} \left(1 - \bar{X}_i \bar{X}_{i+1}\right) \quad \text{and} \quad V_n = \bar{X}_n \prod_{i=1}^{n-1} \left(1 - \bar{X}_i \bar{X}_{i+1}\right), \tag{10.28}$$

and additionally

$$U'_n = \bar{X}_1 \prod_{i=1}^{n-1} \left(1 - \bar{X}_i \bar{X}_{i+1}\right) \quad \text{and} \quad V'_n = \bar{X}_1 \bar{X}_n \prod_{i=1}^{n-1} \left(1 - \bar{X}_i \bar{X}_{i+1}\right). \tag{10.29}$$

Now the first expected value of Eq. (10.27) can be written as

$$\begin{aligned}\left\langle \prod_{i=0}^{L-1} (1 - \bar{x}_i \bar{x}_{i+1}) \right\rangle &= \langle U_{L-1} \rangle - (1-p)\langle V_{L-1} \rangle - (1-p)\langle U'_{L-1} \rangle + (1-p)\langle V'_{L-1} \rangle \\ &= \langle U_L \rangle + p\langle U'_{L-1} \rangle - \langle U'_{L-1} \rangle + (1-p)\langle V'_{L-1} \rangle \\ &= \langle U_L \rangle + p\langle U'_{L-1} \rangle - \langle U'_L \rangle,\end{aligned} \tag{10.30}$$

where we used the fact that $\langle U'_n \rangle, \langle V'_n \rangle$ satisfy the same recurrence equations as $\langle U_n \rangle, \langle V_n \rangle$ (see Sect. 8.5 for details).

The second expected value in Eq. (10.27) can be expressed in terms of U' and V' as well,

$$\begin{aligned}&\left\langle \bar{x}_0(\bar{x}_{L-2} - 1) \prod_{i=0}^{L-1} (1 - \bar{x}_i \bar{x}_{i+1}) \right\rangle \\ &= \left\langle \bar{x}_0(\bar{x}_{L-2} - 1)(1 - \bar{x}_{L-2}\bar{x}_{L-1})(1 - \bar{x}_{L-1}\bar{x}_0) \prod_{i=0}^{L-3} (1 - \bar{x}_i \bar{x}_{i+1}) \right\rangle \\ &= \left\langle \bar{x}_0(\bar{x}_{L-2} - \bar{x}_{L-2}\bar{x}_{L-1} - 1 + \bar{x}_{L-1}) \prod_{i=0}^{L-3} (1 - \bar{x}_i \bar{x}_{i+1}) \right\rangle \\ &= \left\langle \bar{x}_0 \bar{x}_{L-2} \prod_{i=0}^{L-3} (1 - \bar{x}_i \bar{x}_{i+1}) \right\rangle + \left\langle \bar{x}_0 (1 - \bar{x}_{L-2}\bar{x}_{L-1}) \prod_{i=0}^{L-3} (1 - \bar{x}_i \bar{x}_{i+1}) \right\rangle \\ &\quad - 2\left\langle \bar{x}_0 \prod_{i=0}^{L-3} (1 - \bar{x}_i \bar{x}_{i+1}) \right\rangle + \left\langle \bar{x}_0 \bar{x}_{L-1} \prod_{i=0}^{L-3} (1 - \bar{x}_i \bar{x}_{i+1}) \right\rangle \\ &= \langle V'_{L-1} \rangle + \langle U'_k \rangle - 2\langle U'_{L-1} \rangle + (1-p)\langle U'_{L-1} \rangle.\end{aligned}$$

Combining both expected values computed above we get

$$\begin{aligned}\langle [F^n(x)]_j\rangle &= (1-p)^2 p + \langle U_L\rangle + p\langle U'_{L-1}\rangle - \langle U'_L\rangle + \langle V'_{L-1}\rangle + \langle U'_k\rangle \\ &\quad - 2\langle U'_{L-1}\rangle + (1-p)\langle U'_{L-1}\rangle = (1-p)^2 p + \langle U_L\rangle - \langle U'_{L-1}\rangle + \langle V'_{L-1}\rangle.\end{aligned}$$

Since

$$\langle V'_k\rangle = (1-p)\langle U'_{L-1}\rangle - (1-p)\langle V'_{L-1}\rangle,$$

we obtain

$$\langle U'_{L-1}\rangle - \langle V'_{L-1}\rangle = \frac{1}{1-p}\langle V'_k\rangle,$$

and therefore

$$\langle [F^n(x)]_j\rangle = (1-p)^2 p + \langle U_L\rangle - \frac{1}{1-p}\langle V'_k\rangle.$$

The expected value of U_L is already known, it has been calculated in Sect. 8.5, thus we only need to find $\langle V'_k\rangle$. As noted before, the recurrence equations for $\langle U'_n\rangle$, $\langle V'_n\rangle$ are the same as for $\langle U_n\rangle$, $\langle V_n\rangle$, that is, as in Eq. (8.26). Only the initial conditions for $n=2$ are different,

$$\langle U'_2\rangle = \langle \bar{X}_1(1-\bar{X}_1\bar{X}_2)\rangle = \langle \bar{X}_1 - \bar{X}_1\bar{X}_2\rangle = 1-p-(1-p)^2 = p-p^2,$$

$$\langle V'_2\rangle = \langle \bar{X}_1\bar{X}_2(1-\bar{X}_1\bar{X}_2)\rangle = \langle \bar{X}_1\bar{X}_2 - \bar{X}_1\bar{X}_2\rangle = 0.$$

The solution of the recursion given in Eq. (8.37) for U' and V' thus becomes

$$\begin{bmatrix}\langle U'_n\rangle\\ \langle V'_n\rangle\end{bmatrix} = P\begin{bmatrix}\lambda_1^{n-2} & 0\\ 0 & \lambda_2^{n-2}\end{bmatrix}P^{-1}\begin{bmatrix}p-p^2\\ 0\end{bmatrix}, \tag{10.31}$$

where P and P^{-1} are defined in eq. (8.36). After simplification this yields

$$\langle V'_n\rangle = \frac{p(1-p)^2}{\lambda_1-\lambda_2}\left(\lambda_1^{n-2} - \lambda_2^{n-2}\right), \tag{10.32}$$

leading to the final result

$$\begin{aligned}\langle [F^n(x)]_j\rangle &= (1-p)^2 p + \frac{p}{\lambda_2-\lambda_1}\left(a_1\lambda_1^{L-2} + a_2\lambda_2^{L-2}\right)\\ &\quad - \frac{p(1-p)}{\lambda_1-\lambda_2}\left(\lambda_1^{L-2} - \lambda_2^{L-2}\right),\end{aligned}$$

which further simplifies to

$$\langle [F^n(x)]_j \rangle = (1-p)^2 p + \frac{p}{\lambda_2 - \lambda_1}\left((a_1 + 1 - p)\lambda_1^{L-2} + (a_2 - 1 + p)\lambda_2^{L-2}\right), \quad (10.33)$$

where $\lambda_{1,2}$ and $a_{1,2}$ are defined in Eq. (8.21). Note that the above expression does not depend on n, which means that $\langle [F^n(x)]_j \rangle$ becomes constant when $n \geq L$. One can also show (we will omit details) that for $n = L-1$ and $n = L-2$, Eq. (10.33) remains valid. Let us then summarize the final result for the periodic configuration with period L as follows,

$$P_n(1) = \langle [F^n(x)]_j \rangle = (1-p)^2 p + \begin{cases} \frac{p^2}{\lambda_2 - \lambda_1}\left(a_1\lambda_1^{n-1} + a_2\lambda_2^{n-1}\right) & \text{if } n \leq L-3, \\ \frac{p}{\lambda_2 - \lambda_1}\left((a_1 + 1 - p)\lambda_1^{L-2} + (a_2 - 1 + p)\lambda_2^{L-2}\right) & \text{if } n > L-3. \end{cases} \quad (10.34)$$

Recall that in the case of infinite configuration we obtained in Sect. 8.5

$$P_n(0) = 1 - p(1-p)^2 + \alpha_1 \lambda_1{}^n + \alpha_2 \lambda_2{}^n. \quad (10.35)$$

Comparing Eqs. (10.34) and (10.35) we can see that they will result in a very different values of $P_\infty(1)$ (the "steady state" values). The difference will be especially significant if p is small, illustrating the danger of using finite lattices with periodic boundaries as somewhat "resembling" infinite ones, as it is sometimes done in simulations of various cellular automata models.

10.6 Equicontinuity

Theorem 1.1 guarantees that the global function of every cellular automaton is continuous. The Heine-Cantor theorem states that any continuous function with compact domain is also uniformly continuous. Since the Cantor space $\mathcal{A}^{\mathbb{Z}}$ is compact, it follows that all cellular automata are uniformly continuous.

For a given CA with global function F, all the functions belonging to the family F^n, $n = 0, 1, 2, \ldots$ are cellular automata as well, therefore all are uniformly continuous. A natural question is to ask whether they are equicontinuous as well.

Definition 10.4 For a given F point $x \in \mathcal{A}^{\mathbb{Z}}$ is *equicontinuous* if for every $\epsilon > 0$ there exists $\delta > 0$ such that for all $y \in \mathcal{A}^{\mathbb{Z}}$ satisfying $d(y, x) < \delta$ and for all $n \geq 0$ we have $d(F^n(y), F^n(x)) < \epsilon$.

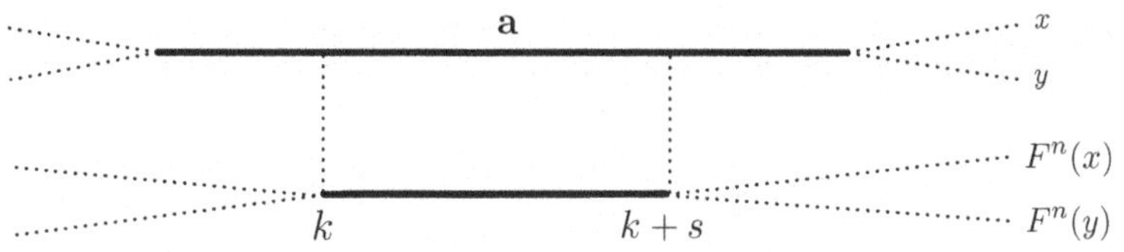

Fig. 10.9 Schematic illustration of the concept of the blocking word

If all points in $\mathcal{A}^{\mathbb{Z}}$ are equicontinuous, we will say that F is *equicontinuous CA*. If the set of equicontinuous points is not empty, we will say that F is *almost equicontinuous CA*. P. Kurka characterized equicontinuous CA as follows.

Theorem 10.1 *(Kůrka [9]) Cellular automaton* $F : \mathcal{A}^{\mathbb{Z}} \to \mathcal{A}^{\mathbb{Z}}$ *is equicontinuous if and only if there exist* $q \geq 0$ *and* $p > 0$ *such that*

$$F^{q+p} = F^q.$$

Using the above theorem together with Tables 2.1 and 5.1 we can conclude that there are 15 minimal ECA which are equicontinuous, namely rules 0, 1, 4, 5, 8, 12, 19, 29, 36, 51, 72, 76, 108, 200 and 204. They can be divided into five different groups,

- rules 51 and 204 satisfying $F^2 = F^0$,
- rules 0, 4, 12, 76 and 200 satisfying $F^2 = F^1$,
- rules 8, 36 and 72 satisfying $F^3 = F^2$,
- rules 1, 5 and 29 satisfying $F^3 = F^1$,
- rules 19 and 108 satisfying $F^4 = F^2$.

Following [9] yet again, almost equicontinuous CA can be characterized using the notion of a blocking word.

Definition 10.5 A word $\mathbf{a} \in \mathcal{A}^\star$ with $|\mathbf{a}| \geq s \geq 0$ is s-*blocking* for F, if there exists an offset $k \in [0, |\mathbf{a}| - s]$ such that for all x, y belonging to the cylinder set $[\mathbf{a}]_0$ and for all $n \geq 0$,

$$F^n(x)_{[k,k+s)} = F^n(y)_{[k,k+s)}.$$

Theorem 10.2 *(Kůrka [9]) Let* $F : \mathcal{A}^{\mathbb{Z}} \to \mathcal{A}^{\mathbb{Z}}$ *be a CA with radius* $r \geq 0$*. Then* F *is almost equicontinuous if and only if it has an* r*-blocking word.*

Figure 10.9 schematically illustrates the concept of the blocking word [10]. We can see that if x and y agree on the word $\mathbf{a}$ extending from indices $i = 0$ to $i = |\mathbf{a}| - 1$ then $F^n(x)$ and $F^n(y)$ must agree on sites with indices from $i = k$ to $i = k + s - 1$, for all n.

In addition to equicontinuous rules listed above, which obviously possess a blocking word, among minimal ECA there exists a significant number (over 20) of rules which are almost equicontinuous but not equicontinuous. Having the deterministic solution of a given rule, it is easy to verify that the rule has a blocking word.

Example 10.4 Consider as an example rule 50, for which the solution of the deterministic initial value problem is given by

$$[F_{50}^n(x)]_j = \frac{1}{2} + \frac{1}{2}(-1)^n + \sum_{i=1}^{n-1}\left((-1)^{i+n}\prod_{p=-i}^{i} x_{p+j}\right) + \sum_{i=0}^{n}\left((-1)^{i+n+1}\prod_{p=-i}^{i} \overline{x}_{p+j}\right).$$

We will show that for this rule the word $\mathbf{a} = 01$ is 1-blocking with offset 0. Configurations x and y belong to the the cylinder set $[\mathbf{a}]_0$ if $x_0 = y_0 = 0$ and $x_1 = y_1 = 1$. For $j = 0$, we have

$$[F_{50}^n(x)]_0 = \frac{1}{2} + \frac{1}{2}(-1)^n + \sum_{i=1}^{n-1}\left((-1)^{i+n}\prod_{p=-i}^{i} x_p\right) + \sum_{i=0}^{n}\left((-1)^{i+n+1}\prod_{p=-i}^{i} \overline{x}_p\right),$$

and since $x_0 = 0$, the first sum will vanish. Similarly, since $x_1 = 1$, the second sum will vanish as well, leaving

$$[F_{50}^n(x)]_0 = \frac{1}{2} + \frac{1}{2}(-1)^n.$$

The above expression does not depend on other components of x at all, thus for y we would obtain exactly the same expression,

$$[F_{50}^n(x)]_0 = [F_{50}^n(y)]_0.$$

This confirms that 01 is indeed 1-blocking, as required by the Definition 10.5. The rule 50, therefore, is almost equicontinuous. Figure 10.10a shows spatiotemporal patterns for this rule, with red line starting at the position of the beginning of the blocking word and passing down through the same cell in subsequent iterations. Note that no information can ever cross this line.

In addition to "regular" equicontinuity one can also consider its following generalization.

Definition 10.6 If there exists $p \in \mathbb{Z}$ and $q \in \mathbb{N}^+$ such that $F^q\sigma^p$ is (almost) equicontinuous, we say that F has *(almost) equicontinuous direction* $\frac{p}{q}$.

Theorem 10.1 implies that if $\frac{p}{q}$ is an equicontinuous direction, then $G = F^q\sigma^p$ must satisfy $G^{s+t} = G^s$ for some $s \geq 0$, $t > 0$, and vice versa. Rules possessing equicontinuous direction must thus satisfy identity

$$(F^q\sigma^p)^{s+t} = (F^q\sigma^p)^s,$$

which simplifies to

$$F^{q(s+t)}\sigma^{pt} = F^{qs}.$$

Tables 2.1 and 5.1 can be used to identify rules of this type. These are:

- rules 0, 2, 10, 34, 42 and 138 satisfying $F^2 = \sigma F$, with $p = -1, q = s = t = 1$;
- rules 8 and 42 satisfying $F^3 = \sigma F^2$, with $p = -1, q = t = 1, s = 2$;
- rules 8 and 24 satisfying $F^3 = \sigma^{-1} F^2$, with $p = 1, q = 1, s = 2, t = 1$;
- rule 38 satisfying $F^4 = \sigma^2 F^2$, with $p = -2, q = 2, s = t = 1$;
- rule 15 satisfying $F^2 = \sigma^{-2} F^0$, with $p = 1, q = 1, s = 0, t = 2$.

The above tables excluded trivial rules, thus we need to add to the above list the rule 170, for which $p = -1, q = 1$. Furthermore, rule 3, not listed in the aforementioned tables, satisfies $F^3 = \sigma^{-1} F$, so it has equicontinuous direction $p = 1, q = 2$.

In the next example we will consider a rule which has almost equicontinuous direction.

Example 10.5 Rule 130 has deterministic solution given by

$$[F_{130}^n(x)]_j = \prod_{i=1}^{2n+1} x_{i+j-n-1} + \sum_{i=0}^{n-1} \left(\left(\frac{1}{2} + \frac{1}{2}(-1)^i - (-1)^i x_{n-2-2i+j} \right) \overline{x}_{n-2i+j-1} x_{n-2i+j} \prod_{p=2n+2-2i}^{2n+1} x_{p+j-n-1} \right)$$

We will show that $\sigma^{-1} F_{130}$ has a 1-blocking word $\mathbf{a} = 0$ with offset $k = 0$. Let us define $G = \sigma^{-1} F_{130}$. Then $[G^n(x)]_j = [F_{130}^n]_{j-n}$, hence

$$[G^n(x)]_0 = \prod_{i=1}^{2n+1} x_{i-2n-1} + \sum_{i=0}^{n-1} \left(\left(\frac{1}{2} + \frac{1}{2}(-1)^i - (-1)^i x_{-2-2i} \right) \overline{x}_{-2i-1} x_{-2i} \prod_{p=2n+2-2i}^{2n+1} x_{p-2n-1} \right).$$

Now suppose that $x_0 = 0$. Both products in the above formula will include x_0, thus both will vanish, therefore

$$[G^n(x)]_0 = 0.$$

This means that if we take two configurations x and y such that $x_0 = y_0 = 0$, we will always have

$$[G^n(x)]_0 = [G^n(y)]_0.$$

The word $\mathbf{a} = 0$ is indeed 1-blocking. By the shift invariance, if $x_1 = y_1 = 0$, we will have

$$[G^n(x)]_1 = [G^n(y)]_1.$$

Fig. 10.10 Spatiotemporal patterns of ECA rules 50 (a), 130 (b) and 27 (c). Arrows indicate almost equicontinuous directions

This means that the word 00 is 2-blocking. The rule $\sigma^{-1}F_{130}$, which is radius 2 rule, has 2-blocking world, thus it is almost equicontinuous. As a result, F_{130} has almost equicontinuous direction with $p = -1, q = 1$. Figure 10.10b shows equicontinuity direction for rule 130 as red arrow. Note that the almost equicontinuous rule 50 considered in the previous example also has almost equicontinuous direction, corresponding to $p = 0, q = 1$. Red arrow in Fig. 10.10a represents this direction.

By a similar method one can show that rule 162, which has a spatiotemporal pattern resembling rule 130, also has almost equicontinuous direction with $p = -1, q = 1$.

Example 10.6 As another example we will consider rule 27. Its solution is given by two slightly different expressions for odd an even iterations,

$$
\begin{aligned}
[F_{27}^n(x)]_j &= \overline{x}_{j-k} + x_{j-k+1}\left(x_{j-k} - x_{j-k-1}\right) \\
&+ \sum_{p=1}^{k} \left(\left(x_{j-k-2+3p} - x_{j-k+1+3p}\right) \prod_{i=0}^{p} x_{j-k-1+3i}\overline{x}_{j-k+3i} \right), \quad n = 2k+1,
\end{aligned}
$$

and

$$[F_{27}^n(x)]_j = x_{j-k} - x_{j-k+1}\overline{x}_{j-k+2}\left(x_{j+3-k}x_{j-k-1} + x_{j-k} - x_{j+3-k} - x_{j-k-1}\right)$$
$$-\sum_{p=1}^{k-1}\left(\left(x_{j-k+3p} - x_{j+3-k+3p}\right)\overline{x}_{j-k-1}\prod_{i=0}^{p} x_{j-k+1+3i}\overline{x}_{j-k+2+3i}\right), \quad n = 2k.$$

Let $G = F_{27}^2\sigma$, then $G^n = F^{2n}\sigma^n$. We will show that G has a blocking word. In order to obtain expressions for $[F^{2n}(x)]_j$, we need to replace k by n in the second formula. For $j = 0$ this yields

$$[F_{27}^{2n}(x)]_0 = x_{-n} - x_{-n+1}\overline{x}_{-n+2}\left(x_{3-n}x_{-n-1} + x_{-n} - x_{3-n} - x_{-n-1}\right)$$
$$-\sum_{p=1}^{n-1}\left(\left(x_{-n+3p} - x_{3-n+3p}\right)\overline{x}_{-n-1}\prod_{i=0}^{p} x_{-n+1+3i}\overline{x}_{-n+2+3i}\right)$$

Multiplication by σ^n means that we add n to each index, yielding

$$[G^n(x)]_0 = x_0 - x_1\overline{x}_2\left(x_3x_{-1} + x_0 - x_3 - x_{-1}\right)$$
$$-\sum_{p=1}^{n-1}\left(\left(x_{3p} - x_{3+3p}\right)\overline{x}_{-1}\prod_{i=0}^{p} x_{1+3i}\overline{x}_{2+3i}\right)$$

Now let us take $x_0 = x_1 = 0$. Since x_1 appears in each product over i, the entire sum over p vanishes. Everything else vanishes as well, leaving

$$[G^n(x)]_0 = 0.$$

By shift invariance, if we take $x_1 = x_2 = 0$, we will obtain $[G^n(x)]_1 = 0$, and $x_2 = x_3 = 0$ will produce $[G^n(x)]_2 = 0$.

If we take two configurations x and y for which $x_0 = y_0 = 0$, $x_1 = y_1 = 0$, $x_2 = y_2 = 0$ and $x_3 = y_3 = 0$, we will have, for all n,

$$[G^n(x)]_0 = [G^n(y)]_0,$$
$$[G^n(x)]_1 = [G^n(y)]_1,$$
$$[G^n(x)]_2 = [G^n(y)]_2,$$

meaning that $\mathbf{a} = 0000$ is 3-blocking for $G = F_{27}^2\sigma$. Since G is of radius 3, it means it is almost equicontinuous. As a result, F_{27} has almost equicontinuous direction with $p = 1, q = 2$, shown in Fig. 10.10c. Note that from that figure one could get a false impression that $p = 1, q = 2$ is not only almost equicontinuous direction, but equicontinuous direction. This, however, is not the case. If this was a fully equicontinuous direction, rule 27 would have to satisfy $F_{27}(x) = F_{27}^2\sigma(x)$ for all x, yet is is easy to verify that this is not true. Configuration $x = \ldots 0101010101\ldots$ produces $F_{27}(x) = \ldots 1111111111\ldots$, $F_{27}^2\sigma(x) = \ldots 0000000000\ldots$, thus $F_{27}(x) \neq F_{27}^2\sigma(x)$.

It is worth mentioning that the almost equicontinuous direction can often be found even if the solution formula is not available.

Example 10.7 For rule 7, the local rule is defined as

$$f_7(x_1, x_2, x_3) = x_1x_2x_3 - x_2x_3 - x_1 + 1,$$

hence

$$[F(x)]_j = x_{j-1}x_jx_{j+1} - x_jx_{j+1} - x_{j-1} + 1.$$

Second iterate of F is easy to construct by computing f^2, yielding

$$\begin{aligned}[F^2(x)]_j = {} & x_{j-2}x_{j-1}x_jx_{j+1}x_{j+2} - x_{j-2}x_jx_{j+1}x_{j+2} - x_{j-2}x_{j-1}x_{j+1}x_{j+2} \\ & - 2x_{j-2}x_{j-1}x_j + x_{j-2}x_{j+1}x_{j+2} + x_{j-2}x_j + x_{j-1}x_j + x_{j-2}x_{j-1}.\end{aligned}$$

Multiplication by σ is equivalent to increasing each index by 1,

$$\begin{aligned}[F^2\sigma(x)]_j = {} & x_{j-1}x_jx_{j+1}x_{j+2}x_{j+3} - x_{j-1}x_{j+1}x_{j+2}x_{j+3} - x_{j-1}x_jx_{j+2}x_{j+3} \\ & - 2x_{j-1}x_jx_{j+1} + x_{j-1}x_{j+2}x_{j+3} + x_{j-1}x_{j+1} + x_jx_{j+1} + x_{j-1}x_j.\end{aligned}$$

It is straightforward to check that if $x_jx_{j+1}x_{j+2} = 000$, then

$$[F^2\sigma(x)]_j = [F^2\sigma(x)]_{j+1} = [F^2\sigma(x)]_{j+2} = 0.$$

This obviously implies that 000 remains in the same position for ever, thus it is a 3-blocking word of the radius-3 rule $F^2\sigma$, so that F has almost equicontinuous direction $p = 1, q = 2$.

References

1. Fukś, H., Boccara, N.: Convergence to equilibrium in a class of interacting particle systems. Phys. Rev. E **64**, 016117 (2001). https://doi.org/10.1103/PhysRevE.64.016117
2. Fukś, H., Boccara, N.: Generalized deterministic traffic rules. Int. J. Mod. Phys. C **9**, 1–12 (1998). https://doi.org/10.1142/S0129183198000029
3. Maerivoet, S., De Moor, B.: Cellular automata models of road traffic. Physics Reports **419**(1), 1–64 (2005)
4. Zabolitzky, J.G.: Critical properties of rule 22 elementary cellular automata. Journal of Statistical Physics **50**, 1255–1262 (1988)

5. Fukś, H.: Remarks on the critical behavior of second order additive invariants in elementary cellular automata. Fundamenta Informaticae **78**, 329–341 (2007)
6. Fukś, H.: Critical behaviour of number-conserving cellular automata with nonlinear fundamental diagrams. J. Stat. Mech.: Theor. Exp. (2004). https://doi.org/10.1088/1742-5468/2004/07/P07005. Art. no. P07005
7. Fukś, H.: Solution of the density classification problem with two cellular automata rules. Phys. Rev. E **55**, 2081R–2084R (1997). https://doi.org/10.1103/PhysRevE.55.R2081
8. Fukś, H.: Explicit solution of the Cauchy problem for cellular automaton rule 172. J. of Cellular Automata **12**(6), 423–444 (2017)
9. Kurka, P.: Languages, equicontinuity and attractors in cellular automata. Ergodic Theory and Dynamical Systems **17**(2), 417–433 (1997)
10. Kůrka, P.: Topological dynamics of cellular automata. In: R.A. Meyers (ed.) Computational Complexity: Theory, Techniques, and Applications, pp. 3212–3233. Springer, New York (2012)

Chapter 11
Approximate Methods

So far we have demonstrated that a significant number of elementary CA are solvable in both deterministic and probabilistic sense. In many cases, however, solution formulae are impossible to obtain (this will be discussed in more detail in Sect. 12.1). In such circumstances, one can resort to approximate methods which will be the subject of this chapter.

We will first demonstrate the method frequently used for construction of approximate values of $P_n(1)$, known as the *mean-field approximation* [1, 2]. It is best explained using an example. Suppose that we want to approximate $P_n(1)$ for rule 44 (this is a solvable rule, but we are using it in order to compare the approximation with the exact solution). The rule is defined as $f_{44}(0,1,0) = f_{44}(0,1,1) = f_{44}(1,0,1) = 1$ and $f_{44}(x_1,x_2,x_3) = 0$ for all other values of x_1, x_2, x_3. This implies that $\mathbf{f}_{132}^{-1}(1) = \{010, 011, 101\}$, thus we can write the RFP equation (Eq. 8.6) for $\mathbf{a} = 1$,

$$P_{n+1}(1) = \sum_{\mathbf{b} \in \mathbf{f}^{-1}(1)} P_n(\mathbf{b}) = P_n(010) + P_n(011) + P_n(101).$$

The idea behind the mean-field approximation is to neglect all correlations between symbols and assume that the probability of a block of symbols is approximately equal to the product of probabilities of individual symbols,

$$\begin{aligned}
P_n(010) &\approx P_n(0) \cdot P_n(1) \cdot P_n(0) = P_n(1) P_n^2(0) = (1 - P_n(1))^2 \, P_n(1), \\
P_n(011) &\approx P_n(0) \cdot P_n(1) \cdot P_n(1) = P_n^2(1) P_n(0) = (1 - P_n(1)) \, P_n^2(1), \\
P_n(101) &\approx P_n(1) \cdot P_n(0) \cdot P_n(1) = P_n^2(1) P_n(0) = (1 - P_n(1)) \, P_n^2(1).
\end{aligned}$$

This yields the difference equation

$$P_{n+1}(1) = (1 - P_n(1))^2 \, P_n(1) + 2\,(1 - P_n(1)) \, P_n^2(1),$$

H. Fukś, *Solvable Cellular Automata*, Understanding Complex Systems,
https://doi.org/10.1007/978-3-031-38700-5_11

which simplifies to

$$P_{n+1}(1) = P_n(1) - P_n^3(1). \tag{11.1}$$

Before we proceed any further, let us note that it is possible to obtain the above equation using the first line of RFP equations for rule 44 given in Appendix D,

$$P_{n+1}(0) = -2P_n(0) + 3P_n(00) + 1 - P_n(000).$$

If we now approximate $P_n(00) \approx P_n^2(0)$, $P_n(000) \approx P_n^3(0)$, we obtain

$$P_{n+1}(0) = -2P_n(0) + 3P_n^2(0) + 1 - P_n^3(0),$$

which, after changing variables to $P_n(1) = 1 - P_n(0)$, yields exactly Eq. (11.1).

The nonlinear difference equation (11.1) is not solvable, but one can easily show that it has stable fixed point at 0, meaning that $P_n(1) \to 0$ as $n \to \infty$. We can say, therefore, that the mean-field approximation predicts

$$P_\infty(1) = 0,$$

whereas inspection of the probabilistic solution for rule 44 given in the Appendix C reveals that

$$P_\infty(1) = \frac{\left(p^2 - p + 1\right) p\,(p-1)}{p^3 - p^2 - 1}.$$

The mean-field prediction is thus very wrong. In order to improve it, one needs to consider not only probabilities of individual symbols, but probabilities of longer blocks. Gutowitz et al. [3, 4] used this idea to develop a method of generalization of the mean-field approximation known as the *local structure theory*. Our presentation of this theory will mostly follow [5].

11.1 Bayesian Extension

In Chap. 7, we have demonstrated that the knowledge of probabilities of blocks of length k (elements of vector $\mathbf{P}^{(k)}$) is sufficient to determine probabilities of all shorter blocks, that is, $\mathbf{P}^{(i)}$ with $i < k$. This can be done using Kolmogorov consistency conditions. Knowledge of $\mathbf{P}^{(k)}$, however, is not enough to determine probabilities $\mathbf{P}^{(i)}$ for $i > k$. This is a result of the fact that the number of independent components in $\mathbf{P}^{(i)}$ is greater than in $\mathbf{P}^{(k)}$ for $i > k$. Nevertheless, it is possible to approximate longer block probabilities by shorter block probabilities using the idea of the so-called *Bayesian extension*.

Let us suppose that we want to approximate $P(a_1a_2\ldots a_{k+1})$ by $P(a_1a_2\ldots a_k)$. One can say that the knowledge of $P(a_1a_2\ldots a_k)$ implies that we know how values of individual symbols in a block of symbols are correlated providing that symbols are not farther apart than k. We do not know, however, anything about correlations between symbols which are separated by distances greater than k. The only thing we can do in this situation is to neglect these higher length correlations, and assume that if a block of length k is extended by appending another symbol to it on the right, then the conditional probability of finding a particular value of that symbol does not significantly depend on the left-most symbol, meaning that

$$\frac{P(a_1a_2\dots a_{k+1})}{P(a_1\dots a_k)} \approx \frac{P(a_2\dots a_{k+1})}{P(a_2\dots a_k)}, \tag{11.2}$$

which yields

$$P(a_1a_2\dots a_{k+1}) \approx \frac{P(a_1\dots a_k)P(a_2\dots a_{k+1})}{P(a_2\dots a_k)}. \tag{11.3}$$

This is the desired approximation of probabilities of blocks of length $k+1$ by probabilities of shorter blocks. Note that since we have the probability $P(a_2\dots a_k)$ in the denominator, we must consider what happens when it becomes zero. We will adopt a simple solution of this problem, namely if the denominator is zero, then we will take $P(a_1a_2\dots a_{k+1}) = 0$. In order to avoid writing separate cases for denominator equal to zero, we define "thick bar" fraction as

$$\genfrac{}{}{2pt}{}{a}{b} := \begin{cases} \dfrac{a}{b} & \text{if } b \neq 0 \\ 0 & \text{if } b = 0. \end{cases} \tag{11.4}$$

Equation (11.3), therefore, should be written as

$$P(a_1a_2\dots a_{k+1}) \approx \genfrac{}{}{2pt}{}{P(a_1\dots a_k)P(a_2\dots a_{k+1})}{P(a_2\dots a_k)}. \tag{11.5}$$

Note that the above only makes sense if $k > 1$. For $k = 1$, the approximation is simply a product of probabilities of both symbols,

$$P(a_1a_2) \approx P(a_1)P(a_2). \tag{11.6}$$

Again, in order to avoid writing the $k = 1$ case separately, we adopt notational convention that

$$P(a_m\dots a_n) = 1 \text{ whenever } n > m, \tag{11.7}$$

and then Eq. (11.5) remains valid even for $k = 1$. Using this convention and the "thick bar" notation, by applying Eq. (11.5) recursively m times, we can express $k+m$ block probabilities in terms of k and $k-1$-block probabilities,

$$P(a_1a_2\dots a_{k+m}) \approx \genfrac{}{}{2pt}{}{\prod_{i=1}^{m+1} P(a_i\dots a_{i+k-1})}{\prod_{i=1}^{m} P(a_{i+1}\dots a_{i+k-1})}. \tag{11.8}$$

Note that if we need to express $k+m$-block probabilities in terms of only k-block probabilities, we can exploit consistency conditions and substitute $P(a_{i+1}\dots a_{i+k-1})$ in the denominator by the sum of longer blocks probabilities,

$$P(a_{i+1}\dots a_{i+k-1}) = \sum_{b\in\mathcal{A}} P(a_{i+1}\dots a_{i+k-1}b). \tag{11.9}$$

For a given k, if the Bayesian extension is constructed for all block probabilities of length greater than k, we obtain a valid probability measure, as the following proposition attests.

Proposition 11.1 *Let $\mathcal{A}$ be a finite alphabet and let $\mu \in \mathfrak{M}(\mathcal{A}^{\mathbb{Z}})$ be a shift invariant measure with associated block probabilities $P : \mathcal{A}^{\star} \to [0, 1]$, $P(\mathbf{b}) = \mu([\mathbf{b}]_i)$ for all $i \in \mathbb{Z}$ and $\mathbf{b} \in \mathcal{A}^{\star}$. For $k > 0$, define probabilities $\widetilde{P} : \mathcal{A}^{\star} \to [0, 1]$ by*

$$\widetilde{P}(a_1a_2\dots a_p) = \begin{cases} P(a_1a_2\dots a_p) & \text{if } p \le k, \\ \dfrac{\prod_{i=1}^{p-k+1} P(a_i\dots a_{i+k-1})}{\prod_{i=1}^{p-k} P(a_{i+1}\dots a_{i+k-1})} & \text{otherwise.} \end{cases} \tag{11.10}$$

Then probabilities $\widetilde{P}$ determine a shift-invariant probability measure $\widetilde{\mu}^{(k)} \in \mathfrak{M}(\mathcal{A}^{\mathbb{Z}})$, to be called Bayesian approximation of μ of order k.

Proof If $\mathbf{b} = b_1b_2\dots b_n$, we will denote subblocks of $\mathbf{b}$ by $\mathbf{b}_{[i,j]} = b_ib_{i+1}\dots b_j$. Using Theorem 7.2, all we need to do is to show that the consistency conditions (7.14) and (7.15) are satisfied by probabilities $\widetilde{P}$. The second one indeed must hold for $\widetilde{P}$, because it already holds for P. For the same reason Eq. (7.14) holds for block $\mathbf{b}$ of length of up to $k-1$. For $\mathbf{b} = b_1b_2\dots b_p$ and $p \ge k$, we have

$$\begin{aligned} \sum_{a\in\mathcal{A}} \widetilde{P}(\mathbf{b}a) &= \sum_{a\in\mathcal{A}} \frac{\prod_{i=1}^{p-k+1} P(\mathbf{b}_{[i,i+k-1]}) \cdot P(\mathbf{b}_{[p-k+2,p]}a)}{\prod_{i=1}^{p-k} P(\mathbf{b}_{[i+1,i+k-1]}) \cdot P(\mathbf{b}_{[p-k+2,p]})} \\ &= \frac{\prod_{i=1}^{p-k+1} P(\mathbf{b}_{[i,i+k-1]})}{\prod_{i=1}^{p-k} P(\mathbf{b}_{[i+1,i+k-1]}) \cdot P(\mathbf{b}_{[p-k+2,p]})} \sum_{a\in\mathcal{A}} P(\mathbf{b}_{[p-k+2,p]}a) \\ &= \frac{\prod_{i=1}^{p-k+1} P(\mathbf{b}_{[i,i+k-1]})}{\prod_{i=1}^{p-k} P(\mathbf{b}_{[i+1,i+k-1]}) \cdot P(\mathbf{b}_{[p-k+2,p]})} P(\mathbf{b}_{[p-k+2,p]}) \\ &= \widetilde{P}(\mathbf{b}). \end{aligned} \tag{11.11}$$

Proof that $\sum_{a\in\mathcal{A}} \widetilde{P}(a\mathbf{b}) = \widetilde{P}(\mathbf{b})$ can be accomplished in a similar way. □

For a given measure μ, if $\mu = \widetilde{\mu}^{(k)}$ for some k, we will call μ a *Markov measure* or a *finite block measure* of order k. The set of Markov measures of order k will be denoted by $\mathfrak{M}^{(k)}(\mathcal{A}^{\mathbb{Z}})$,

$$\mathfrak{M}^{(k)}(\mathcal{A}^{\mathbb{Z}}) = \{\mu \in \mathfrak{M}(\mathcal{A}^{\mathbb{Z}}) : \mu = \widetilde{\mu}^{(k)}\}. \tag{11.12}$$

One of the often mentioned properties of Markov measures, especially in the statistical physics literature, is the maximization of entropy. In order to formulate this property in a rigorous way, we will define *entropy density* of a shift-invariant measure $\mu \in \mathfrak{M}(\mathcal{A}^{\mathbb{Z}})$ as

$$h(\mu) = \lim_{n\to\infty} -\frac{1}{n} \sum_{\mathbf{b}\in\mathcal{A}^n} P(\mathbf{b}) \log P(\mathbf{b}), \tag{11.13}$$

where, as usual, $P(\mathbf{b}) = \mu([\mathbf{b}]_i)$ for all $i \in \mathbb{Z}$ and $\mathbf{b} \in \mathcal{A}^\star$. The following two propositions and the main ideas behind their proofs are due to Fannes and Verbeure [6].

Proposition 11.2 *For any $\mu \in \mathfrak{M}(\mathcal{A}^{\mathbb{Z}})$, the entropy density of the k-th order Bayesian approximation of μ is given by*

$$h(\widetilde{\mu}^{(k)}) = \sum_{\mathbf{a}\in\mathcal{A}^{k-1}} P(\mathbf{a}) \log P(\mathbf{a}) - \sum_{\mathbf{a}\in\mathcal{A}^{k}} P(\mathbf{a}) \log P(\mathbf{a}). \tag{11.14}$$

Proof Since in the definition of the entropy density we need to evaluate the $n \to \infty$ limit, let us assume that $n > k$. Then

$$\begin{aligned}\sum_{\mathbf{b}\in\mathcal{A}^n} \widetilde{P}(\mathbf{b}) \log \widetilde{P}(\mathbf{b}) &= \sum_{\mathbf{b}\in\mathcal{A}^n} \widetilde{P}(\mathbf{b}) \log \frac{\prod_{i=1}^{n-k+1} P(\mathbf{b}_{[i,i+k-1]})}{\prod_{i=1}^{n-k} P(\mathbf{b}_{[i+1,i+k-1]})} \\ &= \sum_{\mathbf{b}\in\mathcal{A}^n} \widetilde{P}(\mathbf{b}) \sum_{i=1}^{n-k+1} \log P(\mathbf{b}_{[i,i+k-1]}) \\ &\quad - \sum_{\mathbf{b}\in\mathcal{A}^n} \widetilde{P}(\mathbf{b}) \sum_{i=1}^{n-k} \log P(\mathbf{b}_{[i+1,i+k-1]})\end{aligned} \tag{11.15}$$

Note that for any $i \in [1, n-k+1]$,

$$\begin{aligned}&\sum_{\mathbf{b}\in\mathcal{A}^n} \widetilde{P}(\mathbf{b}) \log P(\mathbf{b}_{[i,i+k-1]}) \\ &= \sum_{\substack{\mathbf{b}_{[1,i-1]}\\ \in\mathcal{A}^{i-1}}} \sum_{\substack{\mathbf{b}_{[i,i+k-1]}\\ \in\mathcal{A}^{k}}} \sum_{\substack{\mathbf{b}_{[i+k,n]}\\ \in\mathcal{A}^{n-i-k+1}}} \widetilde{P}(\mathbf{b}_{[1,i-1]}\mathbf{b}_{[i,i+k-1]}\mathbf{b}_{[i+k,n]}) \log P(\mathbf{b}_{[i,i+k-1]}) \\ &= \sum_{\mathbf{a}\in\mathcal{A}^k} \widetilde{P}(\mathbf{a}) \log P(\mathbf{a}) = \sum_{\mathbf{a}\in\mathcal{A}^k} P(\mathbf{a}) \log P(\mathbf{a}).\end{aligned}$$

By the same reasoning, for any $i \in [1, n-k]$,

$$\sum_{\mathbf{b}\in\mathcal{A}^n} \widetilde{P}(\mathbf{b}) \log P(\mathbf{b}_{[i+1,i+k-1]}) = \sum_{\mathbf{a}\in\mathcal{A}^{k-1}} P(\mathbf{a}) \log P(\mathbf{a}).$$

Using these results, Eq. (11.15) becomes

$$\sum_{\mathbf{b}\in\mathcal{A}^n} \widetilde{P}(\mathbf{b}) \log \widetilde{P}(\mathbf{b}) = (n-k+1) \sum_{\mathbf{a}\in\mathcal{A}^k} P(\mathbf{a}) \log P(\mathbf{a}) - (n-k) \sum_{\mathbf{a}\in\mathcal{A}^{k-1}} P(\mathbf{a}) \log P(\mathbf{a}). \tag{11.16}$$

Dividing this by $-n$ and taking the limit $n \to \infty$ we obtain the desired expression for the entropy density given in Eq. (11.13). □

Having the formula for the entropy density of $\widetilde{\mu}^{(k)}$, we are now ready to prove that no other measure can have greater entropy density.

Proposition 11.3 *The entropy density of any measure $\mu \in \mathfrak{M}(\mathcal{A}^{\mathbb{Z}})$ does not exceed the entropy density of its k-th order Bayesian approximation,*

$$h(\mu) \le h(\widetilde{\mu}^{(k)}). \tag{11.17}$$

Proof Let us first define

$$H_n(\mu) = -\sum_{\mathbf{b}\in\mathcal{A}^n} P(\mathbf{b}) \log P(\mathbf{b}),$$

$$H_n(\widetilde{\mu}^{(k)}) = -\sum_{\mathbf{b}\in\mathcal{A}^n} \widetilde{P}(\mathbf{b}) \log \widetilde{P}(\mathbf{b}).$$

We will now make use of the convexity of $f(x) = x \log x$. Recall that if f is convex, then its graph lies above all of its tangents,

$$f(y) \ge f(x) + f'(x)(y-x).$$

For $f(x) = x \log x$ this becomes

$$y \log y \ge x \log x + (1 + \log x)(y - x),$$

which results in the inequality

$$x \log x - y \log y \le (x - y)(1 + \log x).$$

Applying this inequality to $H_n(\mu) - H_n(\widetilde{\mu}^{(k)})$ for $n > k$ we obtain

$$\begin{aligned} H_n(\mu) - H_n(\widetilde{\mu}^{(k)}) &= \sum_{\mathbf{b}\in\mathcal{A}^n} \widetilde{P}(\mathbf{b}) \log \widetilde{P}(\mathbf{b}) - \sum_{\mathbf{b}\in\mathcal{A}^n} P(\mathbf{b}) \log P(\mathbf{b}) \\ &\leq \sum_{\mathbf{b}\in\mathcal{A}^n} \left(\widetilde{P}(\mathbf{b}) - P(\mathbf{b})\right)\left(1 + \log \widetilde{P}(\mathbf{b})\right) \\ &= \sum_{\mathbf{b}\in\mathcal{A}^n} \widetilde{P}(\mathbf{b}) \log \widetilde{P}(\mathbf{b}) - \sum_{\mathbf{b}\in\mathcal{A}^n} P(\mathbf{b}) \log \widetilde{P}(\mathbf{b}), \end{aligned}$$

where we used the fact that block probabilities must sum to 1, that is, $\sum_{\mathbf{b}\in\mathcal{A}^n} \widetilde{P}(\mathbf{b}) = \sum_{\mathbf{b}\in\mathcal{A}^n} P(\mathbf{b}) = 1$. Note that we already computed the value of $\sum_{\mathbf{b}\in\mathcal{A}^n} \widetilde{P}(\mathbf{b}) \log \widetilde{P}(\mathbf{b})$ in Eq. (11.16). Nothing would change in the derivation of Eq. (11.16) if we were computing $\sum_{\mathbf{b}\in\mathcal{A}^n} P(\mathbf{b}) \log \widetilde{P}(\mathbf{b})$ instead, hence

$$\sum_{\mathbf{b}\in\mathcal{A}^n} \widetilde{P}(\mathbf{b}) \log \widetilde{P}(\mathbf{b}) = \sum_{\mathbf{b}\in\mathcal{A}^n} P(\mathbf{b}) \log \widetilde{P}(\mathbf{b}). \tag{11.18}$$

We obtain, therefore,

$$H_n(\mu) - H_n(\widetilde{\mu}^{(k)}) \leq 0. \tag{11.19}$$

Dividing this by n and taking the limit $n \to \infty$ we obtain inequality (11.17). □

We will now show that the k-th order Bayesian approximations of μ becomes better as k increases. In order to do this, we need to introduce the notion of convergence of measures. Let $\mu, \mu_n \in \mathfrak{M}(\mathcal{A}^{\mathbb{Z}})$. If $\int_X f d\mu_n \to \int_X f d\mu$ as $n \to \infty$ for every bounded, continuous real function f on $\mathcal{A}^{\mathbb{Z}}$, we say that μ_n *converges weakly* to μ and write $\mu_n \Rightarrow \mu$. There following criterion of the weak convergence is originally due to Kolmogorov and Prohorov [7].

Theorem 11.1 *Let U be a subclass of the smallest σ-algebra containing all open sets of X such that* (i) *U is closed under the formation of finite intersections and* (ii) *each open set in X is a finite or countable union of elements of U. If $\mu_n(A) \to \mu(A)$ for every $A \in U$, then $\mu_n \Rightarrow \mu$.*

Proof of this result can be found in [8], yet since it is beyond the scope of this book, we will not include it here. The subclass U satisfying hypothesis of the above theorem is called *convergence determining class*. It is easy to verify that $Cyl(\mathcal{A}^{\mathbb{Z}})$ is a convergence determining class for measures in $\mathfrak{M}(\mathcal{A}^{\mathbb{Z}})$, hence the following proposition.

Proposition 11.4 *The sequence of k-th order Bayesian approximations of $\mu \in \mathfrak{M}(X)$ weakly converges to μ as $k \to \infty$.*

Proof Let $n > 0$, $\mathbf{b} \in \mathcal{A}^n$ and let $\widetilde{P}_k(\mathbf{b}) = \widetilde{\mu}^{(k)}([\mathbf{b}]_0)$, $P(\mathbf{b}) = \mu([\mathbf{b}]_0)$. Since for $k \geq n$ we have $\widetilde{P}_k(\mathbf{b}) = P(\mathbf{b})$, we obviously must also have $\lim_{k\to\infty} \widetilde{P}_k(\mathbf{b}) = P(\mathbf{b})$. Theorem 11.1, together with the fact that $Cyl(\mathcal{A}^{\mathbb{Z}})$ is a convergence determining class leads to the conclusion that $\widetilde{\mu}^{(k)} \Rightarrow \mu$. □

11.2 The Scramble Operator

Since the Markov measures $\widetilde{\mu}^{(k)}$ weakly converge to μ as $k \to \infty$, we can consider $\widetilde{\mu}^{(k)}$ to be an approximation of μ. Using this idea, as we mentioned at the beginning of this chapter, Gutowitz et al. [3, 4] developed a method of approximating orbits of measures under the action of a CA rules, known as the *local structure theory*.

Following [3], let us define the *scramble operator* of order k, denoted by $\Xi^{(k)}$, to be a map from $\mathfrak{M}(\mathcal{A}^{\mathbb{Z}})$, the set of shift-invariant measures on $\mathcal{A}^{\mathbb{Z}}$, to the set of finite block measures of order k, such that

$$\Xi^{(k)}\mu = \widetilde{\mu}^{(k)}. \tag{11.20}$$

For a given CA rule F (deterministic or probabilistic), the sequence of Markov measures approximating $\{F^n\mu\}_{n=0}^{\infty}$ can be constructed by applying the scramble operator to μ, then applying the rule F to the result, and finally applying the scramble operator again. If repeated recursively, this produces the sequence

$$\left\{\left(\Xi^{(k)} F \,\Xi^{(k)}\right)^n \mu\right\}_{n=0}^{\infty} \tag{11.21}$$

which will be called the *local structure approximation* of level k of the exact orbit $\{F^n\mu\}_{n=0}^{\infty}$. Note that all terms of this sequence are Markov measures, thus the entire local structure approximation of the orbit lies in $\mathfrak{M}^{(k)}(\mathcal{A}^{\mathbb{Z}})$.

The main hypothesis of the local structure theory is that Eq. (11.21) approximates the actual orbit $\{F^n\mu\}_{n=0}^{\infty}$ increasingly well as k increases. Although the formal meaning of this claim is not clearly explained in the original paper of Gutowitz et al. [3], it is straightforward to demonstrate that every point of the approximate orbit weakly converges to the corresponding point of the exact orbit as $k \to \infty$. To do this, we will first prove the following property of the scramble operator.

Proposition 11.5 *Let F be a CA rule of radius r. For $k > 0$ and $\mathbf{b} \in \mathcal{A}^\star$ such that $k \geq |\mathbf{b}| + 2r$, we have*

$$F\mu([\mathbf{b}]) = F\,\Xi^{(k)}\mu([\mathbf{b}]) = \Xi^{(k)} F\mu([\mathbf{b}]). \tag{11.22}$$

Proof Let us first note that $\mu([\mathbf{a}]) = \widetilde{\mu}^{(k)}([\mathbf{a}])$ for all blocks $\mathbf{a}$ of length up to k. The first equality of Eq. (11.22) can be written as

$$\sum_{\mathbf{a}\in\mathcal{A}^{|\mathbf{b}|+2r}} w(\mathbf{b}|\mathbf{a})\mu([\mathbf{a}]) = \sum_{\mathbf{a}\in\mathcal{A}^{|\mathbf{b}|+2r}} w(\mathbf{b}|\mathbf{a})\widetilde{\mu}^{(k)}([\mathbf{a}]), \tag{11.23}$$

and it will hold as long as $|\mathbf{a}| \leq k$, that is, for $|\mathbf{b}| + 2r \leq k$.

Since the scramble operator only modifies probabilities of blocks of length greater than k, the second equality of Eq. (11.22) naturally follows. By the hypothesis of our proposition $k \geq |\mathbf{b}| + 2r$, thus we have $|\mathbf{b}| < k$ and therefore $F\mu([\mathbf{b}]) = \Xi^{(k)} F\mu([\mathbf{b}])$. □

Since F^n is a cellular automaton rule of radius nr, when $k \geq |\mathbf{b}| + 2nr$ the above proposition yields

$$F^n\mu([\mathbf{b}]) = F^n \Xi^{(k)}\mu([\mathbf{b}]) = \Xi^{(k)} F^n \mu([\mathbf{b}]).$$

Combining this results with Eq. (11.22), which obviously also holds for $k \geq |\mathbf{b}| + 2r$, we obtain the following corollary.

Corollary 11.1 *Let k and n be positive integers and $\mathbf{b} \in \mathcal{A}^\star$. If F is a CA rule with radius r and $k \geq |\mathbf{b}| + 2nr$, then*

$$F^n\mu([\mathbf{b}]) = \left(\Xi^{(k)} F \,\Xi^{(k)}\right)^n \mu([\mathbf{b}]).$$

This means that for a given n, measures of cylinder sets in the approximate measure $\left(\Xi^{(k)} F \,\Xi^{(k)}\right)^n \mu$ converge to measures of cylinder sets in $F^n\mu$. By the virtue of Theorem 11.1 we thus obtain the following result.

Theorem 11.2 *Let F be a cellular automaton and μ be a shift-invariant measure in $\mathfrak{M}_\sigma(\mathcal{A}^{\mathbb{Z}})$. Let $\nu_n^{(k)}$ be a local structure approximation of level k of the measure $F^n\mu$, i.e., $\nu_n^{(k)} = \left(\Xi^{(k)} F \,\Xi^{(k)}\right)^n \mu$. Then, for any given $n > 0$, $\nu_n^{(k)}$ weakly converges to $F^n\mu$ as $k \to \infty$.*

11.3 Local Structure Maps

We will now show how to construct finite dimensional maps generating Markov measures belonging to the approximate orbit $\left(\Xi^{(k)} F \,\Xi^{(k)}\right)^n \mu$. If $\nu_n^{(k)} = \left(\Xi^{(k)} F \,\Xi^{(k)}\right)^n \mu$, then $\nu_n^{(k)}$ satisfies recurrence equation

$$\nu_{n+1}^{(k)} = \Xi^{(k)} F \,\Xi^{(k)} \nu_n^{(k)}.$$

On both sides of the above we have Markov measures completely determined by probabilities of blocks of length k. For a cylinder set $[\mathbf{b}]$ such that $|\mathbf{b}| = k$ this yields

$$\nu_{n+1}^{(k)}([\mathbf{b}]) = \Xi^{(k)} F \,\Xi^{(k)} \nu_n^{(k)}([\mathbf{b}]),$$

and, since $\Xi^{(k)}$ does not modify probabilities of blocks of length k, this simplifies to

$$\nu_{n+1}^{(k)}([\mathbf{b}]) = F \,\Xi^{(k)} \nu_n^{(k)}([\mathbf{b}]).$$

Let us further assume that F is a probabilistic cellular automaton of radius r defined by Eq. (9.5). Note that this does not exclude deterministic CA, as these can be viewed as PCA with integer transition probabilities. By the definition of F,

$$\nu_{n+1}^{(k)}([\mathbf{b}]) = \sum_{\mathbf{a} \in \mathcal{A}^{|\mathbf{b}|+2r}} w(\mathbf{b}|\mathbf{a}) \left(\Xi^{(k)} \nu_n^{(k)}\right)([\mathbf{a}]),$$

thus, by using the Bayesian approximation,

$$\nu_{n+1}^{(k)}([\mathbf{b}]) = \sum_{\mathbf{a}\in\mathcal{A}^{|\mathbf{b}|+2r}} w(\mathbf{b}|\mathbf{a}) \frac{\prod_{i=1}^{2r+1} \nu_n^{(k)}([\mathbf{a}_{[i,i+k-1]}])}{\prod_{i=1}^{2r} \nu_n^{(k)}([\mathbf{a}_{[i+1,i+k-1]}])}.$$

To shorten the notation, let us define $Q_n(\mathbf{c}) = \nu_n^{(k)}([\mathbf{c}])$. Using consistency conditions to express the denominator in terms of only probabilities of blocks of length k, we can then rewrite the previous equation as

$$Q_{n+1}(\mathbf{b}) = \sum_{\mathbf{a}\in\mathcal{A}^{|\mathbf{b}|+2r}} w(\mathbf{a}|\mathbf{b}) \frac{\prod_{i=1}^{2r+1} Q_n(\mathbf{a}_{[i,i+k-1]})}{\prod_{i=1}^{2r} \sum_{c\in\mathcal{A}} Q_n(c\mathbf{a}_{[i+1,i+k-1]})}. \tag{11.24}$$

The above equation can be written separately for all $\mathbf{b} \in \mathcal{A}^k$. If we arrange $Q_n(\mathbf{b})$ for all $\mathbf{b} \in \mathcal{A}^k$ in lexicographical order to form a vector $\mathbf{Q}_n$, we will obtain

$$\mathbf{Q}_{n+1} = L^{(k)}(\mathbf{Q}_n),$$

where $L^{(k)} : [0,1]^{|\mathcal{A}|^k} \to [0,1]^{|\mathcal{A}|^k}$ has components defined by Eq. (11.24). $L^{(k)}$ will be called *local structure map* of level k.

Example 11.1 We will construct the map $L^{(2)}$ for ECA 184. As mentioned elsewhere in this book, this deterministic CA can be viewed as a probabilistic rule with transition probabilities

$$\begin{aligned} &w(1|000) = 0, \quad w(1|001) = 0, \quad w(1|010) = 0, \quad w(1|011) = 1, \\ &w(1|100) = 1, \quad w(1|101) = 1, \quad w(1|110) = 0, \quad w(1|111) = 1, \end{aligned} \tag{11.25}$$

and $w(0abc) = 1 - w(abc)$ for all $a, b, c \in \{0, 1\}$. Let $P_n(\mathbf{b}) = F^n\mu([\mathbf{b}])$. For $\mathbf{b} \in \{0,1\}^2$, Eq. (9.5) becomes

$$P_{n+1}(\mathbf{b}) = \sum_{\mathbf{a}\in\mathcal{A}^5} w(\mathbf{b}|\mathbf{a}) P_n(\mathbf{a}).$$

Since there are four blocks of length 2, using definition of $w(\mathbf{b}|\mathbf{a})$ given in Eq. (9.6) and transition probabilities given in Eq. (11.25) we obtain a system of four equations,

$$\begin{aligned} P_{n+1}(00) &= P_n(0000) + P_n(0001) + P_n(0010), \\ P_{n+1}(01) &= P_n(0011) + P_n(0100) + P_n(0101) + P_n(1100) + P_n(1101), \\ P_{n+1}(10) &= P_n(0110) + P_n(1000) + P_n(1001) + P_n(1010) + P_n(1110), \\ P_{n+1}(11) &= P_n(0111) + P_n(1011) + P_n(1111). \end{aligned} \tag{11.26}$$

This set of equations describes exact relationship between block probabilities at step $n+1$ and block probabilities at step n. Note that 3-block probabilities at step $n+1$ are given in terms of 5-blocks probabilities at step n, thus it is not possible to iterate these equations.

On the other hand, local structure map of order 2, given by Eq. (11.24), becomes

$$
\begin{aligned}
Q_{n+1}(00) = {} & \frac{Q_n(00)^3}{(Q_n(00)+Q_n(01))^2} + \frac{Q_n(00)^2 Q_n(01)}{(Q_n(00)+Q_n(01))^2} \\
& + \frac{Q_n(00)Q_n(01)Q_n(10)}{(Q_n(00)+Q_n(01))\,(Q_n(10)+Q_n(11))}, \\
Q_{n+1}(01) = {} & \frac{Q_n(00)Q_n(01)Q_n(11)}{(Q_n(00)+Q_n(01))\,(Q_n(10)+Q_n(11))} \\
& + \frac{Q_n(00)Q_n(01)Q_n(10)}{(Q_n(00)+Q_n(01))\,(Q_n(10)+Q_n(11))} \\
& + \frac{Q_n(01)^2 Q_n(10)}{(Q_n(00)+Q_n(01))\,(Q_n(10)+Q_n(11))} \\
& + \frac{Q_n(11)Q_n(10)Q_n(00)}{(Q_n(00)+Q_n(01))\,(Q_n(10)+Q_n(11))} \\
& + \frac{Q_n(01)Q_n(11)Q_n(10)}{(Q_n(00)+Q_n(01))\,(Q_n(10)+Q_n(11))}, \\
Q_{n+1}(10) = {} & \frac{Q_n(01)Q_n(11)Q_n(10)}{(Q_n(10)+Q_n(11))^2} + \frac{Q_n(10)Q_n(00)^2}{(Q_n(00)+Q_n(01))^2} \\
& + \frac{Q_n(00)Q_n(01)Q_n(10)}{(Q_n(00)+Q_n(01))^2} + \frac{Q_n(01)Q_n(10)^2}{(Q_n(00)+Q_n(01))\,(Q_n(10)+Q_n(11))} \\
& + \frac{Q_n(11)^2 Q_n(10)}{(Q_n(10)+Q_n(11))^2}, \\
Q_{n+1}(11) = {} & \frac{Q_n(01)Q_n(11)^2}{(Q_n(10)+Q_n(11))^2} + \frac{Q_n(01)Q_n(11)Q_n(10)}{(Q_n(00)+Q_n(01))\,(Q_n(10)+Q_n(11))} \\
& + \frac{Q_n(11)^3}{(Q_n(10)+Q_n(11))^2}.
\end{aligned}
\tag{11.27}
$$

These equations contain on both sides only probabilities of blocks of length 2, thus they define a map $[0,1]^4 \to [0,1]^4$. Note that Eq. (11.27) could also be obtained from Eq. (11.26) by replacing P's by Q's and expressing every 5-block probability by its Bayesian approximation of order 3.

We can see that the local structure map given in the example is rather complicated, even though its order is low ($k = 2$). It is possible to simplify it radically if we realize that among block probabilities of length up to k only $N^k - N^{k-1}$ are independent. It is, therefore, sufficient to consider only fundamental probabilities, obtaining the remaining ones from the fundamental ones by using consistency conditions. Let us see how this could be accomplished for rule 184.

Example 11.2 For $k = 2$, among blocks of length of up to 2, only two are fundamental, $P(0)$ and $P(00)$. For rule 184 and these two probabilities, Eq. (9.5) becomes

$$P_{n+1}(0) = P_n(000) + P_n(001) + P_n(010) + P_n(110),$$
$$P_{n+1}(00) = P_n(0000) + P_n(0001) + P_n(0010).$$

Expressing all non-fundamental probabilities on the right hand side by fundamental ones this further simplifies, becoming

$$P_{n+1}(0) = P_n(0)$$
$$P_{n+1}(00) = P_n(000) + P_n(0010).$$

Note that the above is equations are simply RFP equations of order 2 for rule 184, and these could simply be obtained from the Appendix D. As before, this system cannot be iterated because we have a 4-block probability $Q(0010)$ on the right hand side. The corresponding local structure map can be obtained be replacing P by Q and then approximating both $Q(000)$ and $Q(0010)$ by their Bayesian approximations of order 2,

$$Q(000) = \frac{Q(00)Q(00)}{Q(0)} = \frac{Q(00)^2}{Q(0)},$$
$$Q(0010) = \frac{Q(00)Q(01)Q(10)}{Q(0)Q(1)} = \frac{Q(00)\,(Q(0) - Q(00))^2}{Q(0)(1 - Q(0))}.$$

This yields the local structure map

$$Q_{n+1}(0) = Q_n(0) \tag{11.28}$$
$$Q_{n+1}(00) = \frac{Q(00)^2}{Q(0)} + \frac{Q(00)\,(Q(0) - Q(00))^2}{Q(0)(1 - Q(0))}. \tag{11.29}$$

This is the reduced short form of the local structure map for rule 184. Note that this form is dramatically simpler than the 4-dimensional map constructed in Example 11.1.

Let us see how good is the approximation of P_n by Q_n for rule 184 for this map. Recall that in Sect. 8.8 we found that if the initial measure is Bernoulli with $P_0(1) = p \in [0, 1]$, then

$$P_n(0) = 1 - p \tag{11.30}$$

$$P_n(00) = \sum_{j=1}^{n+1} \frac{j}{n+1} \binom{2n+2}{n+1-j} p^{n+1-j}(1-p)^{n+1+j}. \tag{11.31}$$

Furthermore, in Sect. 10.2 we demonstrated that

$$P_\infty(00) = \lim_{n\to\infty} P_n(00) = \begin{cases} 1-2p & \text{if } p < 1/2, \\ 0 & \text{otherwise.} \end{cases} \tag{11.32}$$

It is obvious that if we start with the same Bernoulli measure for Q and P, then $Q_n = Q_0(0) = 1 - p$, thus the local structure predicts the value of the probability of 0 after n steps correctly.

For the probability of the blocks 00 the values predicted by the local structure approximation are different from the exact values. Although $P_1(00) = Q_1(00)$, for higher n the values of $P_n(00)$ and $Q_n(00)$ are not the same. For example, for $n = 2$ we have

$$P_2(00) = 2\,p^6 - 6\,p^5 + 5\,p^4 - 2\,p + 1, \tag{11.33}$$

while the two iterations of the local structure map yield

$$Q_n(00) = p^{10} - 5\,p^9 + 7\,p^8 + 2\,p^7 - 11\,p^6 + 5\,p^5 + 2\,p^4 - 2\,p + 1. \tag{11.34}$$

Nevertheless, it turns out that the limits of $P_n(00)$ and $Q_n(00)$ are actually the same. To determine $Q_\infty(00)$, we can find fixed points of the local structure map, determine their stability, and if there is only one stable fixed point, $Q_n(00)$ should converge to it locally. Since $Q_n(0)$ is constant, let us substitute $Q_n(0) = 1 - p$ in Eq. (11.29), which then becomes

$$Q_{n+1}(00) = \frac{Q_n(00)^2}{1-p} + \frac{Q_n(00)[Q_n(1-p-Q_n(00)]^2}{p(1-p)}. \tag{11.35}$$

This nonlinear difference equation has three fixed points, 0, $1-\rho$ and $1-2\rho$. The second one, $1-\rho$, is always unstable. The first one, 0, is unstable for $\rho < 1/2$, and stable for $\rho > 1/2$. The third one, $1-2\rho$, is is stable for $\rho < 1/2$, and unstable for $\rho > 1/2$. We can, therefore, conclude that

$$Q_\infty(00) = \lim_{n\to\infty} Q_n(00) = \begin{cases} 1-2p & \text{if } p < 1/2, \\ 0 & \text{otherwise,} \end{cases} \tag{11.36}$$

which exactly agrees with the expression for $P_\infty(00)$ given above. We can say that the local structure map in this case "inherits" the limiting value of the probability $P_n(00)$ from the exact orbit.

Example 11.3 As a more complicated example, let us consider local structure approximation of order 3 for ECA rule 14. We have four fundamental probabilities, $P(0)$, $P(00)$, $P(000)$ and $P(010)$, for which Eq. (9.5) becomes

$$\begin{aligned}
P_{n+1}(0) &= P_n(000) + P_n(100) + P_n(101) + P_n(110) + P_n(111),\\
P_{n+1}(00) &= P_n(0000) + P_n(1000) + P_n(1100) + P_n(1101)\\
&\quad + P_n(1110) + P_n(1111),\\
P_{n+1}(000) &= P_n(00000) + P_n(10000) + P_n(11000) + P_n(11100) + P_n(11101)\\
&\quad + P_n(11110) + P_n(11111),\\
P_{n+1}(010) &= P_n(10100) + P_n(10101) + P_n(10110) + P_n(10111),
\end{aligned}$$

Expressing everything in terms of fundamental probabilities, this reduces to

$$\begin{aligned}
P_{n+1}(0) &= 1 - P_n(0) + P_n(000), \qquad (11.37)\\
P_{n+1}(00) &= 1 - 2P_n(0) + P_n(00) + P_n(000),\\
P_{n+1}(000) &= 1 - 3P_n(0) + 2P_n(00) + P_n(000) + P_n(010) - P_n(01000),\\
P_{n+1}(010) &= P_n(0) - 2P_n(00) + P_n(000).
\end{aligned}$$

Again, these are RFP equation of order 3, listed in the Appendix D. Bayesian approximation of $P_n(01000)$ gives

$$P_n(01000) \approx \frac{P_n(010)\,(P_n(00) - P_n(000))\,P_n(000)}{(P_n(0) - P_n(00))\,P_n(00)},$$

thus the local structure map becomes

$$\begin{aligned}
Q_{n+1}(0) &= 1 - Q_n(0) + Q_n(000), \qquad (11.38)\\
Q_{n+1}(00) &= 1 - 2Q_n(0) + Q_n(00) + Q_n(000),\\
Q_{n+1}(000) &= 1 - 3Q_n(0) + 2Q_n(00) + Q_n(000) + Q_n(010) \qquad (11.39)\\
&\quad - \frac{Q_n(010)\,(Q_n(00) - Q_n(000))\,Q_n(000)}{(Q_n(0) - Q_n(00))\,Q_n(00)},\\
Q_{n+1}(010) &= Q_n(0) - 2Q_n(00) + Q_n(000).
\end{aligned}$$

In order to make this a bit more readable, define variables $x_n = P_n(0)$, $y_n = P_n(00)$, $z_n = P_n(000)$, and $v_n = P_n(010)$. Then we can write

$$\begin{aligned}
x_{n+1} &= -x_n + z_n + 1,\\
y_{n+1} &= -2\,x_n + y_n + z_n + 1,\\
z_{n+1} &= 1 + z_n + v_n - 3\,x_n + 2\,y_n - \frac{v_n\,(y_n - z_n)\,z_n}{y_n\,(x_n - y_n)},\\
v_{n+1} &= x_n - 2\,y_n + z_n,
\end{aligned}$$

One can show [9] that the above system has a stable fixed point $(x^\star, y^\star, z^\star, v^\star) = \left(\frac{1}{2}, \frac{1}{4}, 0, 0\right)$. For the initial symmetric Bernoulli measure, the exact values of first four fundamental probabilities, as given in the Appendix C, are

$$
\begin{aligned}
P_n(0) &= \frac{1}{2} + 2^{-2n-2}\binom{2n}{n}, \\
P_n(00) &= \frac{1}{4} + 2^{-2n-2}\binom{2n}{n}, \\
P_n(000) &= \frac{2^{-2n-3}\,(4n+3)}{n+1}\binom{2n}{n}, \\
P_n(010) &= 2^{-2n-2}\binom{2n}{n}.
\end{aligned}
$$

By using the Stirling approximation for the factorials in $\binom{2n}{n}$, it is straightforward to demonstrate that

$$
\lim_{n\to\infty}\big(P_n(0), P_n(00), P_n(000), P_n(010)\big) = \left(\frac{1}{2}, \frac{1}{4}, 0, 0\right) = (x^\star, y^\star, z^\star, v^\star).
$$

Both P's and Q's thus converge to the same values, again demonstrating that the local structure approximation does an excellent job predicting the limiting values of block probabilities.

Example 11.4 As a final example, we will examine probabilistic CA rule investigated in [10], namely α-asynchronous rule 6, for which the transition probabilities are given for all $(b_1, b_2, b_3) \in \{0, 1\}^3$ by

$$
\begin{aligned}
w(1|b_1b_2b_3) &= \alpha f_6(b_1, b_2, b_3) + (1-\alpha)b_2, \\
w(0|b_1b_2b_3) &= 1 - w(1|b_1b_2b_3),
\end{aligned}
$$

and where f_6 is the local function of ECA rule 6,

$$
f_6(x_1, x_2, x_3) = x_3 + x_2 - 2x_2x_3 - x_1x_3 - x_1x_2 + 2x_1x_2x_3.
$$

In the above, $\alpha \in [0, 1]$ is a parameter called the synchrony rate. When $\alpha = 1$, the PCA rule defined this way is equivalent to the deterministic rule with local function f_6, and when $\alpha = 0$, it becomes the identity rule. It is known from Monte Carlo simulations of this rule that $P_\infty(1)$ exhibits phase transition as α varies [11]. When α increases from 0, the value of $P_\infty(1)$ remains non-zero as long as $\alpha < 0.2825$, and remains equal to zero for $\alpha > 0.2825$. We will now demonstrate that local structure approximation can reproduce this behaviour of $P_\infty(1)$ in an approximate way.

We will construct local structure map of order 3 for this rule, using as probabilities $P(1)$, $P(01)$, $P(010)$ and $P(000)$. This is different than fundamental probabilities we normally use in this book, namely $P(0)$, $P(00)$, $P(010)$ and $P(000)$, but it is done for the sake of convenience as in the literature on α-synchronous CA it is typical to consider $P(1)$ instead of $P(0)$. Without going into all tedious details of the construction, we will only show the result. Using variables $X = Q(1)$, $Y = Q(01)$, $Z = Q(010)$ and $V = Q(000)$, with the help of symbolic algebra software, the local structure map of order $k = 3$ for this rule turns out to be

$$\begin{aligned}
X_{n+1} &= X_n - \alpha(2X_n + Y_n - Z_n + V_n - 1) \\
Y_{n+1} &= Y_n + \frac{Z_n}{Y_n}(X_n + V_n - 1)\alpha^2 + (1 - 2X_n + Z_n - V_n)\alpha^2 \\
&\quad + (X_n - 2Y_n + Z_n)\alpha \\
Z_{n+1} &= Z_n + \frac{Z_n}{Y_n}(V_n + X_n - 1)(\alpha^3 + \alpha) - (2X_n - Y_n + V_n - 1)\,\alpha^3 \\
&\quad + (X_n - 4Y_n + 3Z_n)\,\alpha^2 + (2Y_n - Z_n)\,\alpha \\
V_{n+1} &= V_n + \frac{\alpha}{(X_n + Y_n - 1)^2 Y_n^2} D(X_n, Y_n, Z_n, V_n),
\end{aligned}$$

where

$$\begin{aligned}
D(X, Y, Z, V) = {} & (X + Y - 1)(2XY - XZ - Y^2 + YV - ZV - Y + Z)Y\alpha^2 \\
& \times (Y - Z)(3X^2Y - X^2Z + 9XY^2 - 3XYZ + 4XYV - XZV \\
& + 6Y^3 - 2Y^2Z + 4Y^2V - YZV + YV^2 - 6XY + 2XZ - 9Y^2 \\
& + 3YZ - 4YV + ZV + 3Y - Z)\alpha - (X + Y + V - 1) \\
& \times (2XY - 2XZ + 2Y^2 - 2YZ + YV - 2Y + 2Z)Y.
\end{aligned}$$

It is a fairly complicated map, yet it is possible to find its fixed points with symbolic algebra software. The first fixed point is (X_1, Y_1, Z_1, V_1), where

$$\begin{aligned}
X_1 &= \frac{(1-\alpha)}{d_\alpha}(\alpha^6 - 5\alpha^5 + 10\alpha^4 - 7\alpha^3 - 2\alpha^2 + 4\alpha + 2) \\
&\quad \times (\alpha^6 - 7\alpha^5 + 19\alpha^4 - 23\alpha^3 + 8\alpha^2 + 8\alpha - 4), \\
Y_1 &= \frac{1-\alpha}{d_\alpha}\left(\alpha^3 - 2\alpha^2 + \alpha + 2\right)\left(\alpha^6 - 7\alpha^5 + 19\alpha^4 - 23\alpha^3 + 8\alpha^2 + 8\alpha - 4\right), \\
Z_1 &= \frac{1-\alpha}{d_\alpha}\left(\alpha^2 - 2\alpha + 2\right)\left(\alpha^6 - 7\alpha^5 + 19\alpha^4 - 23\alpha^3 + 8\alpha^2 + 8\alpha - 4\right), \\
V_1 &= \frac{\alpha^2}{d_\alpha}\left(\alpha^3 - 2\alpha^2 + \alpha + 2\right)\left(\alpha^2 - 3\alpha + 3\right) \\
&\quad \times \left(2\alpha^6 - 14\alpha^5 + 40\alpha^4 - 55\alpha^3 + 28\alpha^2 + 13\alpha - 16\right),
\end{aligned}$$

and where

$$\begin{aligned}
d_\alpha = {} & 2(\alpha^{12} - 11\alpha^{11} + 54\alpha^{10} - 152\alpha^9 + 259\alpha^8 - 248\alpha^7 + 73\alpha^6 + 95\alpha^5 - 84\alpha^4 \\
& - 15\alpha^3 + 32\alpha^2 + 2\alpha - 8).
\end{aligned}$$

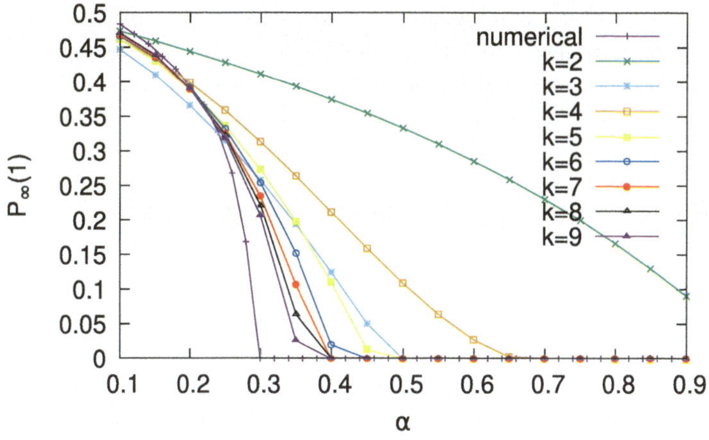

Fig. 11.1 Steady state probability of 1 for α-asynchronous rule 6 obtained by numerical simulations and by local structure approximation of orders from 2 to 9

The second fixed point is $(X_2, Y_2, Z_2, V_2) = (0, 0, 0, 1)$. It is possible to determine the stability of these fixed points, which in [10] has been found to depend on α. The first fixed point is stable for $\alpha < \alpha_c$ and unstable for $\alpha > \alpha_c$, where α_c is the solution of

$$\alpha_c{}^6 - 7\alpha_c{}^5 + 19\alpha_c{}^4 - 23\alpha_c{}^3 + 8\alpha_c{}^2 + 8\alpha_c - 4 = 0,$$

yielding numerical value of $\alpha_c = 0.4828$. The second fixed point, on the other hand, is unstable for $\alpha < \alpha_c$ and stable for $\alpha > \alpha_c$. In the limit of $n \to \infty$ one thus observes that

$$\lim_{n\to\infty} Q_n(1) = \begin{cases} X_1 & \text{if } \alpha < \alpha_c, \\ 0 & \text{otherwise.} \end{cases}$$

Although it is not possible to derive the exact formula for $P_n(1)$ for this rule, numerical simulations reveal that $\lim_{n\to\infty} P_n \neq 0$ for $\alpha < 0.2825$ and $\lim_{n\to\infty} P_n = 0$ for $\alpha > 0.2825$. One can say, therefore, that the local structure approximation of order 3 correctly predicts the existence of the second order phase transition in the α-asynchronous rule 6. The transition point α_c thus predicted is quite different than the numerically obtained value 0.2825, but, as reported in [10], by increasing the order of the local structure approximation one can improve the predicted value of the transition point too. This is illustrated in Fig. 11.1. It shows plots of $Q_\infty(1)$ versus α for local structure approximations of orders 2–9. Values of Q_∞ were approximated by iterating $L^{(k)}$ maps for $k = 2$ up to $k = 9$ for 10^4 steps and recording the final value of $Q_n(1)$. This was done for many different values of α to obtain plots of Q_∞ as a function of α. The leftmost curve represents $P_\infty(1)$ as a function of α after $n = 10^4$ iterations, again approximating $P_\infty(1)$ by $P_n(1)$ with large n. Values of $P_n(1)$ were obtained by Monte Carlo simulations, by iterating the PCA rule for 10^5 steps, using randomly generated initial configurations drawn from symmetric Bernoulli distribution with 4×10^4 sites and periodic boundary conditions. We can see that the these local structure maps not only predict the existence of phase transitions, but also provide a good approximation of the behaviour of $P_\infty(1)$ versus α curves, with increasing accuracy as the order of local structure approximation increases.

11.4 Quality of the Local Structure Approximation

Suppose that for a given rule we construct the local structure map of order k and then, starting from initial values of block probabilities corresponding to the Bernoulli measure ν_p, we iterate it many times. After a large number of iterations we record the values of $Q_n(1)$, call it $Q_\infty(1)$, and then plot $Q_\infty(1)$ versus p. The resulting curve will be called the *LST response curve*. If we plot the actual $P_\infty(1)$ versus p, the resulting curve will be called the *exact response curve*. How well the LST response curve approximates the exact one?

It turns out that for almost all minimal rules excluding two, the two curves match pretty well, the points with the same p differing in their vertical coordinates by not more than a few percent (typically even less) providing that k is high enough. Sometimes one needs $k = 5$ to get this kind of accuracy, but most of the time smaller k is sufficient.

Since for solvable rules $P_n(1)$ is know exactly, we will not discuss their local structure approximation, although we will remark that for many of them the local structure approximation becomes exact—cf. Ref. [12] for more details. We will concentrate instead on rules for which no exact probabilistic solution is given in this book, as listed in Table 12.2. For these, one cannot compare the LST response curves with the exact response curves, thus we will use numerical estimates of $P\{\infty(1)$ instead. In what follows, we report such numerical results obtained by the Monte Carlo method. We iterated the given rule for 10^4 steps on a lattice with 10^4 sites with periodic boundary condition and recorded the final frequency of ones, which then served as the numerical estimate of $P_\infty(1)$.

Among the rules listed in Table 12.2, all with the exception of two exhibit a very good agreement of the numerical response curve and the LST response curve. The top of Fig. 11.2 shows two examples of a such good agreement, for rules 33 and 74.

One can see that for rule 33, already the 4-th order approximation yields very good fit, while for rule 74 one needs 5-th order, and the fit is a bit less satisfactory. Rules 33 and 74 represent quite well the vast majority of minimal ECA.

The two exemption which we mentioned earlier are rules 26 and 154, and their graphs are shown in the bottom of Fig. 11.2. We can see from the graphs that for both rules the local structure approximation of order up to 5 predicts constant $P_\infty(1)$, independent of p, whereas the actual response curve (here represented by numerical data) does depend on p.

It is unfortunately not known which CA rules exhibit response curves well approximated by the local structure theory and which do not. However, in the next section we will consider a related question: is it is possible to construct an alternative approximation which would perform better for the cases when the local structure fails?

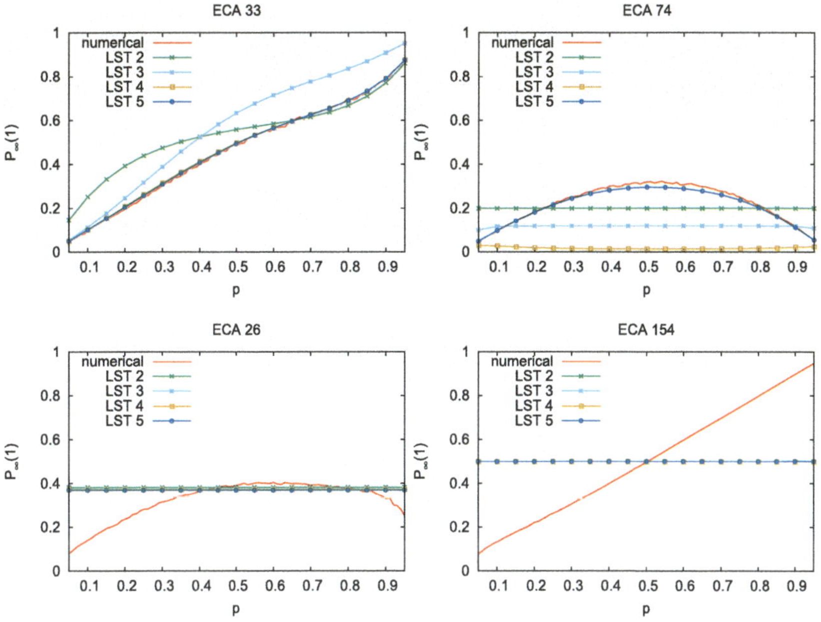

Fig. 11.2 Examples of performance of the local structure approximation for rules 33, 74, 26 and 154

11.5 Minimal Entropy Approximation

Recall that at the heart of the local structure approximation is the Bayesian extension, which, as shown in Sect. 11.1, constructs a probability measure consistent with given block probabilities and maximizing entropy. If instead of maximizing the entropy one constructs the probability measure by some other optimization algorithm, one can develop another method for the construction of approximate orbits. On such possibility has been introduced in [13], where minimization of entropy is used instead of maximization. We will not go into all details of the construction of this approximation, but we will only report the main difference between both methods.

Recall that the Bayesian or maximum entropy approximation is essentially the approximation of $k+1$-block probabilities by k-block probabilities given by the formula

$$P(a_1 a_2 \ldots a_{k+1}) \approx \frac{P(a_1 \ldots a_k) P(a_2 \ldots a_{k+1})}{P(a_2 \ldots a_k)}.$$

If the entropy is minimized instead, as demonstrated in [13], the above approximation must be replaced by

$$P(a_1 a_2 \ldots a_{k+1}) \approx \Upsilon_{a_1, a_{k+1}}\Big(P(0 a_2 \ldots a_k),\, P(a_2 \ldots a_k 0),\, P(a_2 \ldots a_k 1)\Big),$$

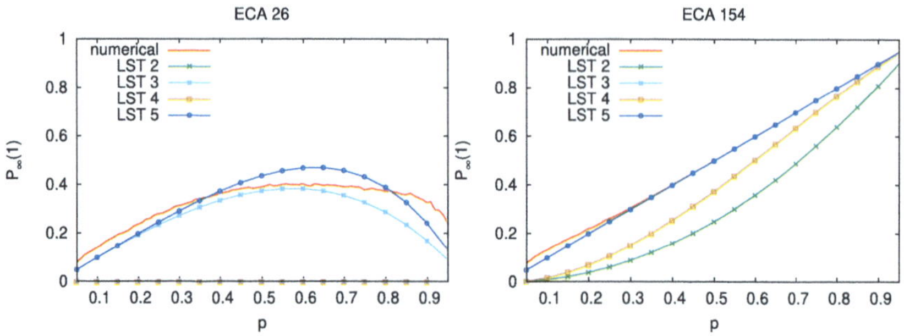

Fig. 11.3 Examples of performance of the minimal entropy approximation for 26 and 154

where

$$\Upsilon(\alpha, \beta, \delta) = \begin{cases} \max\{0, \alpha - \delta\} & \text{if } |\delta - \alpha| < |\beta - \alpha|, \\ \min\{\alpha, \beta\} & \text{otherwise,} \end{cases}$$

and

$$\begin{aligned} \Upsilon_{0,0}(\alpha, \beta, \delta) &= \Upsilon(\alpha, \beta, \delta), \\ \Upsilon_{0,1}(\alpha, \beta, \delta) &= \alpha - \Upsilon(\alpha, \beta, \delta), \\ \Upsilon_{1,0}(\alpha, \beta, \delta) &= \beta - \Upsilon(\alpha, \beta, \delta), \\ \Upsilon_{1,1}(\alpha, \beta, \delta) &= \delta - \alpha + \Upsilon(\alpha, \beta, \delta). \end{aligned}$$

The above approximation will be called the *minimal entropy approximation*. All other details of the construction of finite-dimensional maps (*minimal entropy maps*) remain the same as in the case of the local structure approximation, and a theorem analogous to Theorem 11.2 can be proved [13].

In Fig. 11.3 we show graphs corresponding to those presented in Fig. 11.2 for rules 26 and 154 yet using the minimal entropy maps. We can see that although the agreement between the experimental curves and the minimal entropy curves is not ideal, the fit is certainly much better than in the case of the local structure curves. This demonstrates that the minimal entropy approximation might serve as an alternative approximation, outperforming the local structure theory for some rules. Unfortunately, also for this approximation we do not know which CA rules are amenable to it and which are not.

As a final remark we will mention that some other modifications and improvements to the local structure approximation have been proposed as well. For example, it has been observed [14] that the performance of the approximation for number-conserving rules improves if the block probabilities which are constant are included in the local structure map formulae as constants—that is, if they are not replaced by their Bayesian extension approximations. It is certainly very plausible that further improvements of this type, suited to particular classes of CA rules, are discovered in the future.

References

1. Schulman, L.S., Seiden, P.E.: Statistical mechanics of a dynamical system based on Conway's game of life. Journal of Statistical Physics **19**(3), 293–314 (1978). https://doi.org/10.1007/BF01011727
2. Wolfram, S.: Statistical mechanics of cellular automata. Reviews of Modern Physics **55**(3), 601–644 (1983)
3. Gutowitz, H.A., Victor, J.D., Knight, B.W.: Local structure theory for cellular automata. Physica D **28**, 18–48 (1987)
4. Gutowitz, H.A., Victor, J.D.: Local structure theory in more than one dimension. Complex Systems **1**, 57–68 (1987)
5. Fukś, H.: Construction of local structure maps for cellular automata. J. of Cellular Automata **7**, 455–488 (2013)
6. Fannes, M., Verbeure, A.: On solvable models in classical lattice systems. Commun. Math. Phys. **96**, 115–124 (1984)
7. Kolmogorov, A.N., Prohorov, Y.V.: Zufällige funktionen und grenzverteilungssätze. In: Berichtüber die Tagung Wahrscheinlichkeitsrechnung und Mathematische Statistik, pp. 113–126. Deutscher Verlag der Wissenschaften, Berlin (1954)
8. Bilingsley, P.: Convergence of Probability Measures. John Wiley & Sons, New York (1968)
9. Fukś, H., Combert, F.K.: Evaluating the quality of local structure approximation using elementary rule 14. In: J.M. Baetens, M. Kutrib (eds.) Cellular Automata and Discrete Complex Systems - 24th IFIP WG 1.5 International Workshop, AUTOMATA 2018, Ghent, Belgium, June 20-22, 2018, Proceedings, *Lecture Notes in Computer Science*, vol. 10875, pp. 43–56. Springer (2018). https://doi.org/10.1007/978-3-319-92675-9
10. Fukś, H., Fatès, N.: Local structure approximation as a predictor of second-order phase transitions in asynchronous cellular automata. Natural Computing **14**, 507–522 (2015). https://doi.org/10.1007/s11047-015-9521-6
11. Fatès, N.: Asynchronism induces second order phase transitions in elementary cellular automata. Journal of Cellular Automata **4**(1), 21–38 (2009). http://hal.inria.fr/inria-00138051
12. Gutowitz, H.: Cellular Automata: Theory and Experiment. MIT Press/Bradford Books, Cambridge, Mass. (1991)
13. Fukś, H.: Minimal entropy approximation for cellular automata. J. of Statistical Mechanics: Theory and Experiment **4**, P02009 (2014). https://doi.org/10.1088/1742-5468/2014/02/P02009
14. Fukś, H., Yin, Y.: Approximating dynamics of a number-conserving cellular automaton by a finite-dimensional dynamical system. Int. J. Mod. Phys. C **31**(12), 2050172 (2020). https://doi.org/10.1142/S0129183120501727. https://www.worldscientific.com/doi/10.1142/S0129183120501727

Chapter 12
Beyond Solvable Elementary Rules

Patet omnibus veritas, nondum est occupata. Multum ex illa etiam futuris relictum est.

Seneca

12.1 Solvable Versus Non-solvable Rules

The overview of various methods for obtaining solutions of initial value problems presented in preceding chapters implores the question: which CA rules are solvable? Table 12.1 shows all ECA rules for which deterministic and probabilistic initial value problem has a solution given in this book. These solutions are provided in, respectively, Appendices B and C. The table uses the following abbreviations:

Y	explicit formula for the solution given;
R	recursive formula for the solution given;
Sym	only solution with symmetric initial measure is given ($p = 1/2$);
E	equicontinuous;
AE	almost equicontinuous;
ED	has equicontinuous direction;
AED	has almost equicontinuous direction;
S	surjective.

We can see that 57 out of 88 minimal ECA are solvable in the deterministic sense, which constitutes about 65% of all minimal rules. The percentage of probabilistically solvable rules is even higher if one includes surjective rules, for which the probabilistic solution with uniform initial measure always exists.

It is clear that the vast majority of the solvable rules have some equicontinuity property, ranging from those which are fully equicontinuous to those possessing almost equicontinuous direction. In Sect. 10.6 we found that equicontinuous rules satisfy identities of the form $F^{q+p} = F^q$ and those with equicontinuous direction

H. Fukś, *Solvable Cellular Automata*, Understanding Complex Systems,
https://doi.org/10.1007/978-3-031-38700-5_12

Table 12.1 Table of solvability for all minimal ECA

Rule	Det. sol.	Prob. sol.	Remarks	Rule	Det. sol.	Prob. sol.	Remarks
0	Y	Y	E	56	R		
1	Y	Y	E	57			
2	Y	Y	ED	58			
3	Y	Y	ED	60	Y	Y	S
4	Y	Y	E	62			
5	Y	Y	E	72	Y	Y	E
6				73			AE
7		Y	AED	74			
8	Y	Y	E	76	Y	Y	E
9				77	Y	Y	AE
10	Y	Y	ED	78	Y	Y	AE
11	R			90	Y	Y	S
12	Y	Y	E	94			AE
13	Y	Y	AE	104			AE
14	R	Sym		105	Y		S
15	Y	Y	S, ED	106		Sym	S
18				108	Y	Y	E
19	Y	Y	E	110			
22				122			
23	Y	Y	AE	126			
24	Y	Y	ED	128	Y	Y	AE
25				130	Y	Y	AED
26				132	Y	Y	AE
27	Y	Y	AED	134			
28	Y	Y	AE	136	Y	Y	AE
29	Y	Y	E	138	Y	Y	ED
30		Sym	S	140	Y	Y	AE
32	Y	Y	AE	142	R	Sym	
33			AE	146			
34	Y	Y	ED	150	Y	Y	S
35				152			
36	Y	Y	E	154		Sym	S, AED
37				156	Y	Y	AE
38	Y	Y	ED	160	Y	Y	AE
40	Y	Y	AE	162	Y	Y	AED
41				164	Y	Y	AE
42	Y	Y	ED	168	Y	Y	AE
43	R	Sym		170	Y	Y	S, ED
44	Y	Y	AE	172	Y	Y	AE
45		Sym	S	178	Y	Y	AE
46	Y	Y	ED	184	R	Y	
50	Y	Y	AE	200	Y	Y	E
51	Y	Y	S, E	204	Y	Y	S, E
54				232	Y	Y	AE

identities of the form $F^{q(s+t)}\sigma^{pt} = F^{qs}$. From what we have learned about orbits of rules with such identities in Chap. 5, it is clear that the following general statement can be made.

Proposition 12.1 *Rules which are equicontinuous or have equicontinuous direction are solvable.*

For rules which are almost equicontinuous or have almost equicontinuous direction the situation is less clear. It seems that the existence of the blocking word in the rule itself or in its shifted iterate makes it amenable to solving, because the blocking word severely restricts the orbit's ability to wander "too far" from the initial configuration. In some sense, blocking words play a similar rule as integrals of motion in integrable systems, which restrict orbits to a submanifold within the phase space.

Nevertheless, there are some rules like 7, 33, 73, 94, 104 or 154, which do exhibit equicontinuity property in some form, yet their solution is not given in this book. It is very likely that these rules are actually solvable, and that given enough time and effort their solutions will eventually be found.

Rules with additive invariants (14, 43, 142 and 184) form another distinct group. They are deterministically solvable, but not in an explicit sense: we found their solutions in a recursive form only. Probabilistic solutions of these rules, with the exception of rule 184, are available only for the symmetric initial measure. Rule 56 does not possess any additive invariants, but it emulates rules 184 ($F_{56}^2 = F_{184}F_{56}$), thus it is possible to construct its recursive solution using the solution formula for rule 184.

On the other end of the spectrum there are rules which are most certainly not solvable. It is known that ECA rule 110 is capable of universal computation [1, 2], thus there is no hope that a formula for $[F_{110}]_j$ could ever be obtained. It is conjectured that rule 54, which exhibits a rich system of colliding and interacting gliders [3], is also computationally universal, thus the same comment likely applies to $[F_{54}]_j$.

Another interesting question is the connection between solvability and chaos. In Hamiltonian dynamics chaotic systems are not integrable, thus also not solvable in the sense that their solutions in an explicit functional form cannot be obtained. For cellular automata the situation is a bit different. According to Devaney's definition of chaos, transitivity and density of periodic points are required to call a map chaotic (sensitivity, originally included as one of the condition of chaos, is now known to be implied by transitivity and density of periodic points). Cattaneo et al. demonstrated in [4] that for elementary CA, the only chaotic rules are the non-trivial are the surjective rules, where "non-trivial" excludes single input rules such as 204 and 51. Among minimal ECA, therefore, the only Devaney-chaotic ones are rules 15, 30, 45, 60, 90, 105, 106, 150, 154 and 170. This includes six rules for which the solution of the deterministic IVP is given in this book, namely 15, 60, 90, 105, 150 and 170, demonstrating that chaotic ECA can also be fully solvable. Solvability of the remaining four (30, 45, 106 and 154) remains an open question. Probably the most promising one is rule 154, which has almost equicontinuous direction. Rule 30, on the other hand, is non-linear and it has been used to generate pseudo-random numbers of very good quality [5, 6], thus it is most likely nonsolvable.

For other rules for which the deterministic solution is not known, partial probabilistic solutions may be obtainable nevertheless, and numerical experiments sometimes seem to indicate that this might be possible. For example, for rule 18 we know that

$$P_n(11) \sim n^{-1/2}.$$

The above power law can be explained by the fact that in rule 18 one can view sequences of 0s of even length as "defects" which perform a pseudo-random walk and annihilate upon collision. This has been discovered numerically in [7] and later formally demonstrated in [8]. Another example of a similar power law appears in rule 54, for which Boccara et al. [9] numerically verified that

$$P_n(1) \sim n^{-\gamma},$$

where $\gamma \approx 0.15$. As in the case of rule 18, particle kinematics of rule 54 is now well understood [10], but the above power law has not been formally demonstrated, and the exact value of the exponent γ remains unknown.

Rules 146 and 126 are related to rule 18, by, respectively, emulation $F_{18}^2 = F_{146}F_{18}$ and conjugacy via surjective rule 60, $F_{60}F_{126} = F_{18}F_{60}$ [11]. Their dynamics and solvability properties, therefore, should be somewhat similar [12] as for rule 18, meaning that deterministic solutions are unlikely to be obtainable.

Another complex rule with highly non-trivial dynamics which has been extensively studied is rule 22 [13–15]. Due to its complexity it does not seem to be a likely candidate for a solvable rule either.

All elementary rules for which deterministic solutions are not know at the time of writing are shown in Table 12.2. The last column contains a non-exhaustive list of references investigating dynamical properties of these rules.

12.2 Higher Radius, Number of States and Dimension

Our book was almost exclusively focused on solving binary nearest-neighbour CA. Although not a lot is known about solvability of more general rules, in what follows we will provide some examples of solvable rules with larger neighbourhood, larger number of states and in dimensions higher than one. We will not provide derivations of their solutions—these can be found in references provided. The intention is to demonstrate possibilities of generalization, hoping to stimulate the reader to explore this subject further.

Table 12.2 Elementary rules for which deterministic solution is not known

Rule	Prob. sol.	Remarks	References
6		Emulates rule 134	
7	Y	AED	
9		Interacting gliders	
18		Interacting random walkers	[7, 8, 10]
22		Complex pattern	[13–15]
25		Interacting gliders	
26		Pattern resembling rule 154	[16]
30	Sym	Surjective, chaotic	[5, 6, 17]
33		AE	
35		Interacting gliders, conserves blocks 10	[18, 19]
37		Interacting domains/gliders	
41		Interacting domains/gliders	[20]
45	Sym	Surjective, chaotic	
54		Conjectured universal	[3, 9, 10]
57		Interacting gliders similar to rule 184	[21]
58		Interacting gliders	[22]
62		Interacting gliders	[23]
73		AE	
74		Rather simple interacting domains	[24]
94		AE	
104		AE	
106	Sym	Surjective, chaotic	[25]
110		Universal	[1, 26]
122		Complex pattern resembling rule 22	[27]
126		Conjugate to rule 18 via ECA 60	[12]
134		Interacting domains/gliders	
146		Emulated by rule 18	[12]
152		Simple pattern, monotone nonincreasing	[24, 28]
154	Sym	Surjective, chaotic, AED	[29]

AE almost equicontinuous, *AED* has almost equicontinuous direction, *Sym* probabilistic solution for symmetric (uniform) Bernoulli measure exists

Example 12.1 The first example is the rule with the neighbourhood of 4 sites, one neighbour on the left and two on the right, given by the local function $f : \{0, 1\}^4 \to \{0, 1\}$, so that

$$[F(x)]_i = f(x_{i-1}, x_i, x_{i+1}, x_{i+2}),$$

where

$$\begin{aligned} f(0000) &= f(0001) = f(0010) = f(0011) = f(0101) \\ &= f(1000) = f(1001) = f(1101) = 0, \\ f(0100) &= f(0110) = f(0111) = f(1010) = f(1011) \\ &= f(1100) = f(1110) = f(1111) = 1. \end{aligned}$$

This rule has been investigated in [30]. It conserves the number of 1s, and it is one of 22 number-conserving rules with 4-site neighbourhood. One can interpret it as a particle system, where 1s represent individual particles, and 0s represent empty spaces. In this representation, one can show [31] that the motion of particles will schematically be governed by the following rules,

$$\overset{\curvearrowright}{1}01, \quad \overset{\circlearrowleft}{1}00, \quad \overset{\circlearrowleft}{1}1.$$

This means that only a particle which has a single zero on the right will move to the right (symbol $\curvearrowright$), while particles followed by 1 or by two or more zeros will stay in the same place (symbol $\circlearrowleft$). The structure of preimages of blocks of length 3 under this rule is rather straighforward to understand, and this allows (see [30]) to obtain the following expressions for the fundamental block probabilities of length up to 3,

$$\begin{aligned} P_n(0) &= 1 - p, \\ P_n(00) &= 1 - p - p(1-p)^2 - (1-p)(2-p)^n p^{n+2}, \\ P_n(000) &= 1 - p - (1-p)(2-p)^n p^{n+2} - 2p(1-p)^2, \\ P_n(010) &= p(1-p)^4 + p^2(1-p)^2(p(2-p))^n. \end{aligned} \tag{12.1}$$

In addition to being number-conserving, this rule also has a blocking word 100, meaning that it is almost equicontinuous. The existence of the blocking word is actually exploited in the construction of the solution.

Example 12.2 The next example we will consider is the nearest-neighbour rule with 3 states, $\mathcal{A} = \{0, 1, 2\}$, described in [32, 33]. It is a rule obtained from an elementary CA by "lifting" it to 3-states. What we mean by "lifting" is the following construction. Let $g : \{0, 1\}^3 \to \{0, 1\}$ be a local function of elementary CA satisfying $g(0, 0, 0) = 0$, $g(1, 1, 1) = 1$, and let $f_g : \{0, 1, 2\}^3 \to \{0, 1, 2\}$ be defined by

$$f_g(x_1, x_2, x_3) = \begin{cases} g(x_1, x_2, x_3), & x_1, x_2, x_3 \in \{0, 1\} \\ 2g\left(\frac{x_1}{2}, \frac{x_2}{2}, \frac{x_3}{2}\right), & x_1, x_2, x_3 \in \{0, 2\} \\ g(x_1 - 1, x_2 - 1, x_3 - 1) + 1, & x_1, x_2, x_3 \in \{1, 2\} \\ x_2, & \text{otherwise.} \end{cases} \tag{12.2}$$

This construction ensures that when f_g is restricted to two symbols only, it becomes equivalent to g, otherwise it behaves as identity. Conditions $g(0, 0, 0) = 0, g(1, 1, 1) = 1$ ensure that there are no conflicts, so that, for example, $f(2, 2, 2)$ is the same no matter if we apply the second or the third case of Eq. (12.2). One of the most interesting "lifted" rules is the case of g being elementary CA rule with Wolfram number 140, defined as

$$g(x_1, x_2, x_3) = x_2 - x_1x_2 + x_1x_2x_3, \tag{12.3}$$

where $x_1, x_2, x_3 \in \{0, 1\}$. For this rule we obtain

$$f_g(x_1, x_2, x_3) = \begin{cases} x_3 & \text{for } x_1 = x_2 > x_3, \\ x_2 & \text{otherwise}, \end{cases} \tag{12.4}$$

where $x_1, x_2, x_3 \in \{0, 1, 2\}$. It can be equivalently defined as

$$f_g(x_1, x_2, x_3) = \begin{cases} 0 & \text{for } (x_1, x_2, x_3) = (1, 1, 0) \text{ or } (x_1, x_2, x_3) = (2, 2, 0), \\ 1 & \text{for } (x_1, x_2, x_3) = (2, 2, 1), \\ x_2 & \text{otherwise}. \end{cases} \tag{12.5}$$

By analyzing the FSM of preimages of single symbols 0, 1 and 2, the authors of [32, 33] derived expressions for probabilities of occurrences of these symbols, starting from a Bernoulli measure with probabilities of 0, 1 and 2 equal to, respectively, p, q, and r, where $p + q + r = 1$. These expressions are rather complicated, thus we will give only $P_n(1)$ and $P_n(2)$, and the reader can easily obtain the third one by using $P_n(0) = 1 - P_n(1) - P_n(2)$. The expression for $P_n(1)$ for $q \neq r$ is given by

$$\begin{aligned} P_n(1) = {} & \frac{pq^2\left(-pr + pq + q^2\right)q^n}{(p+r)(q-r)} \\ & + \frac{qr\left(-p^2r + p^2q + pq^2 - 2\,pqr + r^3 - q^2r\right)r^n}{\lambda^2\,(p+q)(q-r)} \\ & + \frac{q\left(p^3 + p^2q + 2\,p^2r + pr^2 + 3\,pqr + r^3 + r^2q + q^2r\right)}{(p+r)(p+q)}, \end{aligned}$$

and for $q = r$ by

$$\begin{aligned} P_n(1) = {} & \frac{pq^3\,(n+1)\,q^n}{q+p} + \frac{q^2\left(2\,p^3 + 4\,p^2q + pq^2 - 2\,q^3\right)q^n}{(q+p)^2} \\ & + \frac{\left(p^3 + 3\,p^2q + 4\,pq^2 + 3\,q^3\right)q}{(q+p)^2}. \end{aligned} \tag{12.6}$$

The expression for $P_n(2)$ is much simpler,

$$P_n(2) = (q + p)\,r + r^{n+2}.$$

The most interesting aspect of the above solution is the fact that for $q = r$ it becomes "degenerate", with linear-exponential convergence to the limit value. One can, for $q = r$, rewrite Eq. (12.6) as

$$P_n(1) = P_\infty(1) - (An + B)\left(\frac{1-p}{2}\right)^n,$$

where

$$\begin{aligned}
A &= \frac{(p-1)^2}{4\,(1+p)^2}\left(p^3 - p\right),\\
B &= \frac{(p-1)^2}{4\,(1+p)^2}\left(1 - p^3 - 5\,p - 3\,p^2\right),\\
P_\infty(1) &= \frac{(1-p)\left(p^3 + 5\,p^2 - p + 3\right)}{4\,(1+p)^2}.
\end{aligned}$$

One can see that for $0 < p < 1$, $P_n(1)$ tends to $P_\infty(1)$ as $n \to \infty$, and that the convergence is linear-exponential in n. No such linear-exponential convergence has been observed for any ECA rule with known solution listed in Appendix C.

Example 12.3 The final example, or rather a set of examples, are CA rules in dimension higher than one. In [34, 35] two-dimension rules with L-shaped neighbourhood were considered. Define the local function for the L-shaped neighbourhood as

$$f : \left(\begin{smallmatrix} x & \\ y & z \end{smallmatrix}\right) \to \{0, 1\}$$

where $x, y, z \in \{0, 1\}$. Similarly as in one dimension, f has a corresponding global mapping, $F : \{0, 1\}^{\mathbb{Z}^2} \to \{0, 1\}^{\mathbb{Z}^2}$ such that

$$[F(x)]_{i,j} = f\left(\begin{smallmatrix} x_{i,j+1} & \\ x_{i,j} & x_{i+1,j} \end{smallmatrix}\right) \tag{12.7}$$

for any $x \in \{0, 1\}^{\mathbb{Z}^2}$. Local rules can be defined using the same Wolfram numbering scheme as for ECA, with

$$W(f) = \sum_{x_0, x_1, x_2 \in \{0,1\}} f\left(\begin{smallmatrix} x_0 & \\ x_1 & x_2 \end{smallmatrix}\right) 2^{4x_0 + 2x_1 + x_2}.$$

Reference [35] derived expressions for $P_n(1)$ for over 20 of such rules. Interestingly, they often exhibit different behaviour than a corresponding ECA rule in one dimension. For instance, for the L-shaped rule 138 we have, according to [35],

$$P_n(1) = \frac{p^{2n+2} + p}{p+1},$$

whereas for the ECA 138 in one dimension Appendix C gives

$$P_n(1) = 1 - (1-p)(p^2+1).$$

This means that in one dimension the probability is constant for a given p, while in two dimensions it decays exponentially toward the limiting value as n increases.

For the L-shaped rule 130, it is possible to obtain an explicit expression for $P_n(1)$ for the symmetric measure [35], that is, for $p = 1/2$:

$$P_n(1) = \frac{1}{2} - \frac{1}{4}\sum_{i=0}^{n-1} 2^{-\frac{i(i+3)}{2}}. \quad (12.8)$$

No similar sum has been encountered in one-dimensional solvable ECA. For ECA rule 130 in one dimension, the expression for $P_n(0)$ given in Appendix C yields for $p = 1/2$

$$P_n(1) = \frac{1}{6} + \frac{2^{-2n}}{3},$$

which is again very different (and much simpler) than in the L-shaped case.

References

1. Cook, M.: Universality in elementary cellular automata. Complex Systems **15**, 1–40 (2004)
2. Wolfram, S.: A New Kind of Science. Wolfram Media (2002)
3. Martinez, G.J., Adamatzky, A., McIntosh, H.: Phenomenology of glider collisions in cellular automaton rule 54 and associated logical gates. Chaos Solitons and Fractals **28**, 100–111 (2006)
4. Cattaneo, G., Finelli, M., Margara, L.: Investigating topological chaos by elementary cellular automata dynamics. Theoretical Computer Science **244**, 219–241 (2000)
5. Wolfram, S.: Random sequence generation by cellular automata. Adv. Appl. Math. **7**(2), 123–169 (1986)
6. Shin, S.H., Yoo, K.Y.: Analysis of 2-state, 3-neighborhood cellular automata rules for cryptographic pseudorandom number generation. In: 2009 International Conference on Computational Science and Engineering, vol. 1, pp. 399–404 (2009)
7. Grassberger, P.: Chaos and diffusion in deterministic cellular automata. Physica D **10**(1), 52–58 (1984)
8. Eloranta, K., Nummelin, E.: The kink of cellular automaton rule 18 performs a random walk. Journal of Statistical Physics **69**(5), 1131–1136 (1992)
9. Boccara, N., Nasser, J., Roger, M.: Particlelike structures and their interactions in spatiotemporal patterns generated by one-dimensional deterministic cellular-automaton rules. Phys. Rev. A **44**, 866–875 (1991)

10. Pivato, M.: Defect particle kinematics in one-dimensional cellular automata. Theoretical Computer Science **377**(1), 205–228 (2007)
11. Boccara, N.: Transformations of one-dimensional cellular automaton rules by translation-invariant local surjective mappings. Physica D **68**, 416–426 (1993)
12. Jiang, Z.: The relationships of three elementary cellular automata. Journal of Systems Science and Complexity **19**, 128–136 (2006)
13. Grassberger, P.: Long-range effects in an elementary cellular automaton. Journal of Statistical Physics **45**, 27–39 (1986)
14. Zabolitzky, J.G.: Critical properties of rule 22 elementary cellular automata. Journal of Statistical Physics **50**, 1255–1262 (1988)
15. Martinez, G.J., Adamatzky, A., Hoffmann, R., Désérable, D., Zelinka, I.: On patterns and dynamics of rule 22 cellular automaton. Complex Systems **28**, 125–174 (2019)
16. Guan, J., Chen, F.: Complex dynamics in elementary cellular automaton rule 26. Journal of cellular automata **10**, 137–147 (2015)
17. Gage, D., Laub, E., McGarry, B.: Cellular automata: is rule 30 random. In: Proceedings of the Midwest NKS Conference, Indiana University (2005)
18. Fukś, H.: Remarks on the critical behavior of second order additive invariants in elementary cellular automata. Fundamenta Informaticae **78**, 329–341 (2007)
19. Han, Q., Liao, X., Li, C., Feng, L.: Complex dynamics behaviors in cellular automata rule 35. Journal of Cellular Automata **6**(6), 487–504 (2011)
20. Guan, J., Wang, K.: Complex shift dynamics of some elementary cellular automaton rules. Complex Systems **20**(1), 31 (2011)
21. Sato, T.: On behaviors of cellular automata with rule number 57. Bulletin of informatics and cybernetics **38** (2006)
22. Pei, Y., Han, Q., Liu, C., Tang, D., Huang, J.: Chaotic behaviors of symbolic dynamics about rule 58 in cellular automata. Mathematical Problems in Engineering **2014**, art. no. 834268 (2014)
23. Shi, L., Chen, F., Jin, W.: Gliders, collisions and chaos of cellular automata rule 62. In: 2009 International Workshop on Chaos-Fractals Theories and Applications, pp. 221–225 (2009)
24. Boccara, N.: Eventually number-conserving cellular automata. International Journal of Modern Physics C **18**(01), 35–42 (2007)
25. Zhao, G., Chen, F., Jin, W., et al.: Infinite number of disjoint chaotic subsystems of cellular automaton rule 106. Applied Mathematics **5**(20), 3256 (2014)
26. Neary, T., Woods, D.: P-completeness of cellular automaton rule 110. In: Automata, Languages and Programming: 33rd International Colloquium, ICALP 2006, Venice, Italy, July 10–14, 2006, Proceedings, Part I 33, pp. 132–143. Springer (2006)
27. Jiang, Z.: A complexity analysis of the elementary cellular automaton of rule 122. Chinese Science Bulletin **46**, 600–603 (2001)
28. Goles, E., Moreira, A., Rapaport, I.: Communication complexity in number-conserving and monotone cellular automata. Theoretical Computer Science **412**(29), 3616–3628 (2011)
29. Gutowitz, H.A., Victor, J.D., Knight, B.W.: Local structure theory for cellular automata. Physica D **28**, 18–48 (1987)
30. Fukś, H., Yin, Y.: Approximating dynamics of a number-conserving cellular automaton by a finite-dimensional dynamical system. Int. J. Mod. Phys. C **31**(12), 2050172 (2020). https://doi.org/10.1142/S0129183120501727. https://www.worldscientific.com/doi/10.1142/S0129183120501727
31. Boccara, N., Fukś, H.: Cellular automaton rules conserving the number of active sites. J. Phys. A: Math. Gen. **31**, 6007–6018 (1998). https://doi.org/10.1088/0305-4470/31/28/014
32. Fukś, H., Midgley-Volpato, J.: An example of degenerate hyperbolicity in cellular automaton with 3 states. In: J. Kari, I. Törmä, M. Szabados (eds.) 21st International Workshop on Cellular Automata and Discrete Complex Systems, *TUCS Lecture Notes*, vol. 24, pp. 47–55. Turku, Finland (2015)
33. Fukś, H., Midgley-Volpato, J.: An example of a deterministic cellular automaton exhibiting linear-exponential convergence to the steady state. Acta Phys. Pol. B **9**(1), 49–62 (2016)

34. Fukś, H., Skelton, A.: Response curves for cellular automata in one and two dimensions – an example of rigorous calculations. International Journal of Natural Computing Research **1**, 85–99 (2010)
35. Fukś, H., Skelton, A.: Response curves and preimage sequences of two-dimensional cellular automata. In: Proceedings of the 2011 International Conference on Scientific Computing: CSC-2011, pp. 165–171. CSERA Press (2011)

Correction to: Solvable Cellular Automata

Correction to:
H. Fukś, *Solvable Cellular Automata*, Understanding Complex Systems, https://doi.org/10.1007/978-3-031-38700-5

In the original version of chapters 6 and 10, below were updated:

1. Equation 6.3 has been corrected.
2. Figure 10.6 has been replaced.
3. Text of p. 183 has been replaced.

The updated versions of these chapters can be found at
https://doi.org/10.1007/978-3-031-38700-5_6
https://doi.org/10.1007/978-3-031-38700-5_10

H. Fukś, *Solvable Cellular Automata*, Understanding Complex Systems,
https://doi.org/10.1007/978-3-031-38700-5_13

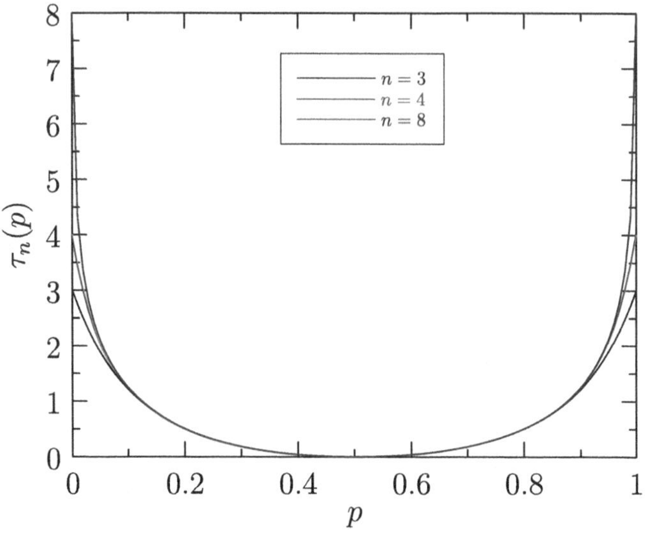

Fig. 10.6 Critical slowing down near $p = 0$ and $p = 1$ for ECA rule 60

Appendix A
Polynomial Representation of Minimal Rules

$$
\begin{aligned}
f_0(x_1, x_2, x_3) &= 0 \\
f_1(x_1, x_2, x_3) &= 1 - x_3 - x_2 - x_1 + x_2x_3 + x_1x_3 + x_1x_2 - x_1x_2x_3 \\
f_2(x_1, x_2, x_3) &= x_3 - x_2x_3 - x_1x_3 + x_1x_2x_3 \\
f_3(x_1, x_2, x_3) &= 1 - x_2 - x_1 + x_1x_2 \\
f_4(x_1, x_2, x_3) &= x_2 - x_2x_3 - x_1x_2 + x_1x_2x_3 \\
f_5(x_1, x_2, x_3) &= 1 - x_3 - x_1 + x_1x_3 \\
f_6(x_1, x_2, x_3) &= x_3 + x_2 - 2\,x_2x_3 - x_1x_3 - x_1x_2 + 2\,x_1x_2x_3 \\
f_7(x_1, x_2, x_3) &= 1 - x_1 - x_2x_3 + x_1x_2x_3 \\
f_8(x_1, x_2, x_3) &= x_2x_3 - x_1x_2x_3 \\
f_9(x_1, x_2, x_3) &= 1 - x_3 - x_2 - x_1 + 2\,x_2x_3 + x_1x_3 + x_1x_2 - 2\,x_1x_2x_3 \\
f_{10}(x_1, x_2, x_3) &= x_3 - x_1x_3 \\
f_{11}(x_1, x_2, x_3) &= 1 - x_2 - x_1 + x_2x_3 + x_1x_2 - x_1x_2x_3 \\
f_{12}(x_1, x_2, x_3) &= x_2 - x_1x_2 \\
f_{13}(x_1, x_2, x_3) &= 1 - x_3 - x_1 + x_2x_3 + x_1x_3 - x_1x_2x_3 \\
f_{14}(x_1, x_2, x_3) &= x_3 + x_2 - x_2x_3 - x_1x_3 - x_1x_2 + x_1x_2x_3 \\
f_{15}(x_1, x_2, x_3) &= 1 - x_1 \\
f_{18}(x_1, x_2, x_3) &= x_3 + x_1 - x_2x_3 - 2\,x_1x_3 - x_1x_2 + 2\,x_1x_2x_3 \\
f_{19}(x_1, x_2, x_3) &= 1 - x_2 - x_1x_3 + x_1x_2x_3 \\
f_{22}(x_1, x_2, x_3) &= x_3 + x_2 + x_1 - 2\,x_2x_3 - 2\,x_1x_3 - 2\,x_1x_2 + 3\,x_1x_2x_3 \\
f_{23}(x_1, x_2, x_3) &= 1 - x_2x_3 - x_1x_3 - x_1x_2 + 2\,x_1x_2x_3 \\
f_{24}(x_1, x_2, x_3) &= x_1 + x_2x_3 - x_1x_3 - x_1x_2 \\
f_{25}(x_1, x_2, x_3) &= 1 - x_3 - x_2 + 2\,x_2x_3 - x_1x_2x_3
\end{aligned}
$$

H. Fukś, *Solvable Cellular Automata*, Understanding Complex Systems,
https://doi.org/10.1007/978-3-031-38700-5

$$f_{26}(x_1, x_2, x_3) = x_3 + x_1 - 2\,x_1x_3 - x_1x_2 + x_1x_2x_3$$
$$f_{27}(x_1, x_2, x_3) = 1 - x_2 + x_2x_3 - x_1x_3$$
$$f_{28}(x_1, x_2, x_3) = x_2 + x_1 - x_1x_3 - 2\,x_1x_2 + x_1x_2x_3$$
$$f_{29}(x_1, x_2, x_3) = 1 - x_3 + x_2x_3 - x_1x_2$$
$$f_{30}(x_1, x_2, x_3) = x_3 + x_2 + x_1 - x_2x_3 - 2\,x_1x_3 - 2\,x_1x_2 + 2\,x_1x_2x_3$$
$$f_{32}(x_1, x_2, x_3) = x_1x_3 - x_1x_2x_3$$
$$f_{33}(x_1, x_2, x_3) = 1 - x_3 - x_2 - x_1 + x_2x_3 + 2\,x_1x_3 + x_1x_2 - 2\,x_1x_2x_3$$
$$f_{34}(x_1, x_2, x_3) = x_3 - x_2x_3$$
$$f_{35}(x_1, x_2, x_3) = 1 - x_2 - x_1 + x_1x_3 + x_1x_2 - x_1x_2x_3$$
$$f_{36}(x_1, x_2, x_3) = x_2 - x_2x_3 + x_1x_3 - x_1x_2$$
$$f_{37}(x_1, x_2, x_3) = 1 - x_3 - x_1 + 2\,x_1x_3 - x_1x_2x_3$$
$$f_{38}(x_1, x_2, x_3) = x_3 + x_2 - 2\,x_2x_3 - x_1x_2 + x_1x_2x_3$$
$$f_{40}(x_1, x_2, x_3) = x_2x_3 + x_1x_3 - 2\,x_1x_2x_3$$
$$f_{41}(x_1, x_2, x_3) = 1 - x_3 - x_2 - x_1 + 2\,x_2x_3 + 2\,x_1x_3 + x_1x_2 - 3\,x_1x_2x_3$$
$$f_{42}(x_1, x_2, x_3) = x_3 - x_1x_2x_3$$
$$f_{43}(x_1, x_2, x_3) = 1 - x_2 - x_1 + x_2x_3 + x_1x_3 + x_1x_2 - 2\,x_1x_2x_3$$
$$f_{44}(x_1, x_2, x_3) = x_2 + x_1x_3 - x_1x_2 - x_1x_2x_3$$
$$f_{45}(x_1, x_2, x_3) = 1 - x_3 - x_1 + x_2x_3 + 2\,x_1x_3 - 2\,x_1x_2x_3$$
$$f_{46}(x_1, x_2, x_3) = x_3 + x_2 - x_2x_3 - x_1x_2$$
$$f_{50}(x_1, x_2, x_3) = x_3 + x_1 - x_2x_3 - x_1x_3 - x_1x_2 + x_1x_2x_3$$
$$f_{51}(x_1, x_2, x_3) = 1 - x_2$$
$$f_{54}(x_1, x_2, x_3) = x_3 + x_2 + x_1 - 2\,x_2x_3 - x_1x_3 - 2\,x_1x_2 + 2\,x_1x_2x_3$$
$$f_{56}(x_1, x_2, x_3) = x_1 + x_2x_3 - x_1x_2 - x_1x_2x_3$$
$$f_{57}(x_1, x_2, x_3) = 1 - x_3 - x_2 + 2\,x_2x_3 + x_1x_3 - 2\,x_1x_2x_3$$
$$f_{58}(x_1, x_2, x_3) = x_3 + x_1 - x_1x_3 - x_1x_2$$
$$f_{60}(x_1, x_2, x_3) = x_2 + x_1 - 2\,x_1x_2$$
$$f_{62}(x_1, x_2, x_3) = x_3 + x_2 + x_1 - x_2x_3 - x_1x_3 - 2\,x_1x_2 + x_1x_2x_3$$
$$f_{72}(x_1, x_2, x_3) = x_2x_3 + x_1x_2 - 2\,x_1x_2x_3$$
$$f_{73}(x_1, x_2, x_3) = 1 - x_3 - x_2 - x_1 + 2\,x_2x_3 + x_1x_3 + 2\,x_1x_2 - 3\,x_1x_2x_3$$
$$f_{74}(x_1, x_2, x_3) = x_3 - x_1x_3 + x_1x_2 - x_1x_2x_3$$
$$f_{76}(x_1, x_2, x_3) = x_2 - x_1x_2x_3$$
$$f_{77}(x_1, x_2, x_3) = 1 - x_3 - x_1 + x_2x_3 + x_1x_3 + x_1x_2 - 2\,x_1x_2x_3$$

$$f_{78}(x_1, x_2, x_3) = x_3 + x_2 - x_2x_3 - x_1x_3$$
$$f_{90}(x_1, x_2, x_3) = x_3 + x_1 - 2\,x_1x_3$$
$$f_{94}(x_1, x_2, x_3) = x_3 + x_2 + x_1 - x_2x_3 - 2\,x_1x_3 - x_1x_2 + x_1x_2x_3$$
$$f_{104}(x_1, x_2, x_3) = x_2x_3 + x_1x_3 + x_1x_2 - 3\,x_1x_2x_3$$
$$f_{105}(x_1, x_2, x_3) = 1 - x_3 - x_2 - x_1 + 2\,x_2x_3 + 2\,x_1x_3 + 2\,x_1x_2 - 4\,x_1x_2x_3$$
$$f_{106}(x_1, x_2, x_3) = x_3 + x_1x_2 - 2\,x_1x_2x_3$$
$$f_{108}(x_1, x_2, x_3) = x_2 + x_1x_3 - 2\,x_1x_2x_3$$
$$f_{110}(x_1, x_2, x_3) = x_3 + x_2 - x_2x_3 - x_1x_2x_3$$
$$f_{122}(x_1, x_2, x_3) = x_3 + x_1 - x_1x_3 - x_1x_2x_3$$
$$f_{126}(x_1, x_2, x_3) = x_3 + x_2 + x_1 - x_2x_3 - x_1x_3 - x_1x_2$$
$$f_{128}(x_1, x_2, x_3) = x_1x_2x_3$$
$$f_{130}(x_1, x_2, x_3) = x_3 - x_2x_3 - x_1x_3 + 2\,x_1x_2x_3$$
$$f_{132}(x_1, x_2, x_3) = x_2 - x_2x_3 - x_1x_2 + 2\,x_1x_2x_3$$
$$f_{134}(x_1, x_2, x_3) = x_3 + x_2 - 2\,x_2x_3 - x_1x_3 - x_1x_2 + 3\,x_1x_2x_3$$
$$f_{136}(x_1, x_2, x_3) = x_2x_3$$
$$f_{138}(x_1, x_2, x_3) = x_3 - x_1x_3 + x_1x_2x_3$$
$$f_{140}(x_1, x_2, x_3) = x_2 - x_1x_2 + x_1x_2x_3$$
$$f_{142}(x_1, x_2, x_3) = x_3 + x_2 - x_2x_3 - x_1x_3 - x_1x_2 + 2\,x_1x_2x_3$$
$$f_{146}(x_1, x_2, x_3) = x_3 + x_1 - x_2x_3 - 2\,x_1x_3 - x_1x_2 + 3\,x_1x_2x_3$$
$$f_{150}(x_1, x_2, x_3) = x_3 + x_2 + x_1 - 2\,x_2x_3 - 2\,x_1x_3 - 2\,x_1x_2 + 4\,x_1x_2x_3$$
$$f_{152}(x_1, x_2, x_3) = x_1 + x_2x_3 - x_1x_3 - x_1x_2 + x_1x_2x_3$$
$$f_{154}(x_1, x_2, x_3) = x_3 + x_1 - 2\,x_1x_3 - x_1x_2 + 2\,x_1x_2x_3$$
$$f_{156}(x_1, x_2, x_3) = x_2 + x_1 - x_1x_3 - 2\,x_1x_2 + 2\,x_1x_2x_3$$
$$f_{160}(x_1, x_2, x_3) = x_1x_3$$
$$f_{162}(x_1, x_2, x_3) = x_3 - x_2x_3 + x_1x_2x_3$$
$$f_{164}(x_1, x_2, x_3) = x_2 - x_2x_3 + x_1x_3 - x_1x_2 + x_1x_2x_3$$
$$f_{168}(x_1, x_2, x_3) = x_2x_3 + x_1x_3 - x_1x_2x_3$$
$$f_{170}(x_1, x_2, x_3) = x_3$$
$$f_{172}(x_1, x_2, x_3) = x_2 + x_1x_3 - x_1x_2$$
$$f_{178}(x_1, x_2, x_3) = x_3 + x_1 - x_2x_3 - x_1x_3 - x_1x_2 + 2\,x_1x_2x_3$$
$$f_{184}(x_1, x_2, x_3) = x_1 + x_2x_3 - x_1x_2$$
$$f_{200}(x_1, x_2, x_3) = x_2x_3 + x_1x_2 - x_1x_2x_3$$
$$f_{204}(x_1, x_2, x_3) = x_2$$
$$f_{232}(x_1, x_2, x_3) = x_2x_3 + x_1x_3 + x_1x_2 - 2\,x_1x_2x_3$$

Appendix B
Deterministic Solution Formulae

Special notation used in solution formulae:

$$\overline{x} = 1 - x$$
$$\mathrm{ev}(n) = \frac{1}{2}(-1)^n + \frac{1}{2},$$
$$I_k(n) = \begin{cases} 1 & n \leq k, \\ 0 & \text{otherwise.} \end{cases}$$

Rule 0

$$[F_0^n(x)]_j = 0$$

Rule 1

$$[F_1^n(x)]_j = \begin{cases} \overline{x}_{j-1}\overline{x}_j\overline{x}_{j+1} & n \text{ odd,} \\ (1 - \overline{x}_{j-2}\overline{x}_{j-1}\overline{x}_j)(1 - \overline{x}_{j-1}\overline{x}_j\overline{x}_{j+1})(1 - \overline{x}_j\overline{x}_{j+1}\overline{x}_{j+2}) & n \text{ even.} \end{cases}$$

Rule 2

$$[F_2^n(x)]_j = x_{j+n}x_{j+n-2}x_{j+n-1} - x_{j+n}x_{j+n-2} - x_{j+n}x_{j+n-1} + x_{j+n}$$

Rule 3

$$[F_3^n(x)]_j = \begin{cases} 1 - \overline{x}_{j-k-1}\overline{x}_{j-k} - \overline{x}_{j-k}\overline{x}_{j-k+1} + \overline{x}_{j-k-1}\overline{x}_{j-k}\overline{x}_{j-k+1} & n = 2k, \\ x_{j-k-1}x_{j-k} - x_{j-k} - x_{j-k-1} + 1 & n = 2k + 1. \end{cases}$$

Rule 4

$$[F_4^n(x)]_j = x_{j-1}x_jx_{j+1} - x_{j-1}x_j - x_jx_{j+1} + x_j$$

H. Fukś, *Solvable Cellular Automata*, Understanding Complex Systems,
https://doi.org/10.1007/978-3-031-38700-5

Rule 5

$$[F_5^n(x)]_j = \begin{cases} 1 - x_{j-1} - x_{j+1} + x_{j-1}x_{j+1} & n \text{ odd}, \\ x_j + x_{j-2}x_{j+2} - x_{j-2}x_jx_{j+2} & n \text{ even}. \end{cases}$$

Rule 8

$$[F_8^n(x)]_i = \begin{cases} -x_{j-1}x_jx_{j+1} + x_jx_{j+1} & n = 1 \\ 0 & \text{otherwise} \end{cases}$$

Rule 10

$$[F_{10}^n(x)]_j = -x_{j+n}x_{j+n-2} + x_{j+n}$$

Rule 11

$$[F_{11}^n(x)]_j = [F_{43}^{n-1}(y)]_j$$

where

$$y_j = 1 + x_{j-1}x_j - x_{j-1} - x_{j+1}x_{j-1}x_j - x_j + x_{j+1}x_j.$$

Rule 12

$$[F_{12}^n(x)]_j = -x_{j-1}x_j + x_j$$

Rule 13

$$[F_{13}^n(x)]_j = x_j + \sum_{r=1}^{n} (-1)^r \prod_{i=0}^{r} x_{j-i} + \sum_{r=1}^{n} (-1)^{r+1} \prod_{i=-1}^{r} \overline{x}_{j-i}$$
$$+ x_{j+2}\overline{x}_{j+1}\overline{x}_j \sum_{r=2}^{n} (-1)^r \prod_{i=1}^{r} x_{j-i}$$

Rule 14

$$[F_{14}^n(x)]_j = \sum_{\substack{\mathbf{a}\in\{0,1\}^{2n+1} \\ M_{2n-1}(\mathbf{a})\leq n-1 \\ M_{2n+1}(\mathbf{a})=n}} \Psi_{\mathbf{a}}(x_{j-n}, x_{j-n+1}, \ldots, x_{j+n}),$$

where

$$M_0 = 0,$$
$$M_i = M_{i-1} + \left[a_i = \frac{1}{2} - \frac{1}{2}(-1)^{\max\{i-1-M_{i-1}, M_{i-1}\}} \vee i = M_{i-1} + n + 2 \right].$$

Rule 15

$$[F_{15}^n(x)]_j = \frac{1}{2}(-1)^{n+1} + \frac{1}{2} + (-1)^n x_{j-n}$$

Rule 19

$$[F^n_{19}(x)]_j = \begin{cases} x_{j-1}x_jx_{j+1} - x_{j-1}x_{j+1} - x_j + 1 & n = 1, \\ \frac{1}{2}\left(-1\right)^{n+1} + \frac{1}{2} + (-1)^n A_j & \text{otherwise,} \end{cases}$$

$$\begin{aligned} A_j = &- x_jx_{j-2}x_{j-1}x_{j+1}x_{j+2} + x_jx_{j-2}x_{j-1}x_{j+1} + x_jx_{j-2}x_{j-1}x_{j+2} \\ &+ x_jx_{j-2}x_{j+1}x_{j+2} + x_jx_{j-1}x_{j+1}x_{j+2} - x_{j-2}x_{j-1}x_j - x_jx_{j-2}x_{j+1} \\ &- x_jx_{j-2}x_{j+2} - 2x_{j-1}x_jx_{j+1} - x_jx_{j-1}x_{j+2} - x_jx_{j+1}x_{j+2} + x_{j-2}x_j \\ &+ x_jx_{j-1} + x_jx_{j+1} + x_jx_{j+2} + x_{j-1}x_{j+1}. \end{aligned}$$

Rule 23

$$\begin{aligned} [F^n_{23}(x)]_j = &\ x_{-n+j}\prod_{i=1}^{n}\overline{x}_{2i-n-1+j}x_{2i-n+j} \\ &+ \sum_{k=0}^{n-1}\Big(x_{k+j}\left(x_{-k-1+j}\overline{x}_{k+1+j} + x_{k+1+j}\right) \\ &\times \prod_{i=0}^{k-1}x_{-k+2i+j}\overline{x}_{-k+2i+1+j}\Big) \quad \text{for } n \text{ even,} \end{aligned}$$

$$\begin{aligned} [F^n_{23}(x)]_j = &\ 1 - x_{-n+j}\prod_{i=1}^{n}\overline{x}_{2i-n-1+j}x_{2i-n+j} \\ &- \sum_{k=0}^{n-1}\Big(x_{k+j}\left(x_{-k-1+j}\overline{x}_{k+1+j} + x_{k+1+j}\right) \\ &\times \prod_{i=0}^{k-1}x_{-k+2i+j}\overline{x}_{-k+2i+1+j}\Big) \quad \text{for } n \text{ odd.} \end{aligned}$$

Rule 24

$$[F^n_{24}(x)]_j = \begin{cases} -x_{j-1}x_j + x_jx_{j+1} - x_{j-1}x_{j+1} + x_{j-1} & n = 1 \\ B_n & \text{otherwise} \end{cases}$$

$$\begin{aligned} B_n = &\ x_{j-n} - x_{j-n}x_{j-n+1}x_{j-n+2}x_{j-n+3} - x_{j-n}x_{j-n+1}x_{j-n+2}x_{j-n+4} \\ &+ x_{j-n}x_{j-n+1}x_{j-n+3}x_{j-n+4} + x_{j-n}x_{j-n+2}x_{j-n+3}x_{j-n+4} \\ &- x_{j-n+1}x_{j-n+2}x_{j-n+3}x_{j-n+4} + x_{j-n}x_{j-n+1}x_{j-n+2} \\ &- x_{j-n}x_{j-n+3}x_{j-n+4} + x_{j-n+1}x_{j-n+2}x_{j-n+3} + x_{j-n+1}x_{j-n+2}x_{j-n+4} \\ &- x_{j-n}x_{j-n+1} - x_{j-n}x_{j-n+2} \end{aligned}$$

Rule 27

$$
\begin{aligned}
[F_{27}^n(x)]_j &= \overline{x}_{j-k} + x_{j-k+1}\left(x_{j-k} - x_{j-k-1}\right) \\
&\quad + \sum_{p=1}^{k}\Bigg(\left(x_{j-k-2+3p} - x_{j-k+1+3p}\right) \\
&\quad \times \prod_{i=0}^{p} x_{j-k-1+3i}\overline{x}_{j-k+3i}\Bigg), \quad n = 2k+1, \\
[F_{27}^n(x)]_j &= x_{j-k} - x_{j-k+1}\overline{x}_{j-k+2}\left(x_{j+3-k}x_{j-k-1} + x_{j-k} - x_{j+3-k} - x_{j-k-1}\right) \\
&\quad - \sum_{p=1}^{k-1}\Bigg(\left(x_{j-k+3p} - x_{j+3-k+3p}\right)\overline{x}_{j-k-1} \\
&\quad \times \prod_{i=0}^{p} x_{j-k+1+3i}\overline{x}_{j-k+2+3i}\Bigg), \quad n = 2k.
\end{aligned}
$$

Rule 28

$$
[F_{28}^n(x)]_j = [F_{156}^{n-1}(y)]_j,
$$

where

$$
y_j = -2x_{j-1}x_j + x_{j-1} + x_j - x_{j-1}x_{j+1} + x_{j-1}x_jx_{j+1}.
$$

Rule 29

$$
[F_{29}^n(x)]_j = \begin{cases} 1 - x_{j-1}x_j + x_jx_{j+1} - x_{j+1} & n \text{ odd}, \\ x_j + \left(x_{j+1} - 1\right)x_{j-1}\left(x_{j-2}x_j - x_jx_{j+2} - x_{j-2} + x_j\right) & n \text{ even}. \end{cases}
$$

Rule 32

$$
[F_{32}^n(x)]_j = x_{j-n}\prod_{i=1}^{n}\overline{x}_{2i-n-1+j}x_{2i-n+j}
$$

Rule 34

$$
[F_{34}^n(x)]_j = -x_{j+n}x_{j+n-1} + x_{j+n}
$$

Rule 36

$$
[F_{36}^n(x)]_j = \begin{cases} -x_{j-1}x_j - x_jx_{j+1} + x_{j-1}x_{j+1} + x_j & n = 1, \\ C_n & \text{otherwise}, \end{cases}
$$

$$
\begin{aligned}
C_n = x_j &- x_jx_{j-2}x_{j-1}x_{j+1} - x_jx_{j-2}x_{j-1}x_{j+2} - x_jx_{j-2}x_{j+1}x_{j+2} \\
&- x_jx_{j-1}x_{j+1}x_{j+2} + x_{j-2}x_{j-1}x_{j+1}x_{j+2} + x_jx_{j-2}x_{j-1} + x_jx_{j-2}x_{j+1} \\
&+ x_jx_{j-2}x_{j+2} + x_jx_{j-1}x_{j+1} + x_jx_{j-1}x_{j+2} + x_jx_{j+1}x_{j+2} \\
&- x_{j-2}x_j - x_{j-1}x_j - x_jx_{j+1} - x_jx_{j+2}
\end{aligned}
$$

Rule 38

$$[F_{38}^n(x)]_j = \begin{cases} x_j x_{j-1} x_{j+1} - x_j x_{j-1} - 2 x_j x_{j+1} + x_j + x_{j+1} & n = 1, \\ A_n & n \text{ even}, \\ B_n & \text{otherwise}, \end{cases}$$

$$\begin{aligned} A_n =& - x_{j+n} x_{j+n-4} x_{j+n-3} x_{j+n-2} x_{j+n-1} + x_{j+n} x_{j+n-4} x_{j+n-3} x_{j+n-1} \\ &+ x_{j+n} x_{j+n-3} x_{j+n-2} - x_{j+n} x_{j+n-3} x_{j+n-1} - x_{j+n} x_{j+n-2} + x_{j+n}, \\ B_n =& \, x_{j+n} x_{j+n-4} x_{j+n-3} x_{j+n-2} x_{j+n-1} - x_{j+n} x_{j+n-4} x_{j+n-3} x_{j+n-1} \\ &- 2 x_{j+n} x_{j+n-3} x_{j+n-2} x_{j+n-1} - x_{j+n-4} x_{j+n-3} x_{j+n-2} x_{j+n-1} \\ &+ x_{j+n} x_{j+n-3} x_{j+n-2} + x_{j+n} x_{j+n-3} x_{j+n-1} + 2 x_{j+n} x_{j+n-2} x_{j+n-1} \\ &+ x_{j+n-4} x_{j+n-3} x_{j+n-1} + x_{j+n-3} x_{j+n-2} x_{j+n-1} - x_{j+n} x_{j+n-2} \\ &- 2 x_{j+n} x_{j+n-1} - x_{j+n-3} x_{j+n-1} - x_{j+n-2} x_{j+n-1} + x_{j+n} + x_{j+n-1} \end{aligned}$$

Rule 40

$$[F_{40}^n(x)]_j = [F_{168}^{n-1}(y)]_j$$

where

$$y_j = - 2x_{j+1} x_{j-1} x_j + x_{j+1} x_j + x_{j+1} x_{j-1}$$

Rule 42

$$[F_{42}^n(x)]_j = - x_{j+n} x_{j+n-2} x_{j+n-1} + x_{j+n}$$

Rule 43

$$[F_{43}^n(x)]_j = \sum_{\substack{\mathbf{a} \in \{0,1\}^{2n+1} \\ M_{2n+1}(\mathbf{a}) = n}} \Psi_{\mathbf{a}}(x_{j-n}, x_{j-n+1}, \ldots, x_{j+n}),$$

where

$$\begin{aligned} M_0 &= 0, \\ M_i &= M_{i-1} + \left[a_i = \frac{1}{2} - \frac{1}{2}(-1)^{\min\{i-1-M_{i-1}, M_{i-1}\}} \vee i = M_{i-1} + n + 2 \right]. \end{aligned}$$

Rule 44

$$\begin{aligned} [F_{44}^n(x)]_j =& \, \overline{x}_{j-2} \overline{x}_{j-1} x_j + \sum_{r=1}^{n-1} \left(C_{r-1}\left(x_{j-r-2}\right) \prod_{k=1}^{2+2r} C_{k+r-1}\left(\overline{x}_{j+k-r-2}\right) \right) \\ &+ \prod_{i=-n}^{n} C_{i-n+1}\left(\overline{x}_{i+j}\right) + \prod_{i=-n}^{n} C_{i-n+1}\left(\overline{x}_{i+j-1}\right), \end{aligned}$$

where

$$C_k(x) = \begin{cases} x & \text{if } k \text{ divisible by 3,} \\ 1-x & \text{otherwise.} \end{cases}$$

Rule 46

$$[F^n_{46}(x)]_j = \begin{cases} -x_{j-1}x_j - x_jx_{j+1} + x_j + x_{j+1} & n = 1, \\ D_n & \text{otherwise,} \end{cases}$$

$$\begin{aligned} D_n = & -x_{j+n}x_{j+n-3}x_{j+n-2}x_{j+n-1} - x_{j+n-4}x_{j+n-3}x_{j+n-2}x_{j+n-1} \\ & + x_{j+n}x_{j+n-3}x_{j+n-2} + x_{j+n}x_{j+n-2}x_{j+n-1} + x_{j+n-4}x_{j+n-3}x_{j+n-1} \\ & + x_{j+n-3}x_{j+n-2}x_{j+n-1} - x_{j+n}x_{j+n-2} - x_{j+n}x_{j+n-1} - x_{j+n-3}x_{j+n-1} \\ & - x_{j+n-2}x_{j+n-1} + x_{j+n} + x_{j+n-1} \end{aligned}$$

Rule 50

$$\begin{aligned} [F^n_{50}(x)]_j = & \frac{1}{2} + \frac{1}{2}(-1)^n + \sum_{i=1}^{n-1} \left((-1)^{i+n} \prod_{p=-i}^{i} x_{p+j} \right) \\ & + \sum_{i=0}^{n} \left((-1)^{i+n+1} \prod_{p=-i}^{i} \overline{x}_{p+j} \right) \end{aligned}$$

Rule 51

$$[F^n_{51}(x)]_j = \frac{1}{2} + \frac{1}{2}(-1)^{n+1} + (-1)^n x_j$$

Rule 56

$$[F^n_{56}(x)]_j = [F^{n-1}_{184}(y)]_j$$

where

$$y_j = x_{j+1}x_j + x_{j-1} - x_{j+1}x_{j-1}x_j - x_{j-1}x_j.$$

Rule 60

$$[F^n_{60}(x)]_j = \sum_{i=0}^{n} \binom{n}{i} x_{i-n+j} \mod 2$$

Rule 72

$$[F^n_{72}(x)]_j = \begin{cases} -2x_{j-1}x_jx_{j+1} + x_{j-1}x_j + x_jx_{j+1} & n = 1 \\ E_n & \text{otherwise} \end{cases}$$

$$\begin{aligned} E_n = & \, x_jx_{j-2}x_{j-1}x_{j+1} + x_jx_{j-1}x_{j+1}x_{j+2} - x_{j-2}x_{j-1}x_j \\ & - 2x_{j-1}x_jx_{j+1} - x_jx_{j+1}x_{j+2} + x_{j-1}x_j + x_jx_{j+1} \end{aligned}$$

Rule 76

$$[F_{76}^n(x)]_j = -x_{j-1}x_jx_{j+1} + x_j$$

Rule 77

$$[F_{77}^n(x)]_j = x_j + \sum_{r=1}^{n}(-1)^r \prod_{i=-r}^{r} x_{j+i} + \sum_{r=1}^{n}(-1)^{r+1} \prod_{i=-r}^{r} \overline{x}_{j+i}$$

Rule 78

$$[F_{78}^n(x)]_j = 1 - x_{j-1}\overline{x}_j - \prod_{i=n-1}^{2n} \overline{x}_{i-n+j} - S_{n,j}^{(A)} - S_{n,j}^{(B)} + C_{n,j},$$

where

$$\begin{aligned}
S_{n,j}^{(A)} &= \sum_{k=2}^{n-1}\left(\mathrm{ev}(k)x_{j+k-1}\overline{x}_{j+k}\prod_{j-2}^{j-2+k}\overline{x}_i\right)\\
&\quad+\sum_{k=1}^{n-1}\left(\mathrm{ev}(k+1)x_{j+k-1}x_{j+k}\overline{x}_{j+k+1}\prod_{j-2}^{j-2+k}\overline{x}_i\right)\\
&\quad+\sum_{k=2}^{n-1}\left(\mathrm{ev}(k)x_{j+k-1}x_{j+k}x_{j+k+1}\prod_{j-2}^{j-2+k}\overline{x}_i\right),\\
S_{n,j}^{(B)} &= x_{j-1}x_jx_{j+1},\\
C_{n,j} &= x_{j-3}x_{j-2}x_{j-1}x_jx_{j+1}\sum_{r=2}^{2\lfloor n/2\rfloor-2}\left((-1)^r\prod_{k=2}^{r}x_{j+k}\right) - \prod_{i=-2}^{n}x_{j+i}\\
&\quad+\overline{x}_{j-3}x_{j-2}x_{j-1}x_jx_{j+1} + I_3(n)x_{j-3}x_{j-2}x_{j-1}x_jx_{j+1}.
\end{aligned}$$

Rule 90

$$[F_{90}^n(x)]_j = \sum_{i=0}^{n}\binom{n}{i}x_{2i-n+j} \mod 2$$

Rule 105

$$[F_{105}^n(x)]_j = \begin{cases} \sum_{i=0}^{2n}\binom{n}{i-n}_2 x_{i+j-n} \mod 2 & n \text{ even},\\ 1-\sum_{i=0}^{2n}\binom{n}{i-n}_2 x_{i+j-n} \mod 2 & n \text{ odd}. \end{cases}$$

Rule 108

$$[F_{108}^n(x)]_j = \begin{cases} -2\,x_j x_{j-1} x_{j+1} + x_{j-1} x_{j+1} + x_j & n = 1, \\ A & n \text{ even}, \\ B & \text{otherwise}, \end{cases}$$

$$\begin{aligned} A = {} & x_j - x_j \left(x_{j-2} x_{j+1} + x_{j-2} x_{j+2} + x_{j-1} x_{j+2}\right) \\ & - 2\,x_j x_{j-2} x_{j+2} \left(x_{j-1} x_{j+1} - x_{j-1} - x_{j+1}\right) \\ B = {} & -4 x_j x_{j-3} x_{j-2} x_{j-1} x_{j+1} x_{j+2} x_{j+3} + 2\,x_j x_{j-3} x_{j-2} x_{j-1} x_{j+1} x_{j+2} \\ & + 2\,x_j x_{j-3} x_{j-2} x_{j-1} x_{j+1} x_{j+3} + 2\,x_j x_{j-3} x_{j-1} x_{j+1} x_{j+2} x_{j+3} \\ & + 2\,x_j x_{j-2} x_{j-1} x_{j+1} x_{j+2} x_{j+3} + 4\,x_{j-3} x_{j-2} x_{j-1} x_{j+1} x_{j+2} x_{j+3} \\ & - 2\,x_j x_{j-3} x_{j-2} x_{j-1} x_{j+1} - x_j x_{j-3} x_{j-1} x_{j+1} x_{j+2} \\ & - x_j x_{j-3} x_{j-1} x_{j+1} x_{j+3} - 4\,x_j x_{j-2} x_{j-1} x_{j+1} x_{j+2} \\ & - x_j x_{j-2} x_{j-1} x_{j+1} x_{j+3} - 2\,x_j x_{j-1} x_{j+1} x_{j+2} x_{j+3} \\ & - 2\,x_{j-3} x_{j-2} x_{j-1} x_{j+1} x_{j+2} - 2\,x_{j-3} x_{j-2} x_{j-1} x_{j+1} x_{j+3} \\ & - 2\,x_{j-3} x_{j-1} x_{j+1} x_{j+2} x_{j+3} - 2\,x_{j-2} x_{j-1} x_{j+1} x_{j+2} x_{j+3} \\ & + x_j x_{j-3} x_{j-1} x_{j+1} + 2\,x_j x_{j-2} x_{j-1} x_{j+1} + 2\,x_j x_{j-2} x_{j-1} x_{j+2} \\ & + 2\,x_j x_{j-2} x_{j+1} x_{j+2} + 2\,x_j x_{j-1} x_{j+1} x_{j+2} + x_j x_{j-1} x_{j+1} x_{j+3} \\ & + 2\,x_{j-3} x_{j-2} x_{j-1} x_{j+1} + x_{j-3} x_{j-1} x_{j+1} x_{j+2} + x_{j-3} x_{j-1} x_{j+1} x_{j+3} \\ & + x_{j-2} x_{j-1} x_{j+1} x_{j+2} + x_{j-2} x_{j-1} x_{j+1} x_{j+3} + 2\,x_{j-1} x_{j+1} x_{j+2} x_{j+3} \\ & - x_j x_{j-2} x_{j+1} - x_j x_{j-2} x_{j+2} - 2\,x_j x_{j-1} x_{j+1} - x_j x_{j-1} x_{j+2} \\ & - x_{j-3} x_{j-1} x_{j+1} - x_{j-2} x_{j-1} x_{j+1} - x_{j-1} x_{j+1} x_{j+2} \\ & - x_{j-1} x_{j+1} x_{j+3} + x_{j-1} x_{j+1} + x_j \end{aligned}$$

Rule 128

$$[F_{128}^n(x)]_j = \prod_{i=-n}^{n} x_{i+j}$$

Rule 130

$$\begin{aligned} [F_{130}^n(x)]_j = {} & \prod_{i=1}^{2n+1} x_{i+j-n-1} \\ & + \sum_{i=0}^{n-1} \left(\left(\frac{1}{2} + \frac{1}{2} (-1)^i - (-1)^i x_{n-2-2i+j} \right) \overline{x}_{n-2i+j-1} x_{n-2i+j} \right. \\ & \left. \times \prod_{p=2n+2-2i}^{2n+1} x_{p+j-n-1} \right) \end{aligned}$$

Rule 132

$$[F^n_{132}(x)]_j = x_j - \sum_{r=1}^{n} \left(\left(x_{j-r} - x_{j+r}\right)^2 \prod_{i=-r+1}^{r-1} x_{i+j} \right)$$

Rule 136

$$[F^n_{136}(x)]_j = \prod_{i=0}^{n} x_{i+j}$$

Rule 138

$$[F^n_{138}(x)]_j = x_{j+n}x_{j+n-2}x_{j+n-1} - x_{j+n}x_{j+n-2} + x_{j+n}$$

Rule 140

$$[F^n_{140}(x)]_j = \left(1 - x_{j-1}\right) x_j + \prod_{i=n-1}^{2n} x_{i-n+j}$$

Rule 142

$$[F^n_{142}(x)]_j = \sum_{\substack{\mathbf{a}\in\{0,1\}^{2n+1} \\ M_{2n+1}(\mathbf{a})=n}} \Psi_{\mathbf{a}}(x_{j-n}, x_{j-n+1}, \ldots, x_{j+n}),$$

where

$$M_0 = 0,$$

$$M_i = M_{i-1} + \left[a_i = \frac{1}{2} - \frac{1}{2}(-1)^{\max\{i-1-M_{i-1}, M_{i-1}\}} \vee i = M_{i-1} + n + 2 \right].$$

Rule 150

$$[F^n_{150}(x)]_j = \sum_{i=0}^{2n} T(n,i)\, x_{i+j-n} \mod 2$$

$$T(n,i) = \sum_{p=0}^{n} \binom{n}{p} \binom{p}{i-p}$$

Rule 156

$$[F^n_{156}(x)]_j = \overline{x}_{j-1} x_j + \prod_{i=n-1}^{2n} x_{i-n+j} + D_{n,j} + B^{(0)}_{n,j} + B^{(0)}_{n,j} + S^{(0)}_{n,j} + S^{(1)}_{n,j},$$

where

$$D_{n,j} = x_{j-n} \prod_{i=n-1}^{2n-1} \overline{x}_{-i+n+j}.$$

$$B_{n,j}^{(0)} = \mathrm{ev}\,(n) \sum_{r=2}^{n} \left(\mathrm{ev}\,(r+1)\, x_{j-r} x_{j-r+1} x_{2+j} \prod_{m=j-r+2}^{1+j} \overline{x}_m \right)$$
$$+ \mathrm{ev}\,(n+1) \sum_{r=2}^{n} \left(\mathrm{ev}\,(r)\, \overline{x}_{j-r} x_{j-r+1} x_{2+j} \prod_{m=j-r+2}^{1+j} \overline{x}_m \right),$$
$$B_{n,j}^{(1)} = \mathrm{ev}\,(n+1) \sum_{m=0}^{n-1} \left(\mathrm{ev}\,(m+1)\, \overline{x}_{j-2} \overline{x}_{j+m+1} \overline{x}_{j+m+2} \prod_{k=-1}^{m} x_{j+k} \right)$$
$$+ \mathrm{ev}\,(n) \sum_{m=0}^{n-1} \left(\mathrm{ev}\,(m)\, \overline{x}_{j-2} \overline{x}_{j+m+1} x_{j+m+2} \prod_{k=-1}^{m} x_{j+k} \right),$$
$$S_{n,j}^{(0)} = \sum_{k=1}^{n} \left(\mathrm{ev}\,(k)\, x_{j-k} x_{j-k+1} \prod_{m=j-k+2}^{1+j} \overline{x}_m \right)$$
$$+ \sum_{k=2}^{n} \left(\mathrm{ev}\,(k+1)\, \overline{x}_{j-k} x_{j-k+1} \prod_{m=j-k+2}^{1+j} \overline{x}_m \right),$$
$$S_{n,j}^{(1)} = \sum_{k=2}^{n} \left((-1)^k \left(x_{j+k} + \mathrm{ev}\,(k+1) \left(x_{j-1+k} - 1 \right) \right) \prod_{i=j-2}^{j-2+k} x_i \right)$$
$$- \sum_{k=2}^{n} \left(x_j x_{j-2} x_{j-1} (-1)^k \prod_{m=1}^{k} x_{j+m} \right).$$

Rule 160

$$[F_{160}^n(x)]_j = \prod_{i=0}^{n} x_{2i+j-n}$$

Rule 162

$$[F_{162}^n(x)]_j = x_{j+n} + \sum_{p=-n+1}^{n} \left((-1)^{p+n} \prod_{i=j-p}^{j+n} x_i \right)$$

Rule 164

$$
\begin{aligned}
[F^n_{164}]_j = &\prod_{p=0}^{n} x_{j-n+2p} - \prod_{p=-n}^{n} x_{j+p} + \left(1+\sum_{p=0}^{n}\binom{n}{p}x_{j-n+2p} \mod 2\right) \\
&\times \prod_{p=0}^{n-1} x_{j-n+2p+1} + \sum_{i=1}^{n-1}\left[\left(1-\prod_{p=0}^{i} x_{j-i+2p}\right)\overline{x}_{j-1-i}\overline{x}_{j+i+1}\right. \\
&\times \left.\left(\prod_{p=0}^{i-1} x_{j-i+2p+1}\right)\left(1+\sum_{p=0}^{i}\binom{i}{p}x_{j-i+2p} \mod 2\right)\right]
\end{aligned}
$$

Rule 168

$$
\begin{aligned}
[F^n_{168}(x)]_j = &\sum_{k=n+2}^{2n}\Big(H_k\,\overline{x}_{n-k+j}\overline{x}_{n-k+1+j}x_{n-k+2+j}x_{n+j} \\
&\times \prod_{i=2n-k+4}^{2n}(1-\overline{x}_{i-n-1+j}\overline{x}_{i-n+j})\Big) \\
&+ x_{n+j}\prod_{i=1}^{2n-1}(1-\overline{x}_{i-n-1+j}\overline{x}_{i-n+j})
\end{aligned}
$$

$$
H_k = \begin{cases} 1 & \text{if} \displaystyle\sum_{i=2n-k+4}^{2n} x_{i-n-1+j} \geq n-1, \\ 0 & \text{otherwise.} \end{cases}
$$

Rule 170

$$
[F^n_{170}(x)]_j = x_{j+n}
$$

Rule 172

$$
[F^n_{172}(x)]_j = \overline{x}_{j-2}\overline{x}_{j-1}x_j + \left(\overline{x}_{j+n-2}x_{j+n-1} + x_{j+n-2}x_{j+n}\right)\prod_{i=j-2}^{j+n-3}(1-\overline{x}_i\overline{x}_{i+1})
$$

Rule 178

$$
[F^n_{178}(x)]_j = \begin{cases} x_j + \displaystyle\sum_{r=1}^{n}(-1)^r\prod_{i=-r}^{r}x_{j+i} + \sum_{r=1}^{n}(-1)^{r+1}\prod_{i=-r}^{r}\overline{x}_{j+i} & n \text{ even}, \\ 1 - x_j - \displaystyle\sum_{r=1}^{n}(-1)^r\prod_{i=-r}^{r}x_{j+i} - \sum_{r=1}^{n}(-1)^{r+1}\prod_{i=-r}^{r}\overline{x}_{j+i} & n \text{ odd}. \end{cases}
$$

Rule 184

$$[F^n_{184}(x)]_j = \sum_{\substack{\mathbf{a}\in\{0,1\}^{2n+1}\\ M_{2n+1}(\mathbf{a})=n}} \Psi_{\mathbf{a}}(x_{j-n}, x_{j-n+1}, \ldots, x_{j+n}),$$

where Ψ is the density polynomial and

$$M_i(\mathbf{a}) = \begin{cases} 0 & \text{if}\, i = 0, \\ M_{i-1}(\mathbf{a}) + 1 & \text{if}\, a_i = 0 \,\text{and}\, i \leq 2M_{i-1} + 1, \\ M_{i-1}(\mathbf{a}) + 1 & \text{if}\, a_i = 1 \,\text{and}\, i > 2M_{i-1} + 1, \\ M_{i-1}(\mathbf{a}) + 1 & \text{if}\, i = M_{i-1} + n + 2, \\ M_{i-1}(\mathbf{a}) & \text{otherwise}. \end{cases}$$

Rule 200

$$[F^n_{200}(x)]_j = -x_{j-1}x_jx_{j+1} + x_{j-1}x_j + x_jx_{j+1}$$

Rule 204

$$[F^n_{204}(x)]_j = x_j$$

Rule 232

$$\begin{aligned} [F^n_{232}(x)]_j = {} & x_{j-n} \prod_{i=1}^{n} \overline{x}_{j+2i-n-1} x_{j+2i-n} \\ & + \sum_{k=1}^{n} x_{j-k} x_{j-k+1} \left(\prod_{i=1}^{k-1} \overline{x}_{j-k+2i} x_{j-k+2i+1} \right) \overline{x}_{j+k} \\ & + \sum_{k=1}^{n} \left(\prod_{i=0}^{k-2} x_{j-k+2i+1} \overline{x}_{j-k+2i+2} \right) x_{j+k-1} x_{j+k} \end{aligned}$$

Appendix C
Probabilistic Solution Formulae

This appendix lists expressions for probabilities of the first four fundamental blocks ($P_n(0)$ $P_n(00)$, $P_n(000)$ and $P_n(010)$) after n iterations of a given ECA rule, assuming that one starts from the initial Bernoulli measure. The remaining 10 probabilities of blocks of length up to 3 can be obtained using the following formulae [cf. Eq. (7.41)]:

$$
\begin{aligned}
P(1) &= 1 - P(0), \\
P(01) &= P(0) - P(00), \\
P(10) &= P(0) - P(00), \\
P(11) &= 1 - 2P(0) + P(00), \\
P(001) &= P(00) - P(000), \\
P(011) &= P(0) - P(00) - P(010), \\
P(100) &= P(00) - P(000), \\
P(101) &= P(0) - 2P(00) + P(000), \\
P(110) &= P(0) - P(00) - P(010), \\
P(111) &= 1 - 3P(0) + 2P(00) + P(010).
\end{aligned}
$$

The expressions are generally valid for $p \in (0, 1)$ and $n > 0$. Although many of them remain valid for $p = 0$, $p = 1$ or $n = 0$ as well, the expressions given below should be used for $p = 0$ and $p = 1$, depending on the rule:

f	$f(000) = 0$ $f(111) = 0$		$f(000) = 0$ $f(111) = 1$		$f(000) = 1$ $f(111) = 0$		$f(000) = 1$ $f(111) = 1$	
p	$p = 0$	$p = 1$	$p = 0$	$p = 1$	$p = 0$	$p = 1$	$p = 0$	$p = 1$
$P_n(0)$	1	1	0	1	$\text{ev}(n)$	$\text{ev}(n+1)$	0	0
$P_n(00)$	1	1	0	1	$\text{ev}(n)$	$\text{ev}(n+1)$	0	0
$P_n(000)$	1	1	0	1	$\text{ev}(n)$	$\text{ev}(n+1)$	0	0
$P_n(010)$	0	0	0	0	0	0	0	0

H. Fukś, *Solvable Cellular Automata*, Understanding Complex Systems,
https://doi.org/10.1007/978-3-031-38700-5

For $n = 0$, expressions corresponding to the Bernoulli initial measure should be used,

$$\begin{aligned} P_0(0) &= 1 - p, \\ P_0(00) &= (1 - p)^2, \\ P_0(000) &= (1 - p)^3, \\ P_0(010) &= p(1 - p)^2. \end{aligned}$$

If only the $p = 1/2$ solution is known, the probabilities are denoted by $P^{(s)}$.

Rule 0

$$\begin{aligned} P_n(0) &= 1, \\ P_n(00) &= 1, \\ P_n(000) &= 1, \\ P_n(010) &= 0. \end{aligned}$$

Rule 1

$$\begin{aligned} P_n(0) &= \frac{1}{2}\,(-1)^{1+n}\left(2\,p^4 - 4\,p^3 + 4\,p - 1\right) - p^4 + 3\,p^3 - 3\,p^2 + p + \frac{1}{2}, \\ P_n(00) &= \frac{1}{2} - p\left(p^3 - 2\,p^2 + 2\right)(-1)^n + \frac{1}{2}\,(-1)^n, \\ P_n(000) &= \frac{1}{2}\,(-1)^{1+n}\left(2\,p^4 - 4\,p^3 + 4\,p - 1\right) + p^4 - 3\,p^3 + 3\,p^2 - p + \frac{1}{2}, \\ P_n(010) &= \frac{1}{2}\,p\,(-1)^n\left(p^3 - 3\,p^2 + 4\,p - 1\right)(-1 + p)^3 \\ &\quad + \frac{1}{2}\left(p^3 - 3\,p^2 + 2\,p - 1\right)(-1 + p)^3\,p. \end{aligned}$$

Rule 2

$$\begin{aligned} P_n(0) &= -\,p^3 + 2\,p^2 - p + 1, \\ P_n(00) &= -\,2\,p^3 + 4\,p^2 - 2\,p + 1, \\ P_n(000) &= -\,3\,p^3 + 6\,p^2 - 3\,p + 1, \\ P_n(010) &= p\,(p - 1)^2\,. \end{aligned}$$

Rule 3

$$P_n(0) = \begin{cases} (p+1)(p-1)^2 & \text{if} n \text{even}, \\ p(2-p) & \text{otherwise}, \end{cases}$$

$$P_n(00) = \begin{cases} (p-1)^2 & \text{if } n \text{ even}, \\ -p\left(p^2-p-1\right) & \text{otherwise}, \end{cases}$$

$$P_n(000) = \begin{cases} -(p-1)^3 & \text{if } n \text{ even}, \\ -p^2(2p-3) & \text{otherwise}, \end{cases}$$

$$P_n(010) = \begin{cases} p(p-1)^4 & \text{if } n \text{ even}, \\ p^2(p-1)^2 & \text{otherwise}. \end{cases}$$

Rule 4

$$\begin{aligned} P_n(0) &= -p^3 + 2p^2 - p + 1, \\ P_n(00) &= -2p^3 + 4p^2 - 2p + 1, \\ P_n(000) &= -p^5 + 3p^4 - 6p^3 + 7p^2 - 3p + 1, \\ P_n(010) &= p^3 - 2p^2 + p. \end{aligned}$$

Rule 5

$$P_n(0) = \begin{cases} (p+1)(p-1)^2 & \text{if} n \text{even}, \\ -p(p-2) & \text{otherwise}, \end{cases}$$

$$P_n(00) = \begin{cases} (p+1)^2(p-1)^4 & \text{if } n \text{ even}, \\ p^2(p-2)^2 & \text{otherwise}, \end{cases}$$

$$P_n(000) = \begin{cases} (p+1)(p-1)^4 & \text{if } n \text{ even}, \\ p^2(p-2)\left(p^2-p-1\right) & \text{otherwise}, \end{cases}$$

$$P_n(010) = -p\left(p^2-p-1\right)(p-1)^2.$$

Rule 7

$$P_n(0) = \begin{cases} \frac{(p-1)(p+1)\left(p^3-3p^2+p\right)\lambda^n}{p^2-p+1} + \frac{(p-1)(p+1)\left(-p^4+p^3+p-1\right)}{p^2-p+1} & \text{if } n \text{ even}, \\ \frac{\left(2p^3-p^2-2p+1\right)\lambda^{n+1}}{p^2-p+1} + \frac{p^5-p^4-2p^3+3p^2}{p^2-p+1} & \text{otherwise}, \end{cases}$$

$$P_n(00) = \begin{cases} -\frac{(p+1)(p-1)^3\left(p^2-2p\right)\lambda^n}{p^2-p+1} - \frac{(p+1)^2(p-1)^3}{p^2-p+1} & \text{if } n \text{ even}, \\ \frac{p^2\left(2p^2-2\right)\lambda^{n+1}}{p^2-p+1} + \frac{p^2\left(p^4-p^3-p^2+2\right)}{p^2-p+1} & \text{otherwise}, \end{cases}$$

$$P_n(000) = \begin{cases} \frac{(p+1)(p-1)^3\left(p^3-3\,p^2+3\,p\right)\lambda^n}{p^2-p+1} + \frac{(p+1)(p-1)^3\left(p^2-p-1\right)}{p^2-p+1} & \text{if } n \text{ even}, \\ \frac{p^2\left(p^3+2\,p^2-p-2\right)\lambda^{n+1}}{p^2-p+1} + \frac{p^2\left(2\,p^4-3\,p^3-p^2+2\,p+1\right)}{p^2-p+1} & \text{otherwise}, \end{cases}$$

$$P_n(010) = \begin{cases} p\,(p+1)\,(p-1)^2\,\lambda^n & \text{if } n \text{ even}, \\ -\,p\,(p-1)\,(p+1)\,\lambda^{n+1} & \text{otherwise}, \end{cases}$$

where $\lambda = \sqrt{p(1-p)}$.

Rule 8

$$P_n(0) = \begin{cases} p^3 - p^2 + 1 & \text{if } n = 1, \\ 1 & \text{otherwise}, \end{cases}$$

$$P_n(00) = \begin{cases} 2\,p^3 - 2\,p^2 + 1 & \text{if } n = 1, \\ 1 & \text{otherwise}, \end{cases}$$

$$P_n(000) = \begin{cases} 3\,p^3 - 3\,p^2 + 1 & \text{if } n = 1, \\ 1 & \text{otherwise}, \end{cases}$$

$$P_n(010) = \begin{cases} -p^2\,(p-1) & \text{if } n = 1, \\ 0 & \text{otherwise}. \end{cases}$$

Rule 10

$$\begin{aligned} P_n(0) &= p^2 - p + 1, \\ P_n(00) &= \left(p^2 - p + 1\right)^2, \\ P_n(000) &= \left(p^2 - p + 1\right)\left(2\,p^2 - 2\,p + 1\right), \\ P_n(010) &= -\,p\,(p-1)\left(2\,p^2 - 2\,p + 1\right). \end{aligned}$$

Rule 12

$$\begin{aligned} P_n(0) &= p^2 - p + 1, \\ P_n(00) &= 2\,p^2 - 2\,p + 1, \\ P_n(000) &= p^4 - 2\,p^3 + 4\,p^2 - 3\,p + 1, \\ P_n(010) &= -\,p\,(p-1). \end{aligned}$$

Rule 13

$$\begin{aligned} P_n(0) = &-\frac{p^6 - 4\,p^5 + 6\,p^4 - 4\,p^3 + p + 1}{(p+1)\,(p-2)} - \frac{\left(p^2 - 2\,p + 2\right)p^2\,(-p)^n}{p+1} \\ &+ \frac{\left(p^3 - 3\,p^2 + 3\,p - 1\right)(-1+p)^n}{p-2}, \end{aligned}$$

$$
\begin{aligned}
P_n(00) = & -\frac{p\left(2\,p^5-8\,p^4+12\,p^3-8\,p^2+p+1\right)}{(p+1)\,(p-2)} \\
& +\frac{1}{2}\,\frac{\left(p^3-5\,p^2+8\,p-6\right)p^2\,(-p)^n}{p+1} \\
& +\left(-\frac{1}{2}\,p^3+\frac{3}{2}\,p^2-\frac{3}{2}\,p+\frac{1}{2}\right)(1-p)^n \\
& +\frac{1}{2}\,\frac{\left(p^4-p^3-3\,p^2+5\,p-2\right)(-1+p)^n}{p-2}+\frac{1}{2}\,p^2\left(p^2-2\,p+2\right)p^n, \\
P_n(000) = & -\frac{1}{2}\left(p^2-2\,p+2\right)p^2\,(-p)^n+\left(-\frac{1}{2}\,p^3+\frac{3}{2}\,p^2-\frac{3}{2}\,p+\frac{1}{2}\right)(1-p)^n \\
& +\left(-\frac{1}{2}\,p^3+\frac{3}{2}\,p^2-\frac{3}{2}\,p+\frac{1}{2}\right)(-1+p)^n+\frac{1}{2}\,p^2\left(p^2-2\,p+2\right)p^n, \\
P_n(010) = & \frac{p^6-4\,p^5+6\,p^4-4\,p^3+p^2-1}{(p+1)\,(p-2)} \\
& +\frac{1}{2}\,\frac{\left(p^3-3\,p^2+4\,p-2\right)p^3\,(-p)^n}{p+1} \\
& -\frac{1}{2}\,\frac{p\left(p^4-4\,p^3+6\,p^2-4\,p+1\right)(-1+p)^n}{p-2} \\
& +\left(\frac{1}{2}\,p^4-p^3+p-\frac{1}{2}\right)(1-p)^n+\frac{1}{2}\,p^2\left(p^3-4\,p^2+6\,p-4\right)p^n.
\end{aligned}
$$

Rule 14

$$
\begin{aligned}
P_n^{(s)}(0) &= \frac{1}{2}+2^{-2\,n-2}\binom{2\,n}{n}, \\
P_n^{(s)}(00) &= \frac{1}{4}+2^{-2\,n-2}\binom{2\,n}{n}, \\
P_n^{(s)}(000) &= \frac{2^{-2\,n-3}\,(4\,n+3)}{n+1}\binom{2\,n}{n}, \\
P_n^{(s)}(010) &= 2^{-2\,n-2}\binom{2\,n}{n}.
\end{aligned}
$$

Rule 15

$$
\begin{aligned}
P_n(0) &= \frac{1}{2}-\frac{1}{2}\,(-1)^{n+1}-(-1)^n\,p, \\
P_n(00) &= \frac{1}{4}\left(2\,(-1)^n\,p-(-1)^n-1\right)^2, \\
P_n(000) &= -\frac{1}{8}\left(2\,(-1)^n\,p-(-1)^n-1\right)^3, \\
P_n(010) &= \frac{1}{2}\,p\left(2\,(-1)^n\,p^2+3\,(-1)^{n+1}\,p+(-1)^n-p+1\right).
\end{aligned}
$$

Rule 19

$$P_n(0) = \begin{cases} -p\left(p^2-p-1\right) & \text{if } n=1, \\ \frac{1}{2}-\frac{1}{2}(-1)^{n+1}-(-1)^n\left(-p^5+4p^4-7p^3+5p^2\right) & \text{otherwise,} \end{cases}$$

$$P_n(00) = \begin{cases} -p^2(2p-3) & \text{if } n=1, \\ 7(-1)^n p^3+5p^3+4(-1)^{n+1}p^4-4p^4+(-1)^n p^5 & \\ \quad +p^5+\frac{1}{2}+\frac{1}{2}(-1)^n+5(-1)^{n+1}p^2-2p^2 & \text{otherwise,} \end{cases}$$

$$P_n(000) = \begin{cases} p^2\left(p^3-4p^2+3p+1\right) & \text{if } n=1, \\ \frac{1}{2}+\frac{1}{2}(-1)^n-8p^4+7(-1)^n p^3+10p^3-4p^2 & \\ \quad +4(-1)^{n+1}p^4+5(-1)^{n+1}p^2+(-1)^n p^5+2p^5 & \text{otherwise,} \end{cases}$$

$$P_n(010) = 0.$$

Rule 23

$$P_n(0) = \frac{1}{2}+\frac{1}{2}\frac{\left(2p^3-3p^2-p+1\right)(-1)^n}{p^2-p+1}-\frac{p\left(2p^2-3p+1\right)\left(p^2-p\right)^n}{p^2-p+1},$$

$$\begin{aligned} P_n(00) = &-\frac{1}{2}\frac{2p^4-4p^3+p^2+p-1}{p^2-p+1}+\frac{1}{2}\frac{\left(2p^3-3p^2-p+1\right)(-1)^n}{p^2-p+1} \\ &-\frac{p\left(2p^2-3p+1\right)\left(p^2-p\right)^n}{p^2-p+1} \\ &+\frac{p\left(2p^3-4p^2+3p-1\right)\left(-p^2+p\right)^n}{p^2-p+1}, \end{aligned}$$

$$\begin{aligned} P_n(000) = &-\frac{1}{2}\frac{4p^4-8p^3+3p^2+p-1}{p^2-p+1}+\frac{1}{2}\frac{\left(2p^3-3p^2-p+1\right)(-1)^n}{p^2-p+1} \\ &-\frac{1}{2}\frac{p\left(2p^4-5p^3+10p^2-10p+3\right)\left(p^2-p\right)^n}{p^2-p+1} \\ &+\frac{1}{2}\frac{p\left(7p^3-14p^2+10p-3\right)\left(-p^2+p\right)^n}{p^2-p+1}, \end{aligned}$$

$$P_n(010) = \frac{1}{2}p\left(2p^2-3p+1\right)\left(p^2-p\right)^n-\frac{1}{2}(p-1)p\left(-p^2+p\right)^n.$$

Rule 24

$$P_n(0) = \begin{cases} p^2-p+1 & \text{if } n=1, \\ p^4-2p^3+2p^2-p+1 & \text{otherwise,} \end{cases}$$

$$P_n(00) = \begin{cases} 2p^2-2p+1 & \text{if } n=1, \\ 2p^4-4p^3+4p^2-2p+1 & \text{otherwise,} \end{cases}$$

$$P_n(000) = \begin{cases} p^4 - 2\,p^3 + 4\,p^2 - 3\,p + 1 & \text{if}\, n = 1, \\ 3\,p^4 - 6\,p^3 + 6\,p^2 - 3\,p + 1 & \text{otherwise}, \end{cases}$$

$$P_n(010) = \begin{cases} -p\,(p-1) & \text{if}\, n = 1, \\ -p\,(p-1)\left(p^2 - p + 1\right) & \text{otherwise}. \end{cases}$$

Rule 27

$$P_n(0) = \frac{1}{2} - \frac{1}{2}\,p^4 + p^3 - \frac{1}{2}\,p^2 + \left(-\frac{1}{2}\,p^4 + p^3 - 1/2\,p^2 - p + \frac{1}{2}\right)(-1)^n,$$

$$P_n(00) = \frac{1}{2} + \frac{1}{2}\,p^4 - p^3 + \frac{3}{2}\,p^2 - p + \left(-\frac{1}{2}\,p^4 + p^3 - \frac{1}{2}\,p^2 - p + \frac{1}{2}\right)(-1)^n,$$

$$P_n(000) = \frac{1}{2} + \frac{1}{2}\,p^4 + 2\,p^2 - \frac{3}{2}\,p - p^3 + \left(-\frac{1}{2}\,p^4 + p^2 - \frac{3}{2}\,p + \frac{1}{2}\right)(-1)^n,$$

$$P_n(010) = 2\,p^3 - \frac{3}{2}\,p^2 + \frac{1}{2}\,p - p^4 + \left(p^3 - \frac{3}{2}\,p^2 + \frac{1}{2}\,p\right)(-1)^n.$$

Rule 28

$$\begin{aligned} P_n(0) = {} & \frac{1}{2} - \frac{(p-2)\,p^{n+3}}{1+p} \\ & - \frac{1}{2}\,\frac{p\,(-1)^n\,(-1+p)\left(2\,p^4 - 6\,p^3 + 4\,p^2 + 2\,p - 1\right)}{(1+p)\,(p-2)} + \frac{(1-p)^{n+3}}{2-p}, \\ P_n(00) = {} & \frac{1}{2}\,\frac{p\,(-1+p)\left(2\,p^4 - 4\,p^3 + 2\,p^2 + 1\right)}{(1+p)\,(p-2)} - \frac{1}{2}\,\frac{(3\,p+1)\,(p-2)\,p^{n+2}}{1+p} \\ & + \frac{1}{2}\,\frac{(-1+p)\,(p-2)\,(-p)^{n+2}}{1+p} + \frac{1}{2}\,\frac{(3\,p-4)\,(1-p)^{n+2}}{p-2} \\ & - \frac{1}{2}\,\frac{p\,(-1)^n\,(-1+p)\left(2\,p^4 - 6\,p^3 + 4\,p^2 + 2\,p - 1\right)}{(1+p)\,(p-2)} \\ & - \frac{1}{2}\,\frac{p\,(2\,p-1)\,(-1+p)^{n+2}}{p-2}, \\ P_n(000) = {} & (1-p)^{n+3} - (p-2)\,p^{n+3}, \\ P_n(010) = {} & -\frac{1}{2}\,\frac{4\,p^6 - 12\,p^5 + 12\,p^4 - 4\,p^3 + p^2 - p + 2}{(1+p)\,(p-2)} + \frac{(p-2)\,p^{n+2}}{1+p} \\ & - \frac{(-1+p)\,(p-2)\,(-p)^{n+2}}{1+p} + \frac{(1-p)^{n+2}}{p-2} \\ & + \frac{p\,(2\,p-1)\,(-1+p)^{n+2}}{p-2} \\ & - \frac{1}{2}\,\frac{p\,(-1)^n\,(-1+p)\left(2\,p^4 - 6\,p^3 + 4\,p^2 + 2\,p - 1\right)}{(1+p)\,(p-2)}. \end{aligned}$$

Rule 29

$$P_n(0) = \begin{cases} 1-p & \text{if } n \text{ even,} \\ p & \text{otherwise,} \end{cases}$$

$$P_n(00) = \begin{cases} -(p+1)(p-1)^3 & \text{if } n \text{ even,} \\ -p^3(p-2) & \text{otherwise,} \end{cases}$$

$$P_n(000) = \begin{cases} (p+1)(p-1)^4 & \text{if } n \text{ even,} \\ -p^4(p-2) & \text{otherwise,} \end{cases}$$

$$P_n(010) = \begin{cases} -p\left(p^2-2p-1\right)(p-1)^2 & \text{if } n \text{ even,} \\ p^2(p-1)\left(p^2-2\right) & \text{otherwise.} \end{cases}$$

Rule 30

$$P_n^{(s)}(\mathbf{a}) = 2^{-|\mathbf{a}|}$$

Rule 32

$$\begin{aligned} P_n(0) &= -p\left(-p^2+p\right)^n+1, \\ P_n(00) &= -2p\left(-p^2+p\right)^n+1, \\ P_n(000) &= 1-p\left(p^2-p+3\right)\left(-p^2+p\right)^n, \\ P_n(010) &= p\left(-p^2+p\right)^n. \end{aligned}$$

Rule 34

$$\begin{aligned} P_n(0) &= p^2-p+1, \\ P_n(00) &= 2p^2-2p+1, \\ P_n(000) &= p^4-2p^3+4p^2-3p+1, \\ P_n(010) &= -p(p-1). \end{aligned}$$

Rule 36

$$P_n(0) = \begin{cases} p^2-p+1 & \text{if } n=1, \\ 3p^4-6p^3+4p^2-p+1 & \text{otherwise,} \end{cases}$$

$$P_n(00) = \begin{cases} 2p^4-4p^3+4p^2-2p+1 & \text{if } n=1, \\ 6p^4-12p^3+8p^2-2p+1 & \text{otherwise,} \end{cases}$$

$$P_n(000) = \begin{cases} 4p^4-8p^3+7p^2-3p+1 & \text{if } n=1, \\ 9p^4-18p^3+12p^2-3p+1 & \text{otherwise,} \end{cases}$$

$$P_n(010) = -p(p-1)\left(3p^2-3p+1\right).$$

Rule 38

$$P_n(0) = \begin{cases} -p^3 + 3p^2 - 2p + 1 & \text{if} n = 1, \\ 1 + \frac{3}{2}p^4 - 3p^3 + 3p^2 - \frac{3}{2}p \\ \quad + \frac{1}{2}p\left(2p^4 - 5p^3 + 6p^2 - 4p + 1\right)(-1)^n & \text{otherwise,} \end{cases}$$

$$P_n(00) = \begin{cases} p^4 - 3p^3 + 5p^2 - 3p + 1 & \text{if} n = 1, \\ 1 + \frac{5}{2}p^4 - 5p^3 + 5p^2 - \frac{5}{2}p \\ \quad + \frac{1}{2}p\left(2p^4 - 5p^3 + 6p^2 - 4p + 1\right)(-1)^n & \text{otherwise,} \end{cases}$$

$$P_n(000) = \begin{cases} -p^5 + 5p^4 - 8p^3 + 8p^2 - 4p + 1 & \text{if} n = 1, \\ 1 + \frac{1}{2}p^6 - \frac{3}{2}p^5 + \frac{11}{2}p^4 - \frac{17}{2}p^3 + \frac{15}{2}p^2 - \frac{7}{2}p + \frac{1}{2}p \\ \quad (2p^6 - 7p^5 + 13p^4 - 15p^3 + 11p^2 - 5p + 1)(-1)^n & \text{otherwise,} \end{cases}$$

$$P_n(010) = -\frac{1}{2}p^4 + p^3 - p^2 + \frac{1}{2}p + \frac{1}{2}p\left(2p^4 - 5p^3 + 6p^2 - 4p + 1\right)(-1)^n.$$

Rule 40

$$\begin{aligned} P_n(0) &= 1 + \alpha_1 {\lambda_1}^n + \alpha_2 {\lambda_2}^n, \\ P_n(00) &= 1 + \beta_1 {\lambda_1}^n + \beta_2 {\lambda_2}^n, \\ P_n(000) &= 1 + \delta_1 {\lambda_1}^n + \delta_2 {\lambda_2}^n, \\ P_n(010) &= \eta_1 {\lambda_1}^n + \eta_2 {\lambda_2}^n, \end{aligned}$$

where

$$\begin{aligned} \Delta &= \sqrt{p^2 - 6p + 5}, \\ \lambda_{1,2} &= -\frac{2p^2 - 2p}{p - 1 \pm \Delta}, \\ \alpha_{1,2} &= \frac{p(\Delta - p \pm 5)(p + \Delta \pm 1)}{4p - 20}, \\ \beta_{1,2} &= -\frac{p(\Delta - p \pm 5)\left(p^2\Delta + p^3 - 3p^2 - 2\Delta \mp 2\right)}{4p - 20}, \\ \delta_{1,2} &= -\frac{p(\Delta - p \pm 5)\left(p^2\Delta \pm p^3 + p\Delta \mp 6p^2 - 3\Delta \pm 2p - 3\right)}{4p - 20}, \\ \eta_{1,2} &= \frac{p(\Delta - p + \pm 5)\left(2p^2\Delta + 2p^3 \mp 6p^2 - \Delta \pm 3p - 1\right)}{4p - 20}. \end{aligned}$$

Rule 42

$$\begin{aligned} P_n(0) &= p^3 - p + 1, \\ P_n(00) &= p^3 + p^2 - 2\,p + 1, \\ P_n(000) &= p^5 - p^4 + 3\,p^2 - 3\,p + 1, \\ P_n(010) &= p\,(p-1)^2. \end{aligned}$$

Rule 43

$$\begin{aligned} P_n^{(s)}(0) &= \frac{1}{2}, \\ P_n^{(s)}(00) &= \frac{1}{4}, \\ P_n^{(s)}(000) &= 2^{-2\,n-3}\binom{2\,n+1}{n+1}, \\ P_n^{(s)}(010) &= 2^{-2\,n-3}\binom{2\,n+1}{n+1}. \end{aligned}$$

Rule 44

$$\begin{aligned} P_n(0) = {} & \frac{\left((1+p)\left(p^2-p-1\right)+D_3\,(n+2)\,(2\,p-1)+D_3\,(n+1)\left(-2\,p^2+3\,p-1\right)\right)}{1+p^2\,(1-p)} \\ & \times p^{1+n+\lfloor n/3\rfloor}\,(1-p)^{n-\lfloor n/3\rfloor} - \frac{\left(p^2-p+1\right)p\,(-1+p)}{p^3-p^2-1} + 1 \end{aligned}$$

$$\begin{aligned} P_n(00) &= p^{1+\lfloor 4n/3\rfloor}\,(1-p)^{2\,n-\lfloor 4n/3\rfloor}\,(2-p+D_3\,(n)\,p-D_3\,(n)\,(1-p)) \\ &\quad +2\,P_n(0)-1, \\ P_n(000) &= 4P_n(0)+3\,p^{1+\lfloor 4n/3\rfloor}\,(1-p)^{2\,n-\lfloor 4n/3\rfloor} \\ &\quad \times(-p+2+D_3\,(n)\,p-D_3\,(n)\,(1-p))-P_{n+1}(0)-2, \\ P_n(010) &= p^{2+4\,\lfloor n/3\rfloor+D_3(n+1)}\,(1-p)^{2n-4\lfloor n/3\rfloor-D_3(n+1)}+I_1(n)p^3\,(1-p)^2+1 \\ &\quad -P_n(0)-2\,p^{1+\lfloor 4n/3\rfloor}\,(1-p)^{2\,n-\lfloor 4n/3\rfloor} \\ &\quad \times(2-p+D_3\,(n)\,p-D_3\,(n)\,(1-p)), \\ D_3(n) &= \begin{cases} 1 & \text{if } n \text{ divisible by } 3, \\ 0 & \text{otherwise,} \end{cases} \\ I_1(n) &= \begin{cases} 1 & n=1, \\ 0 & \text{otherwise.} \end{cases} \end{aligned}$$

Rule 45

$$P_n^{(s)}(\mathbf{a}) = 2^{-|\mathbf{a}|}$$

Rule 46

$$P_n(0) = \begin{cases} 2\,p^2 - 2\,p + 1 & \text{if}\, n = 1, \\ 2\,p^4 - 4\,p^3 + 4\,p^2 - 2\,p + 1 & \text{otherwise}, \end{cases}$$

$$P_n(00) = \begin{cases} p^4 - 2\,p^3 + 4\,p^2 - 3\,p + 1 & \text{if}\, n = 1, \\ 3\,p^4 - 6\,p^3 + 6\,p^2 - 3\,p + 1 & \text{otherwise}, \end{cases}$$

$$P_n(000) = \begin{cases} \left(p^2 - p + 1\right)\left(3\,p^2 - 3\,p + 1\right) & \text{if}\, n = 1, \\ p^6 - 3\,p^5 + 8\,p^4 - 11\,p^3 + 9\,p^2 - 4\,p + 1 & \text{otherwise}, \end{cases}$$

$$P_n(010) = 0.$$

Rule 50

$$\begin{aligned}
P_n(0) &= \frac{1}{2} + \frac{p^{2n+1}}{p^2+1} + \frac{1}{2}\,\frac{p^2\left(2\,p^3 - 5\,p^2 + 4\,p - 1\right)(-1)^n}{\left(p^2 - 2\,p + 2\right)\left(p^2+1\right)} \\
&\quad - \frac{\left(p^3 - 3\,p^2 + 3\,p - 1\right)(1-p)^{2n}}{p^2 - 2\,p + 2}, \\
P_n(00) &= -\frac{1}{2}\,\frac{p\left(2\,p^3 - 4\,p^2 + p + 1\right)}{(p-2)(p+1)} + \frac{p\left(p^2 + p + 2\right)p^{2n}}{p^3 + p^2 + p + 1} \\
&\quad + \frac{1}{2}\,\frac{p^2\left(2\,p^3 - 5\,p^2 + 4\,p - 1\right)(-1)^n}{\left(p^2 - 2\,p + 2\right)\left(p^2+1\right)} \\
&\quad + \frac{\left(p^5 - 6\,p^4 + 16\,p^3 - 22\,p^2 + 15\,p - 4\right)(1-p)^{2n}}{p^3 - 4\,p^2 + 6\,p - 4}, \\
P_n(000) &= p^{2n+1} + \left(-p^3 + 3\,p^2 - 3\,p + 1\right)(1-p)^{2n}, \\
P_n(010) &= \frac{1}{2}\,\frac{4\,p^4 - 8\,p^3 + 3\,p^2 + p - 2}{(p-2)(p+1)} - \frac{p\left(2\,p^2 - p + 1\right)p^{2n}}{p^3 + p^2 + p + 1} \\
&\quad + \frac{1}{2}\,\frac{p^2\left(2\,p^3 - 5\,p^2 + 4\,p - 1\right)(-1)^n}{\left(p^2 - 2\,p + 2\right)\left(p^2+1\right)} \\
&\quad - \frac{\left(2\,p^5 - 9\,p^4 + 17\,p^3 - 17\,p^2 + 9\,p - 2\right)(1-p)^{2n}}{p^3 - 4\,p^2 + 6\,p - 4}.
\end{aligned}$$

Rule 51

$$
\begin{aligned}
P_n(0) &= \frac{1}{2} - \frac{1}{2}\,(-1)^{n+1} - (-1)^n\,p,\\
P_n(00) &= \frac{1}{4}\left(2\,(-1)^n\,p - (-1)^n - 1\right)^2,\\
P_n(000) &= -\frac{1}{8}\left(2\,(-1)^n\,p - (-1)^n - 1\right)^3,\\
P_n(010) &= \frac{1}{2}\,p\left(2\,(-1)^n\,p^2 + 3\,(-1)^{n+1}\,p + (-1)^n - p + 1\right).
\end{aligned}
$$

Rule 60

$$
\begin{aligned}
P_n(0) &= \frac{1}{2} + \frac{1}{2}\,(1-2\,p)^{G(n)},\\
P_n(00) &= \frac{1}{4}\,(1-2\,p)^{G(n+1)} + \frac{1}{2}\,(1-2\,p)^{G(n)} + \frac{1}{4},\\
P_n(000) &= \frac{1}{8} + \frac{3}{8}\,(1-2\,p)^{G(n)} + \frac{1}{4}\,(1-2\,p)^{G(n+1)}\\
&\quad + \frac{1}{8}\,(1-2\,p)^{G(n+2)} + \frac{1}{8}\,(1-2\,p)^{H(n)},\\
P_n(010) &= \frac{1}{8} + \frac{1}{8}\,(1-2\,p)^{G(n)} - \frac{1}{4}\,(1-2\,p)^{G(n+1)}\\
&\quad + \frac{1}{8}\,(1-2\,p)^{G(n+2)} - \frac{1}{8}\,(1-2\,p)^{H(n)}.
\end{aligned}
$$

where

$$
\begin{aligned}
G(n) &= \sum_{k=0}^{n}\left(\binom{n}{k} \mod 2\right).\\
H(n) &= \sum_{k=0}^{n+2}\left(\binom{n+1}{k} + \binom{n}{k-2} \mod 2\right).
\end{aligned}
$$

Rule 72

$$
\begin{aligned}
P_n(0) &= \begin{cases} 2\,p^3 - 2\,p^2 + 1 & \text{if}\, n = 1,\\ -2\,p^4 + 4\,p^3 - 2\,p^2 + 1 & \text{otherwise,}\end{cases}\\
P_n(00) &= \begin{cases} p^4 + 2\,p^3 - 3\,p^2 + 1 & \text{if}\, n = 1,\\ -3\,p^4 + 6\,p^3 - 3\,p^2 + 1 & \text{otherwise,}\end{cases}\\
P_n(000) &= \begin{cases} p^4 + 3\,p^3 - 4\,p^2 + 1 & \text{if}\, n = 1,\\ -p^7 + 3\,p^6 - 3\,p^5 - 3\,p^4 + 8\,p^3 - 4\,p^2 + 1 & \text{otherwise,}\end{cases}
\end{aligned}
$$

$$P_n(010) = \begin{cases} -2\,p^3\,(p-1) & \text{if } n = 1, \\ 0 & \text{otherwise.} \end{cases}$$

Rule 76

$$\begin{aligned} P_n(0) &= p^3 - p + 1, \\ P_n(00) &= p^4 + p^2 - 2\,p + 1, \\ P_n(000) &= p^5 - p^3 + 3\,p^2 - 3\,p + 1, \\ P_n(010) &= -\,p\,(p-1)\left(2\,p^2 - p + 1\right). \end{aligned}$$

Rule 77

$$\begin{aligned} P_n(0) &= 1 - p + \frac{(-1)^{n+1}\,p^{2n+3} + p^3}{p^2+1} + \frac{(-1)^n\,(1-p)^{2n+3} - (1-p)^3}{p^2 - 2\,p + 2}, \\ P_n(00) &= -\frac{p\left(p^7 - 5\,p^6 + 11\,p^5 - 11\,p^4 + 4\,p^3 - p + 1\right)}{\left(p^2+1\right)\left(p^2 - 2\,p + 2\right)(p+1)(p-2)} - \frac{p^3\left(-p^2\right)^n}{p^2+1} \\ &\quad - \frac{\left(p^3 - 3\,p^2 + 3\,p - 1\right)\left(-p^2 + 2\,p - 1\right)^n}{p^2 - 2\,p + 2} + \frac{p^{2n+2}}{p+1} \\ &\quad + \frac{\left(p^3 - 3\,p^2 + 3\,p - 1\right)(p-1)^{2n}}{p-2}, \\ P_n(000) &= -\frac{1}{2}\,p^3\left(-p^2\right)^n + \left(-\frac{1}{2}\,p^3 + \frac{3}{2}\,p^2 - \frac{3}{2}\,p + \frac{1}{2}\right)\left(-p^2 + 2\,p - 1\right)^n \\ &\quad + \frac{1}{2}\,p^{2n+3} + \left(-\frac{1}{2}\,p^3 + \frac{3}{2}\,p^2 - \frac{3}{2}\,p + \frac{1}{2}\right)(p-1)^{2n}, \\ P_n(010) &= \frac{2\,p^8 - 7\,p^7 + 12\,p^6 - 16\,p^5 + 17\,p^4 - 11\,p^3 + 3\,p - 2}{\left(p^2+1\right)\left(p^2 - 2\,p + 2\right)(p+1)(p-2)} \\ &\quad + \frac{1}{2}\,\frac{\left(p^2 - 1\right)p^3\left(-p^2\right)^n}{p^2+1} \\ &\quad + \frac{1}{2}\,\frac{p\left(p^4 - 5\,p^3 + 9\,p^2 - 7\,p + 2\right)\left(-p^2 + 2\,p - 1\right)^n}{p^2 - 2\,p + 2} \\ &\quad + \frac{1}{2}\,\frac{(p-3)\,p^{2n+3}}{p+1} - \frac{1}{2}\,\frac{\left(p^4 - p^3 - 3\,p^2 + 5\,p - 2\right)(p-1)^{2n}}{p-2}. \end{aligned}$$

Rule 78

$$\begin{aligned} P_n(0) &= 1 - \frac{p^6 - 2\,p^5 + p^4 - p^2 + p + 1}{(p+1)(-p+2)} \\ &\quad + \frac{1}{2}\,\frac{\left(2\,p^2 - 2\,p + 1\right)p\,(-1+p)^{n+1}}{-p+2} \end{aligned}$$

$$
\begin{aligned}
& + \frac{1}{2}\,(2\,p-1)\,p^{n+2} + \frac{1}{2}\,\frac{(1-p)\,(-p)^{n+2}}{p+1} - \frac{1}{2}\,(2\,p-1)\,(1-p)^{n+1} \\
& + I_1\,(n)\,(1-p)\,p^3, \\
P_n(00) &= (1-p)^{n+2} + p^{n+3}, \\
P_n(000) &= (1-p)^{n+3} + p^{n+4}, \\
P_n(010) &= \frac{3\,p^6 - 6\,p^5 + 3\,p^4 - p^2 + p - 1}{(p+1)\,(p-2)} - \frac{1}{2}\,\frac{\left(2\,p^3 - p + 1\right)p\,(-1+p)^{n+1}}{p-2} \\
& + \left(\frac{1}{2}\,p - 1\right)p^{n+2} - \frac{1}{2}\,\frac{(-p+2)\,(-p)^{n+2}\,(-1+p)}{p+1} \\
& - \frac{1}{2}\,(p+1)\,(1-p)^{n+1} + 2\,I_1\,(n)\,(1-p)\,p^3.
\end{aligned}
$$

where

$$
I_1(n) = \begin{cases} 1 & \text{if}\, n = 1, \\ 0 & \text{otherwise.} \end{cases}
$$

Rule 90

$$
\begin{aligned}
P_n(0) &= \frac{1}{2} + \frac{1}{2}\,(1-2\,p)^{G(n)}, \\
P_n(00) &= \left(\frac{1}{2} + \frac{1}{2}\,(1-2\,p)^{G(n)}\right)^2, \\
P_n(000) &= \left(\frac{1}{4} + \frac{1}{2}\,(1-2\,p)^{G(n)} + \frac{1}{4}\,(1-2\,p)^{G(n+1)}\right)\left(\frac{1}{2} + \frac{1}{2}\,(1-2\,p)^{G(n)}\right), \\
P_n(010) &= \frac{1}{8} + \frac{1}{8}\,(1-2\,p)^{G(n)} + \frac{1}{8}\,(1-2\,p)^{G(n+1)} - \frac{1}{4}\,(1-2\,p)^{2\,G(n)} \\
& - \frac{1}{8}\,(1-2\,p)^{G(n+1)+G(n)}.
\end{aligned}
$$

where

$$
G(n) = \sum_{k=0}^{n} \left(\binom{n}{k} \mod 2\right).
$$

Rule 106

$$
P_n^{(s)}(\mathbf{a}) = 2^{-|\mathbf{a}|}
$$

Rule 108

$$P_n(0) = \begin{cases} 2\,p^3 - p^2 - p + 1 & \text{if } n = 1, \\ 2\,p^5 - 4\,p^4 + 3\,p^3 - p + 1 & \text{if } n \text{ even}, \\ 4\,p^7 - 12\,p^6 + 19\,p^5 - 18\,p^4 + 9\,p^3 - p^2 - p + 1 & \text{otherwise}, \end{cases}$$

$$P_n(00) = \begin{cases} p^4 + p^2 - 2\,p + 1 & \text{if } n = 1, \\ p^6 + 2\,p^5 - 6\,p^4 + 4\,p^3 + p^2 - 2\,p + 1 & \text{if } n \text{ even}, \\ p^6 + 2\,p^5 - 6\,p^4 + 4\,p^3 + p^2 - 2\,p + 1 & \text{otherwise}, \end{cases}$$

$$P_n(000) = \begin{cases} p^5 - p^3 + 3\,p^2 - 3\,p + 1 & \text{if } n = 1, \\ -4\,p^7 + 14\,p^6 - 14\,p^5 + 4\,p^4 + 3\,p^2 - 3\,p + 1 & \text{if } n \text{ even}, \\ -4\,p^7 + 14\,p^6 - 14\,p^5 + 4\,p^4 + 3\,p^2 - 3\,p + 1 & \text{otherwise}, \end{cases}$$

$$P_n(010) = \begin{cases} p\left(3\,p^2 - 2\,p + 1\right)(p - 1)^2 & \text{if } n = 1, \\ -p\,(p - 1)\left(2\,p^4 - p^3 + p^2 - p + 1\right) & \text{if } n \text{ even}, \\ p\left(8\,p^4 - 10\,p^3 + 9\,p^2 - 2\,p + 1\right)(p - 1)^2 & \text{otherwise}. \end{cases}$$

Rule 128

$$P_n(0) = -\,p^{2n+1} + 1,$$
$$P_n(00) = 1 + p^{2n+1}\,(p - 2)\,,$$
$$P_n(000) = p^{2n+1}\,(2\,p - 3) + 1,$$
$$P_n(010) = \left(p^2 - 2\,p + 1\right) p^{2n+1}.$$

Rule 130

$$P_n(0) = -\frac{p^4 - p^3 - 2\,p^2 - 1}{p^3 + p^2 + p + 1} + \frac{1}{2}\,\frac{(-1)^n\,p^{2n+1}\,(-1 + p)\,(2\,p - 1)}{p^2 + 1} - \frac{1}{2}\,\frac{p^{2n+1}\,(2\,p + 1)}{p + 1},$$

$$P_n(00) = -\frac{2\,p^4 - p^3 - 3\,p^2 + p - 1}{p^3 + p^2 + p + 1} + \frac{p^{2n+1}\left(p^2 - p - 1\right)}{p + 1} + \frac{(-1)^n\,p^{2n+1}\,(-1 + p)\,(2\,p - 1)}{p^2 + 1},$$

$$P_n(000) = -\frac{3\,p^4 - p^3 - 4\,p^2 + 2\,p - 1}{p^3 + p^2 + p + 1} + \frac{1}{2}\,\frac{p^{2n+1}\left(3\,p^2 - 2\,p - 2\right)}{p + 1} + \frac{1}{2}\,\frac{(-1)^{n+1}\,p^{2n+1}\,(-1 + p)\,(2\,p - 1)\left(p^2 - 2\right)}{p^2 + 1},$$

$$P_n(010) = \frac{p\left(p^3 - p + 1\right)}{p^3 + p^2 + p + 1} + \frac{1}{2}\,\frac{p^{2n+1}\left(2\,p^3 - 2\,p^2 - 2\,p + 1\right)}{p + 1}$$

$$-\frac{1}{2}\frac{(-1)^n\, p^{2n+1}\left(2\,p^2-3\,p+1\right)}{p^2+1}.$$

Rule 132

$$\begin{aligned}
P_n(0) &= -\frac{2p^{2n+2}}{p+1}+\frac{p^2+1}{p+1},\\
P_n(00) &= \frac{p^{2n+2}\,(p-3)}{p+1}+\frac{2\,p^2-p+1}{p+1},\\
P_n(000) &= \frac{2p^{2n+2}\,(p-2)}{p+1}-\frac{\left(p^2-p+1\right)\left(p^4-p^3-2\,p^2+p-1\right)}{p+1},\\
P_n(010) &= \frac{p^{2n+3}\,(p-1)}{p+1}+\frac{p\,(1-p)}{p+1}.
\end{aligned}$$

Rule 136

$$\begin{aligned}
P_n(0) &= 1-p^{n+1},\\
P_n(00) &= p^{2+n}-2\,p^{n+1}+1,\\
P_n(000) &= 2\,p^{2+n}-3\,p^{n+1}+1,\\
P_n(010) &= p^{n+1}\,(p-1)^2.
\end{aligned}$$

Rule 138

$$\begin{aligned}
P_n(0) &= -\,(p-1)\left(p^2+1\right),\\
P_n(00) &= (p-1)\left(p^3-2\,p^2+p-1\right),\\
P_n(000) &= \left(2\,p^2-p+1\right)(p-1)^2,\\
P_n(010) &= p\left(p^2-p+1\right)(p-1)^2.
\end{aligned}$$

Rule 140

$$\begin{aligned}
P_n(0) &= p^2-p+1-p^{2+n},\\
P_n(00) &= 2\,p^2-2\,p+1-p^{2+n},\\
P_n(000) &= p^4-2\,p^3+4\,p^2-3\,p+1-p^{2+n},\\
P_n(010) &= p^{2+n}\,(p-1)-p\,(p-1).
\end{aligned}$$

Rule 142

$$\begin{aligned}
P_n^{(s)}(0) &= \frac{1}{2},\\
P_n^{(s)}(00) &= \frac{1}{4},
\end{aligned}$$

$$P_n^{(s)}(000) = \frac{(2n+1)}{4^{n+1}(2n+2)}\binom{2n}{n},$$
$$P_n^{(s)}(010) = \frac{(2n+1)}{4^{n+1}(2n+2)}\binom{2n}{n}.$$

Rule 150

$$P_n(0) = \frac{1}{2} + \frac{1}{2}(1-2p)^{R(n)}.$$

where

$$R(n) = \sum_{i=0}^{2n}\left(\binom{n}{i-n}_2 \mod 2\right).$$

Rule 154

$$P_n^{(s)}(\mathbf{a}) = 2^{-|\mathbf{a}|}$$

Rule 156

$$P_n(0) = \frac{1}{2} - \frac{p^{n+3}}{1+p} - \frac{(1-p)^{n+3}}{p-2} - \frac{1}{2}\frac{(p-1)p(2p-1)(-1)^n}{(1+p)(p-2)},$$
$$P_n(00) = \frac{1}{2}\frac{p(p-1)\left(2p^4-4p^3+2p^2+1\right)}{(1+p)(p-2)} - \frac{1}{2}\frac{(p-1)p^{n+2}}{1+p}$$
$$+\frac{1}{2}\frac{(3p-4)(1-p)^{n+2}}{p-2} + \frac{1}{2}\frac{(2p-1)(p-1)(-p)^{n+2}}{1+p}$$
$$-\frac{1}{2}\frac{p(2p-1)(p-1)^{n+2}}{p-2} - \frac{1}{2}\frac{p(p-1)(2p-1)(-1)^n}{(1+p)(p-2)},$$
$$P_n(000) = (1-p)^{n+3},$$
$$P_n(010) = -\frac{1}{2}\frac{4p^6-12p^5+12p^4-4p^3+p^2-p+2}{(1+p)(p-2)} + \frac{\left(p^2-p-1\right)p^{n+2}}{1+p}$$
$$+\frac{(1-p)^{n+2}}{p-2} - \frac{(2p-1)(p-1)(-p)^{n+2}}{1+p} + \frac{p(2p-1)(p-1)^{n+2}}{p-2}$$
$$-\frac{1}{2}\frac{p(p-1)(2p-1)(-1)^n}{(1+p)(p-2)}.$$

Rule 160

$$P_n(0) = 1 - p^{n+1},$$
$$P_n(00) = 1 - 2p^{n+1} + p^{2n+2},$$
$$P_n(000) = 1 - (p-2)p^{2n+2} + p(p-3)p^n,$$
$$P_n(010) = (p-2)p^{2n+2} + p^{n+1}.$$

Rule 162

$$P_n(0) = \frac{1 - p^{2n+2}}{p+1},$$
$$P_n(00) = -\frac{p-1}{p+1} + \frac{(p-1)\,p^{2n+2}}{p+1},$$
$$P_n(000) = \frac{p^2 - 2p + 1}{p+1} - \frac{\left(p^2 - 2p + 1\right) p^{2n+2}}{p+1},$$
$$P_n(010) = \frac{p}{p+1} + \frac{\left(p^2 - p - 1\right) p^{2n+2}}{p+1}.$$

Rule 164

$$P_n(0) = -\frac{1}{2}\,\frac{2p^4 - 3p^3 - p - 2}{1+p} - \frac{3}{2}\,p^{1+n} + 2\,\frac{p^{2+2n}}{1+p} - \frac{1}{2}\,p^n\,(1-2p)^{G(n)}$$
$$-\frac{1}{2}\,(1-p)^2 \sum_{i=1}^{n-1} p^i\,(1-2p)^{G(i)},$$
$$P_n(00) = -\frac{2p^4 - 3p^3 - 1}{1+p} - 2\,p^{n+1} - \frac{(p-3)\,p^{2n+2}}{1+p}$$
$$+ (-1+p)\,p^n\,(1-2p)^{G(n)} + \left(-p^2 + 2p - 1\right) \sum_{i=1}^{n-1} p^i\,(1-2p)^{G(i)},$$
$$P_n(000) = -\frac{1}{2}\,\frac{6p^4 - 9p^3 + p - 2}{1+p} + \frac{1}{2}\,(p-5)\,p^{n+1} - 2\,\frac{(p-2)\,p^{2n+2}}{1+p}$$
$$+ \left(-\frac{1}{2}\,p^2 + 2p - \frac{3}{2}\right) p^n\,(1-2p)^{G(n)}$$
$$+ \left(3p - \frac{3}{2} - \frac{3}{2}\,p^2\right) \sum_{i=1}^{n-1} p^i\,(1-2p)^{G(i)},$$
$$P_n(010) = \frac{1}{2}\,\frac{p\left(2p^3 - 3p^2 + 1\right)}{1+p} + 1/4\,(-1+2p)\,p^{1+n} - \frac{(-1+p)\,p^{2n+3}}{1+p}$$
$$+ \left(\frac{1}{2}\,p^2 - \frac{3}{2}\,p + \frac{1}{2}\right) p^n\,(1-2p)^{G(n)} + \frac{1}{4}\,p^{1+n}\,(1-2p)^{G(1+n)}$$
$$+ \left(\frac{1}{2}\,p^2 - p + \frac{1}{2}\right) \sum_{i=1}^{n-1} p^i\,(1-2p)^{G(i)},$$

where

$$G(n) = \sum_{k=0}^{n} \left(\binom{n}{k} \mod 2 \right).$$

Rule 168

$$\begin{aligned}
P_n(0) &= 1 - p^{n+1}(2-p)^n\,,\\
P_n(00) &= 1 - p^{n+1}(2-p)^{n+1}\,,\\
P_n(000) &= 1 - p^{n+1}(2-p)^n\left(p^2 - 3\,p + 3\right),\\
P_n(010) &= (p-1)^2\,p^{n+1}(2-p)^n\,.
\end{aligned}$$

Rule 170

$$\begin{aligned}
P_n(0) &= 1 - p,\\
P_n(00) &= (1-p)^2\,,\\
P_n(000) &= (1-p)^3\,,\\
P_n(010) &= p\,(1-p)^2\,.
\end{aligned}$$

Rule 172

$$\begin{aligned}
P_n(0) &= 1 - (1-p)^2\,p + \alpha_1\lambda_1{}^n + \alpha_2\lambda_2{}^n,\\
P_n(00) &= -2\,p^3 + 4\,p^2 - 2\,p + 1 + \beta_1\lambda_1{}^n + \beta_2\lambda_2{}^n,\\
P_n(000) &= -3\,p^3 + 6\,p^2 - 3\,p + 1 + \beta_1\lambda_1{}^n + \beta_2\lambda_2{}^n,\\
P_n(010) &= p\left(p^2 - 2\,p + 1\right) + \delta_1\lambda_1{}^n + \delta_2\lambda_2{}^n.
\end{aligned}$$

where

$$\begin{aligned}
\lambda_{1,2} &= \frac{1}{2}\,p \pm \frac{1}{2}\sqrt{p\,(4-3\,p)}\\
\alpha_{1,2} &= -\frac{1}{2}\,p \mp \frac{1}{2}\,\frac{\sqrt{p\,(4-3\,p)}\,p\,(2\,p-3)}{-4+3\,p}\\
\delta_{1,2} &= -\frac{1}{2}\,p\,(1-p)^2 \pm \frac{1}{2}\,\frac{\sqrt{p\,(4-3\,p)}\,p\,(1-p)^2}{-4+3\,p}\\
\beta_{1,2} &= \frac{1}{2}\left(p^3 - p^2 - 1\right)p \pm \frac{1}{2}\,\frac{\sqrt{p\,(4-3\,p)}\,p\left(p^3 - 3\,p^2 + 3\right)}{-4+3\,p}
\end{aligned}$$

Rule 178

$$\begin{aligned}
P_n(0) = \frac{1}{2} - \frac{p^{2n+3}}{p^2+1} + \frac{1}{2}\,\frac{p^2\left(2\,p^3 - 5\,p^2 + 4\,p - 1\right)(-1)^n}{\left(p^2+1\right)\left(p^2 - 2\,p + 2\right)}\\
- \frac{\left(p^3 - 3\,p^2 + 3\,p - 1\right)(1-p)^{2n}}{p^2 - 2\,p + 2},
\end{aligned}$$

$$
\begin{aligned}
P_n(00) &= -\frac{1}{2}\,\frac{p\left(2\,p^3-4\,p^2+p+1\right)}{(p-2)\,(p+1)}+\frac{(p-1)\,p^{2\,n+4}}{p^3+p^2+p+1}\\
&+\frac{1}{2}\,\frac{p^2\left(2\,p^3-5\,p^2+4\,p-1\right)(-1)^n}{\left(p^2+1\right)\left(p^2-2\,p+2\right)}\\
&+\frac{\left(p^5-6\,p^4+16\,p^3-22\,p^2+15\,p-4\right)(1-p)^{2\,n}}{p^3-4\,p^2+6\,p-4},\\
P_n(000) &= \left(-p^3+3\,p^2-3\,p+1\right)(1-p)^{2\,n},\\
P_n(010) &= \frac{1}{2}\,\frac{4\,p^4-8\,p^3+3\,p^2+p-2}{(p-2)\,(p+1)}+\frac{p^3\left(p^3-p^2-2\right)p^{2\,n}}{p^3+p^2+p+1}\\
&+\frac{1}{2}\,\frac{p^2\left(2\,p^3-5\,p^2+4\,p-1\right)(-1)^n}{\left(p^2+1\right)\left(p^2-2\,p+2\right)}\\
&-\frac{\left(2\,p^5-9\,p^4+17\,p^3-17\,p^2+9\,p-2\right)(1-p)^{2\,n}}{p^3-4\,p^2+6\,p-4}.
\end{aligned}
$$

Rule 184

$$
\begin{aligned}
P_n(0) &= 1-p,\\
P_n(00) &= \sum_{j=1}^{n+1}\frac{j}{n+1}\binom{2\,n+2}{n+1-j}p^{n+1-j}\,(1-p)^{n+1+j},\\
P_n(000) &= \sum_{j=1}^{n+1}\frac{j}{n+1}\binom{2\,n+2}{n+1-j}p^{n+1-j}\,(1-p)^{n+2+j},\\
P_n(010) &= 2\,(1-p)^2-(2-p)\sum_{j=1}^{n+1}\frac{j}{n+1}\binom{2\,n+2}{n+1-j}p^{n+1-j}\,(1-p)^{n+1+j}.
\end{aligned}
$$

Rule 200

$$
\begin{aligned}
P_n(0) &= p^3-2\,p^2+1,\\
P_n(00) &= (2\,p+1)\,(p-1)^2,\\
P_n(000) &= -\left(p^3+p^2-2\,p-1\right)(p-1)^2,\\
P_n(010) &= 0.
\end{aligned}
$$

Rule 204

$$
\begin{aligned}
P_n(0) &= 1-p,\\
P_n(00) &= (1-p)^2,\\
P_n(000) &= (1-p)^3,\\
P_n(010) &= p\,(1-p)^2.
\end{aligned}
$$

Rule 232

$$P_n(0) = -\frac{\left(-p^2+p\right)^n p\left(2p^2-3p+1\right)}{p^2-p+1} - \frac{-p^3+p^2+p-1}{p^2-p+1},$$

$$P_n(00) = \frac{2\left(-p^2+p\right)^n p\,(p-1)^3}{p^2-p+1} - \frac{\left(p^2-p-1\right)(p-1)^2}{p^2-p+1},$$

$$P_n(000) = -\frac{\left(-p^2+p\right)^n p\,(p-3)\,(p-1)^3}{p^2-p+1} - \frac{(2p+1)\,(p-1)^3}{p^2-p+1},$$

$$P_n(010) = (p-1)^2\, p\left(-p^2+p\right)^n.$$

Appendix D
Ruelle-Frobenius-Perron Equation of Order 3 for Minimal ECA

Rule 0

$$\begin{aligned}
P_{n+1}(0) &= 1,\\
P_{n+1}(00) &= 1,\\
P_{n+1}(000) &= 1,\\
P_{n+1}(010) &= 0.
\end{aligned}$$

Rule 1

$$\begin{aligned}
P_{n+1}(0) &= -P_n(000) + 1,\\
P_{n+1}(00) &= 1 - 2P_n(000) + P_n(0000),\\
P_{n+1}(000) &= 1 - 3P_n(000) + 2P_n(0000),\\
P_{n+1}(010) &= P_n(000) - 2P_n(0000) + P_n(00000).
\end{aligned}$$

Rule 2

$$\begin{aligned}
P_{n+1}(0) &= P_n(000) - P_n(00) + 1,\\
P_{n+1}(00) &= 2P_n(000) - 2P_n(00) + 1,\\
P_{n+1}(000) &= 1 + 3P_n(000) - 3P_n(00),\\
P_{n+1}(010) &= -P_n(000) + P_n(00).
\end{aligned}$$

Rule 3

$$\begin{aligned}
P_{n+1}(0) &= -P_n(00) + 1,\\
P_{n+1}(00) &= 1 + P_n(000) - 2P_n(00),\\
P_{n+1}(000) &= 1 + 2P_n(000) - 3P_n(00),\\
P_{n+1}(010) &= P_n(00) - 2P_n(000) + P_n(0000).
\end{aligned}$$

H. Fukś, *Solvable Cellular Automata*, Understanding Complex Systems,
https://doi.org/10.1007/978-3-031-38700-5

Rule 4

$$
\begin{aligned}
P_{n+1}(0) &= -P_n(010) + 1, \\
P_{n+1}(00) &= -2P_n(010) + 1, \\
P_{n+1}(000) &= 1 - 3P_n(010) + P_n(01010), \\
P_{n+1}(010) &= P_n(010).
\end{aligned}
$$

Rule 5

$$
\begin{aligned}
P_{n+1}(0) &= -P_n(000) - P_n(010) + 1, \\
P_{n+1}(00) &= 1 - 2P_n(000) - 2P_n(010) + P_n(0000), \\
P_{n+1}(000) &= 1 - 3P_n(000) - 3P_n(010) + P_n(00010) + P_n(01000) \\
&\quad + P_n(01010) + 2P_n(0000), \\
P_{n+1}(010) &= P_n(000) - 2P_n(0000) + P_n(00000) + P_n(010).
\end{aligned}
$$

Rule 6

$$
\begin{aligned}
P_{n+1}(0) &= P_n(000) - P_n(00) - P_n(010) + 1, \\
P_{n+1}(00) &= 1 + 2P_n(000) - 2P_n(010) - 2P_n(00) + P_n(0010), \\
P_{n+1}(000) &= 1 + 3P_n(000) - 3P_n(010) - P_n(01000) + P_n(01010) \\
&\quad - 3P_n(00) + 2P_n(0010) + P_n(0100), \\
P_{n+1}(010) &= -P_n(000) - 2P_n(0010) + P_n(00) + P_n(010).
\end{aligned}
$$

Rule 7

$$
\begin{aligned}
P_{n+1}(0) &= -P_n(00) - P_n(010) + 1, \\
P_{n+1}(00) &= 1 + P_n(000) - 2P_n(010) - 2P_n(00) + P_n(0010), \\
P_{n+1}(000) &= 1 + 2P_n(000) - 3P_n(010) + P_n(01010) \\
&\quad - 3P_n(00) + 2P_n(0010) + P_n(0100), \\
P_{n+1}(010) &= -2P_n(0010) + P_n(00010) + P_n(00) \\
&\quad - 2P_n(000) + P_n(0000) + P_n(010).
\end{aligned}
$$

Rule 8

$$
\begin{aligned}
P_{n+1}(0) &= P_n(00) + P_n(010) - P_n(0) + 1, \\
P_{n+1}(00) &= 2P_n(00) + 2P_n(010) - 2P_n(0) + 1, \\
P_{n+1}(000) &= 1 + 3P_n(010) - 3P_n(0) + 3P_n(00), \\
P_{n+1}(010) &= -P_n(00) - P_n(010) + P_n(0).
\end{aligned}
$$

Rule 9

$$
\begin{aligned}
P_{n+1}(0) &= P_n(010) + P_n(00) - P_n(0) + 1 - P_n(000),\\
P_{n+1}(00) &= 1 - 2P_n(000) + 2P_n(010) - 2P_n(0) + 2P_n(00) + P_n(0000),\\
P_{n+1}(000) &= 1 - 2P_n(000) + 3P_n(010) - P_n(00010)\\
&\quad - 3P_n(0) + 3P_n(00) + P_n(0000),\\
P_{n+1}(010) &= -P_n(00) + P_n(000) - 2P_n(0000) + P_n(00000) + P_n(0) - P_n(010).
\end{aligned}
$$

Rule 10

$$
\begin{aligned}
P_{n+1}(0) &= P_n(010) - P_n(0) + 1 + P_n(000),\\
P_{n+1}(00) &= 1 + P_n(000) + 2P_n(010) - 2P_n(0) + P_n(00) - P_n(0010),\\
P_{n+1}(000) &= 1 + P_n(000) + 3P_n(010) - 3P_n(0) + 2P_n(00) - 2P_n(0010),\\
P_{n+1}(010) &= 2P_n(0010) + P_n(0) - P_n(010) - 2P_n(00) + P_n(000).
\end{aligned}
$$

Rule 11

$$
\begin{aligned}
P_{n+1}(0) &= P_n(010) - P_n(0) + 1,\\
P_{n+1}(00) &= 1 + 2P_n(010) - 2P_n(0) + P_n(00) - P_n(0010),\\
P_{n+1}(000) &= 1 + 3P_n(010) - 3P_n(0) + 2P_n(00) - 2P_n(0010),\\
P_{n+1}(010) &= 2P_n(0010) - P_n(00010) + P_n(0) - P_n(010) - 2P_n(00) + P_n(000).
\end{aligned}
$$

Rule 12

$$
\begin{aligned}
P_{n+1}(0) &= P_n(00) - P_n(0) + 1,\\
P_{n+1}(00) &= 2P_n(00) - 2P_n(0) + 1,\\
P_{n+1}(000) &= 1 + P_n(010) - 3P_n(0) + 3P_n(00) - P_n(0100),\\
P_{n+1}(010) &= -P_n(00) + P_n(0).
\end{aligned}
$$

Rule 13

$$
\begin{aligned}
P_{n+1}(0) &= P_n(00) - P_n(0) + 1 - P_n(000),\\
P_{n+1}(00) &= 1 - 2P_n(000) - 2P_n(0) + 2P_n(00) + P_n(0000),\\
P_{n+1}(000) &= 1 - 2P_n(000) + P_n(010) + P_n(01000) - 3P_n(0)\\
&\quad + 3P_n(00) + P_n(0000) - P_n(0100),\\
P_{n+1}(010) &= P_n(000) + P_n(00000) + P_n(0) - P_n(00) - 2P_n(0000).
\end{aligned}
$$

Rule 14

$$\begin{aligned}
P_{n+1}(0) &= -P_n(0) + 1 + P_n(000),\\
P_{n+1}(00) &= 1 + P_n(000) - 2P_n(0) + P_n(00),\\
P_{n+1}(000) &= 1 + P_n(000) + P_n(010) - P_n(01000) - 3P_n(0) + 2P_n(00),\\
P_{n+1}(010) &= P_n(0) - 2P_n(00) + P_n(000).
\end{aligned}$$

Rule 15

$$\begin{aligned}
P_{n+1}(0) &= -P_n(0) + 1,\\
P_{n+1}(00) &= P_n(00) - 2P_n(0) + 1,\\
P_{n+1}(000) &= 2P_n(00) - 3P_n(0) + P_n(010) + 1,\\
P_{n+1}(010) &= P_n(0) - 2P_n(00) + P_n(000).
\end{aligned}$$

Rule 18

$$\begin{aligned}
P_{n+1}(0) &= -2P_n(00) + 1 + 2P_n(000),\\
P_{n+1}(00) &= 1 + 2P_n(000) - 3P_n(00) + P_n(0000),\\
P_{n+1}(000) &= 1 + 3P_n(000) + P_n(00000) + P_n(00100) - 4P_n(00),\\
P_{n+1}(010) &= -2P_n(0000) + 2P_n(000).
\end{aligned}$$

Rule 19

$$\begin{aligned}
P_{n+1}(0) &= P_n(000) - 2P_n(00) + 1,\\
P_{n+1}(00) &= 1 + 2P_n(000) - 3P_n(00),\\
P_{n+1}(000) &= 1 + 3P_n(000) + P_n(00100) - 4P_n(00),\\
P_{n+1}(010) &= 0.
\end{aligned}$$

Rule 22

$$\begin{aligned}
P_{n+1}(0) &= -2P_n(00) - P_n(010) + 1 + 2P_n(000),\\
P_{n+1}(00) &= 1 + 2P_n(000) - 2P_n(010) - 3P_n(00) + P_n(0010)\\
&\quad + P_n(0100) + P_n(0000),\\
P_{n+1}(000) &= 1 + 3P_n(000) - 3P_n(010) + P_n(00000) + P_n(01010)\\
&\quad - 4P_n(00) + 2P_n(0010) + 2P_n(0100),\\
P_{n+1}(010) &= -P_n(00010) - 2P_n(0000) + 2P_n(000) + P_n(010)\\
&\quad - P_n(0100) - P_n(0010) + P_n(00100) - P_n(01000).
\end{aligned}$$

Rule 23

$$\begin{aligned}
P_{n+1}(0) &= P_n(000) - 2P_n(00) - P_n(010) + 1,\\
P_{n+1}(00) &= 1 + 2P_n(000) - 2P_n(010) - 3P_n(00) + P_n(0010) + P_n(0100),\\
P_{n+1}(000) &= 1 + 3P_n(000) - 3P_n(010) + P_n(01010)\\
&\quad - 4P_n(00) + 2P_n(0010) + 2P_n(0100),\\
P_{n+1}(010) &= P_n(010) - P_n(0100) - P_n(0010) + P_n(00100).
\end{aligned}$$

Rule 24

$$\begin{aligned}
P_{n+1}(0) &= P_n(000) + P_n(010) - P_n(0) + 1,\\
P_{n+1}(00) &= 2P_n(000) + 2P_n(010) - 2P_n(0) + 1,\\
P_{n+1}(000) &= 1 + P_n(000) + 3P_n(010) + P_n(00010) + P_n(01100)\\
&\quad - 3P_n(0) + P_n(00) + P_n(0000) - P_n(0010),\\
P_{n+1}(010) &= -P_n(000) - P_n(010) + P_n(0).
\end{aligned}$$

Rule 25

$$\begin{aligned}
P_{n+1}(0) &= P_n(010) - P_n(0) + 1,\\
P_{n+1}(00) &= 1 + P_n(000) + 2P_n(010) - 2P_n(0),\\
P_{n+1}(000) &= 1 + P_n(000) + 3P_n(010) + P_n(01100)\\
&\quad - 3P_n(0) + P_n(00) - P_n(0010),\\
P_{n+1}(010) &= -2P_n(000) + P_n(0) - P_n(010) + P_n(0000).
\end{aligned}$$

Rule 26

$$\begin{aligned}
P_{n+1}(0) &= P_n(010) - P_n(0) + 1 - P_n(00) + 2P_n(000),\\
P_{n+1}(00) &= 1 + P_n(000) + 2P_n(010) - 2P_n(0) - P_n(0010) + P_n(0000),\\
P_{n+1}(000) &= 1 + P_n(000) + 3P_n(010) + P_n(00000) + P_n(00100)\\
&\quad + P_n(01100) - 3P_n(0) + P_n(00) - 2P_n(0010),\\
P_{n+1}(010) &= P_n(00010) + P_n(0010) + P_n(0) - P_n(010)\\
&\quad - 2P_n(00) + 2P_n(000) - P_n(0000).
\end{aligned}$$

Rule 27

$$\begin{aligned}
P_{n+1}(0) &= P_n(010) - P_n(0) + 1 - P_n(00) + P_n(000),\\
P_{n+1}(00) &= 1 + P_n(000) + 2P_n(010) - 2P_n(0) - P_n(0010),\\
P_{n+1}(000) &= 1 + P_n(000) + 3P_n(010) + P_n(00100) + P_n(01100)\\
&\quad - 3P_n(0) + P_n(00) - 2P_n(0010),\\
P_{n+1}(010) &= P_n(0010) + P_n(0) - P_n(010) - 2P_n(00) + P_n(000).
\end{aligned}$$

Rule 28

$$\begin{aligned}
P_{n+1}(0) &= P_n(000) - P_n(0) + 1,\\
P_{n+1}(00) &= -2P_n(0) + P_n(0100) + 2P_n(000) + 1,\\
P_{n+1}(000) &= P_n(0000) - 3P_n(0) + P_n(010) + P_n(00)\\
&\quad + P_n(0100) + P_n(01100) + P_n(000) + 1,\\
P_{n+1}(010) &= -P_n(000) + P_n(0) - 2P_n(0100).
\end{aligned}$$

Rule 29

$$\begin{aligned}
P_{n+1}(0) &= -P_n(0) + 1,\\
P_{n+1}(00) &= -2P_n(0) + P_n(0100) + P_n(000) + 1,\\
P_{n+1}(000) &= -3P_n(0) + P_n(010) + P_n(00) + P_n(0100)\\
&\quad + P_n(01100) + P_n(000) + 1,\\
P_{n+1}(010) &= -2P_n(000) + P_n(01000) + P_n(0) + P_n(0000) - 2P_n(0100).
\end{aligned}$$

Rule 30

$$\begin{aligned}
P_{n+1}(0) &= -P_n(0) - P_n(00) + 1 + 2P_n(000),\\
P_{n+1}(00) &= -2P_n(0) + P_n(0100) + P_n(000) + 1 + P_n(0000),\\
P_{n+1}(000) &= P_n(00000) - 3P_n(0) + P_n(010) + P_n(00) + P_n(0100)\\
&\quad + P_n(01100) + P_n(000) + 1,\\
P_{n+1}(010) &= -P_n(0100) + P_n(00100) + P_n(0) - 2P_n(00)\\
&\quad + 2P_n(000) - P_n(0000) - P_n(01000).
\end{aligned}$$

Rule 32

$$\begin{aligned}
P_{n+1}(0) &= -P_n(0) + 2P_n(00) - P_n(000) + 1,\\
P_{n+1}(00) &= -2P_n(0) + 4P_n(00) - 2P_n(000) + 1,\\
P_{n+1}(000) &= 1 - 3P_n(000) + P_n(010) + P_n(00100)\\
&\quad - 3P_n(0) + 6P_n(00) - P_n(0010) - P_n(0100),\\
P_{n+1}(010) &= P_n(0) - 2P_n(00) + P_n(000).
\end{aligned}$$

Rule 33

$$\begin{aligned}
P_{n+1}(0) &= -2P_n(000) - P_n(0) + 2P_n(00) + 1,\\
P_{n+1}(00) &= 1 - 4P_n(000) - 2P_n(0) + 4P_n(00) + P_n(0000),\\
P_{n+1}(000) &= 1 - 6P_n(000) + P_n(010) + P_n(00100) - 3P_n(0)\\
&\quad + 6P_n(00) + 2P_n(0000) - P_n(0010) - P_n(0100),\\
P_{n+1}(010) &= 2P_n(000) - 2P_n(0000) + P_n(00000) + P_n(0) - 2P_n(00).
\end{aligned}$$

Rule 34

$$\begin{aligned}
P_{n+1}(0) &= P_n(00) - P_n(0) + 1,\\
P_{n+1}(00) &= 2P_n(00) - 2P_n(0) + 1,\\
P_{n+1}(000) &= 1 + P_n(010) - 3P_n(0) + 3P_n(00) - P_n(0100),\\
P_{n+1}(010) &= -P_n(00) + P_n(0).
\end{aligned}$$

Rule 35

$$\begin{aligned}
P_{n+1}(0) &= P_n(00) - P_n(0) + 1 - P_n(000),\\
P_{n+1}(00) &= -2P_n(0) + 2P_n(00) - P_n(000) + 1,\\
P_{n+1}(000) &= -3P_n(0) + P_n(010) + 3P_n(00) - P_n(0100) - P_n(000) + 1,\\
P_{n+1}(010) &= -P_n(00) - P_n(000) + P_n(0000) + P_n(0).
\end{aligned}$$

Rule 36

$$\begin{aligned}
P_{n+1}(0) &= -P_n(0) + 2P_n(00) - P_n(000) + 1 - P_n(010),\\
P_{n+1}(00) &= 1 - 2P_n(000) - 2P_n(0) + 4P_n(00) - P_n(0010) - P_n(0100),\\
P_{n+1}(000) &= 1 - 3P_n(000) + P_n(010) - 3P_n(0)\\
&\quad + 6P_n(00) - 2P_n(0010) - 2P_n(0100),\\
P_{n+1}(010) &= P_n(00100) + P_n(0010) + P_n(0)\\
&\quad - 2P_n(010) - 2P_n(00) + P_n(0100) + P_n(000) + P_n(01010).
\end{aligned}$$

Rule 37

$$\begin{aligned}
P_{n+1}(0) &= -2P_n(000) - P_n(0) + 2P_n(00) + 1 - P_n(010),\\
P_{n+1}(00) &= 1 - 4P_n(000) - 2P_n(0) + 4P_n(00) + P_n(0000)\\
&\quad - P_n(0010) - P_n(0100),\\
P_{n+1}(000) &= 1 - 6P_n(000) + P_n(010) + P_n(00010) + P_n(01000)\\
&\quad - 3P_n(0) + 6P_n(00) + 2P_n(0000) - 2P_n(0010) - 2P_n(0100),\\
P_{n+1}(010) &= P_n(00100) + 2P_n(000) - 2P_n(0000) + P_n(00000) + P_n(0010)\\
&\quad + P_n(0) - 2P_n(010) - 2P_n(00) + P_n(0100) + P_n(01010).
\end{aligned}$$

Rule 38

$$\begin{aligned}
P_{n+1}(0) &= P_n(00) - P_n(0) + 1 - P_n(010),\\
P_{n+1}(00) &= 1 - 2P_n(0) + 2P_n(00) - P_n(0100),\\
P_{n+1}(000) &= 1 + P_n(010) - P_n(01000) - 3P_n(0) + 3P_n(00) - P_n(0100),\\
P_{n+1}(010) &= -P_n(00) + P_n(0) - 2P_n(010) + P_n(0100) + P_n(01010).
\end{aligned}$$

Rule 40

$$\begin{aligned}
P_{n+1}(0) &= -2P_n(0) + 3P_n(00) + 1 - P_n(000) + P_n(010),\\
P_{n+1}(00) &= 1 - P_n(000) + P_n(010) - 3P_n(0) + 4P_n(00) + P_n(0010),\\
P_{n+1}(000) &= 1 - P_n(000) + 2P_n(010) + P_n(00100) - P_n(01100)\\
&\quad - 4P_n(0) + 5P_n(00) + P_n(0010) - P_n(0100) + P_n(0110),\\
P_{n+1}(010) &= P_n(00) - P_n(000) - 2P_n(0010) + P_n(010).
\end{aligned}$$

Rule 41

$$\begin{aligned}
P_{n+1}(0) &= -2P_n(0) + 3P_n(00) - 2P_n(000) + 1 + P_n(010),\\
P_{n+1}(00) &= 1 - 3P_n(000) + P_n(010) - 3P_n(0)\\
&\quad + 4P_n(00) + P_n(0000) + P_n(0010),\\
P_{n+1}(000) &= 1 - 3P_n(000) + 2P_n(010) - P_n(00010) + P_n(00100) - P_n(01100)\\
&\quad - 4P_n(0) + 5P_n(00) + P_n(0000) + P_n(0010) - P_n(0100) + P_n(0110),\\
P_{n+1}(010) &= P_n(00) - 2P_n(0010) - 2P_n(0000) + P_n(00000) + P_n(010).
\end{aligned}$$

Rule 42

$$\begin{aligned}
P_{n+1}(0) &= 2P_n(00) - 2P_n(0) + 1 + P_n(010),\\
P_{n+1}(00) &= 3P_n(00) - 3P_n(0) + P_n(010) + 1,\\
P_{n+1}(000) &= 4P_n(00) - P_n(0100) - P_n(01100)\\
&\quad - 4P_n(0) + 2P_n(010) + P_n(0110) + 1,\\
P_{n+1}(010) &= P_n(010).
\end{aligned}$$

Rule 43

$$\begin{aligned}
P_{n+1}(0) &= 2P_n(00) - 2P_n(0) - P_n(000) + 1 + P_n(010),\\
P_{n+1}(00) &= 3P_n(00) - P_n(000) - 3P_n(0) + P_n(010) + 1,\\
P_{n+1}(000) &= 4P_n(00) - P_n(0100) - P_n(01100) - P_n(000) - 4P_n(0)\\
&\quad + 2P_n(010) + P_n(0110) + 1,\\
P_{n+1}(010) &= -P_n(00010) + P_n(010).
\end{aligned}$$

Rule 44

$$\begin{aligned}
P_{n+1}(0) &= -2P_n(0) + 3P_n(00) + 1 - P_n(000),\\
P_{n+1}(00) &= 1 - P_n(000) + P_n(010) - 3P_n(0) + 4P_n(00) - P_n(0100),\\
P_{n+1}(000) &= 1 - P_n(000) + 2P_n(010) - P_n(01100) - 4P_n(0)\\
&\quad + 5P_n(00) - 2P_n(0100) + P_n(0110),\\
P_{n+1}(010) &= P_n(00100) + P_n(00) - P_n(000) - P_n(0010).
\end{aligned}$$

Rule 45

$$\begin{aligned}
P_{n+1}(0) &= -2P_n(0) + 3P_n(00) - 2P_n(000) + 1, \\
P_{n+1}(00) &= 4P_n(00) - 3P_n(000) + P_n(0000) - P_n(0100) \\
&\quad - 3P_n(0) + P_n(010) + 1, \\
P_{n+1}(000) &= 5P_n(00) - 3P_n(000) + P_n(0000) - 2P_n(0100) + P_n(01000) \\
&\quad - P_n(01100) - 4P_n(0) + 2P_n(010) + P_n(0110) + 1, \\
P_{n+1}(010) &= P_n(00100) + P_n(00) - P_n(0010) - 2P_n(0000) + P_n(00000).
\end{aligned}$$

Rule 46

$$\begin{aligned}
P_{n+1}(0) &= 2P_n(00) - 2P_n(0) + 1, \\
P_{n+1}(00) &= 3P_n(00) - P_n(0100) - 3P_n(0) + P_n(010) + 1, \\
P_{n+1}(000) &= -P_n(01000) + 4P_n(00) - P_n(0100) - P_n(01100) \\
&\quad - 4P_n(0) + 2P_n(010) + P_n(0110) + 1, \\
P_{n+1}(010) &= 0.
\end{aligned}$$

Rule 50

$$\begin{aligned}
P_{n+1}(0) &= P_n(000) - P_n(0) + 1, \\
P_{n+1}(00) &= -2P_n(0) + P_n(00) + 1 + P_n(0000), \\
P_{n+1}(000) &= P_n(00000) - 3P_n(0) + P_n(010) + 2P_n(00) + 1, \\
P_{n+1}(010) &= -2P_n(0000) + 3P_n(000) + P_n(0) - 2P_n(00).
\end{aligned}$$

Rule 51

$$\begin{aligned}
P_{n+1}(0) &= -P_n(0) + 1, \\
P_{n+1}(00) &= P_n(00) - 2P_n(0) + 1, \\
P_{n+1}(000) &= 2P_n(00) - 3P_n(0) + P_n(010) + 1, \\
P_{n+1}(010) &= P_n(0) - 2P_n(00) + P_n(000).
\end{aligned}$$

Rule 54

$$\begin{aligned}
P_{n+1}(0) &= P_n(000) - P_n(0) - P_n(010) + 1, \\
P_{n+1}(00) &= -2P_n(0) + P_n(00) + 1 + P_n(0000), \\
P_{n+1}(000) &= P_n(00000) - 3P_n(0) + P_n(010) + 2P_n(00) + 1, \\
P_{n+1}(010) &= -P_n(00010) - 2P_n(0000) + 3P_n(000) - P_n(01000) + P_n(0010) \\
&\quad + P_n(0) - 2P_n(010) - 2P_n(00) + P_n(0100) + P_n(01010).
\end{aligned}$$

Rule 56

$$\begin{aligned}
P_{n+1}(0) &= -2P_n(0) + 1 + 2P_n(00) + P_n(010),\\
P_{n+1}(00) &= -3P_n(0) + P_n(010) + 2P_n(00) + 1 + P_n(000) + P_n(0010),\\
P_{n+1}(000) &= P_{n+1}(0000) + P_n(00010) - 4P_n(0) + 2P_n(010)\\
&\quad + P_n(0110) + 3P_n(00) + 1,\\
P_{n+1}(010) &= 2P_n(00) - 2P_n(000) - 2P_n(0010) + P_n(010).
\end{aligned}$$

Rule 57

$$\begin{aligned}
P_{n+1}(0) &= -P_n(000) - 2P_n(0) + 1 + 2P_n(00) + P_n(010),\\
P_{n+1}(00) &= -3P_n(0) + P_n(010) + 2P_n(00) + 1 + P_n(0010),\\
P_{n+1}(000) &= -4P_n(0) + 2P_n(010) + P_n(0110) + 3P_n(00) + 1,\\
P_{n+1}(010) &= 2P_n(00) - 3P_n(000) - 2P_n(0010) + P_n(0000) + P_n(010).
\end{aligned}$$

Rule 58

$$\begin{aligned}
P_{n+1}(0) &= P_n(00) - 2P_n(0) + 1 + P_n(000) + P_n(010),\\
P_{n+1}(00) &= -3P_n(0) + P_n(010) + 2P_n(00) + 1 + P_n(0000),\\
P_{n+1}(000) &= P_n(00000) - 4P_n(0) + 2P_n(010) + P_n(0110) + 3P_n(00) + 1,\\
P_{n+1}(010) &= P_n(00010) + P_n(000) - P_n(0000) + P_n(010) - P_n(0010).
\end{aligned}$$

Rule 60

$$\begin{aligned}
P_{n+1}(0) &= -2P_n(0) + 1 + 2P_n(00),\\
P_{n+1}(00) &= -3P_n(0) + P_n(010) + 2P_n(00) + 1 + P_n(000),\\
P_{n+1}(000) &= P_n(0000) - 4P_n(0) + 2P_n(010) + P_n(0110) + 3P_n(00) + 1,\\
P_{n+1}(010) &= 2P_n(00) - 2P_n(000) - P_n(0010) - P_n(0100).
\end{aligned}$$

Rule 62

$$\begin{aligned}
P_{n+1}(0) &= P_n(00) - 2P_n(0) + 1 + P_n(000),\\
P_{n+1}(00) &= -3P_n(0) + P_n(010) + 2P_n(00) + 1 + P_n(0000),\\
P_{n+1}(000) &= P_n(00000) - 4P_n(0) + 2P_n(010) + P_n(0110) + 3P_n(00) + 1,\\
P_{n+1}(010) &= P_n(000) - P_n(0000) - P_n(01000).
\end{aligned}$$

Rule 72

$$\begin{aligned}
P_{n+1}(0) &= -2P_n(0) + 1 + 2P_n(010) + 2P_n(00),\\
P_{n+1}(00) &= 1 + 4P_n(010) - 4P_n(0) + 4P_n(00) + P_n(0110),
\end{aligned}$$

$$
\begin{aligned}
P_{n+1}(000) &= 1 + P_n(000) + 4P_n(010) + P_n(01010) + P_n(01110) - 5P_n(0) \\
&\quad + 4P_n(00) + P_n(0010) + P_n(0100) + 2P_n(0110), \\
P_{n+1}(010) &= -2P_n(00) + 2P_n(0) - 2P_n(010) - 2P_n(0110).
\end{aligned}
$$

Rule 73

$$
\begin{aligned}
P_{n+1}(0) &= -P_n(000) - 2P_n(0) + 2P_n(010) + 1 + 2P_n(00), \\
P_{n+1}(00) &= 4P_n(010) + 1 + P_n(0110) - 4P_n(0) + 4P_n(00) \\
&\quad - 2P_n(000) + P_n(0000), \\
P_{n+1}(000) &= P_n(0010) + P_n(0100) - P_n(01000) + P_n(01010) - P_n(00010) \\
&\quad + 4P_n(010) + 1 + 2P_n(0110) + P_n(01110) - 5P_n(0) + 4P_n(00), \\
P_{n+1}(010) &= P_n(000) - 2P_n(010) + P_n(00000) + 2P_n(0) \\
&\quad - 2P_n(00) - 2P_n(0000) - 2P_n(0110).
\end{aligned}
$$

Rule 74

$$
\begin{aligned}
P_{n+1}(0) &= -2P_n(0) + 2P_n(010) + 1 + P_n(00) + P_n(000), \\
P_{n+1}(00) &= 4P_n(010) - P_n(0010) + 1 + P_n(0110) - 4P_n(0) \\
&\quad + 3P_n(00) + P_n(000), \\
P_{n+1}(000) &= P_n(01000) + P_n(01010) + P_n(0000) + 4P_n(010) - P_n(0010) + 1 \\
&\quad + 2P_n(0110) + P_n(01110) - 5P_n(0) + 4P_n(00), \\
P_{n+1}(010) &= 2P_n(0010) + P_n(00110) + 2P_n(0) - 2P_n(010) \\
&\quad - 2P_n(0110) - 3P_n(00) + P_n(000).
\end{aligned}
$$

Rule 76

$$
\begin{aligned}
P_{n+1}(0) &= -2P_n(0) + P_n(010) + 1 + 2P_n(00), \\
P_{n+1}(00) &= 1 + P_n(0110) - 4P_n(0) + 2P_n(010) + 4P_n(00), \\
P_{n+1}(000) &= P_{n+1}(000) + 1 + 2P_n(0110) + P_n(01110) - 5P_n(0) \\
&\quad + 3P_n(010) + 4P_n(00), \\
P_{n+1}(010) &= -P_n(010) + 2P_n(0) - 2P_n(00) - 2P_n(0110).
\end{aligned}
$$

Rule 77

$$
\begin{aligned}
P_{n+1}(0) &= -P_n(000) - 2P_n(0) + P_n(010) + 1 + 2P_n(00), \\
P_{n+1}(00) &= 1 + P_n(0110) - 4P_n(0) + 2P_n(010) + 4P_n(00) \\
&\quad - 2P_n(000) + P_n(0000),
\end{aligned}
$$

$$
\begin{aligned}
P_{n+1}(000) &= 1 + 2P_n(0110) + P_n(01110) - 5P_n(0) + 3P_n(010) + 4P_n(00), \\
P_{n+1}(010) &= P_n(000) - P_n(010) + P_n(00000) + 2P_n(0) \\
&\quad - 2P_n(00) - 2P_n(0000) - 2P_n(0110).
\end{aligned}
$$

Rule 78

$$
\begin{aligned}
P_{n+1}(0) &= -2P_n(0) + P_n(010) + 1 + P_n(00) + P_n(000), \\
P_{n+1}(00) &= 1 + P_n(0110) - 4P_n(0) + 2P_n(010) + 3P_n(00) + P_n(000), \\
P_{n+1}(000) &= P_n(0000) + 1 + 2P_n(0110) + P_n(01110) \\
&\quad - 5P_n(0) + 3P_n(010) + 4P_n(00), \\
P_{n+1}(010) &= -P_n(010) + P_n(00110) + 2P_n(0) - 2P_n(0110) \\
&\quad - 3P_n(00) + P_n(000).
\end{aligned}
$$

Rule 90

$$
\begin{aligned}
P_{n+1}(0) &= -2P_n(0) + 2P_n(010) + 1 + 2P_n(000), \\
P_{n+1}(00) &= 4P_n(010) - P_n(0100) - P_n(0010) + 1 + P_n(0110) \\
&\quad - 4P_n(0) + 3P_n(00) + P_n(0000), \\
P_{n+1}(000) &= P_n(00000) + P_n(01010) + 4P_n(010) - P_n(0100) - P_n(0010) \\
&\quad + P_n(00100) + 1 + 2P_n(0110) + P_n(01110) - 5P_n(0) + 4P_n(00), \\
P_{n+1}(010) &= P_n(00010) + P_n(01000) + P_n(0010) + P_n(00110) \\
&\quad + 2P_n(0) - 2P_n(010) - 2P_n(0110) \\
&\quad - 4P_n(00) + 2P_n(000) + P_n(0100) + P_n(01100).
\end{aligned}
$$

Rule 94

$$
\begin{aligned}
P_{n+1}(0) &= -2P_n(0) + P_n(010) + 1 + 2P_n(000), \\
P_{n+1}(00) &= 1 + P_n(0110) - 4P_n(0) + 2P_n(010) + 3P_n(00) + P_n(0000), \\
P_{n+1}(000) &= P_n(00000) + 1 + 2P_n(0110) + P_n(01110) \\
&\quad - 5P_n(0) + 3P_n(010) + 4P_n(00), \\
P_{n+1}(010) &= -P_n(010) + P_n(00100) + P_n(00110) + 2P_n(0) - 2P_n(0110)
\end{aligned}
$$

Rule 104

$$
\begin{aligned}
P_{n+1}(0) &= -3P_n(0) + 4P_n(00) + 2P_n(010) - P_n(000) + 1, \\
P_{n+1}(00) &= P_{n+1}(0100) + P_n(0010) + 1 + P_n(0110) \\
&\quad - 4P_n(0) + 2P_n(010) + 4P_n(00), \\
P_{n+1}(000) &= P_n(00100) + P_n(0100) + P_n(000) + P_n(0010) + 1 + 2P_n(0110) \\
&\quad + P_n(01110) - 5P_n(0) + 3P_n(010) + 4P_n(00),
\end{aligned}
$$

$$\begin{aligned}P_{n+1}(010) &= 2P_n(00) - 2P_n(000) - P_n(0010) - P_n(00110)\\ &\quad + P_n(01010) - P_n(0100) - P_n(01100).\end{aligned}$$

Rule 105

$$\begin{aligned}P_{n+1}(0) &= 1 - 2P_n(000) + 2P_n(010) - 3P_n(0) + 4P_n(00),\\ P_{n+1}(00) &= P_n(0100) + P_n(0010) + 1 + P_n(0110) - 4P_n(0) + 2P_n(010)\\ &\quad + 4P_n(00) - 2P_n(000) + P_n(0000),\\ P_{n+1}(000) &= P_n(00100) + P_n(0100) - P_n(01000) + P_n(0010) - P_n(00010) + 1\\ &\quad + 2P_n(0110) + P_n(01110) - 5P_n(0) + 3P_n(010) + 4P_n(00),\\ P_{n+1}(010) &= 2P_n(00) - P_n(000) - P_n(0010) - P_n(00110) + P_n(01010)\\ &\quad - 2P_n(0000) + P_n(00000) - P_n(0100) - P_n(01100).\end{aligned}$$

Rule 106

$$\begin{aligned}P_{n+1}(0) &= 3P_n(00) - 3P_n(0) + 2P_n(010) + 1,\\ P_{n+1}(00) &= 1 + P_n(0110) - 4P_n(0) + 2P_n(010)\\ &\quad + 3P_n(00) + P_n(0100) + P_n(000),\\ P_{n+1}(000) &= P_n(01000) + P_n(0000) + 1 + 2P_n(0110) + P_n(01110)\\ &\quad - 5P_n(0) + 3P_n(010) + 4P_n(00),\\ P_{n+1}(010) &= P_n(01010) + P_n(0010) + P_n(00)\\ &\quad - P_n(0100) - P_n(01100) - P_n(000).\end{aligned}$$

Rule 108

$$\begin{aligned}P_{n+1}(0) &= -3P_n(0) + P_n(010) + 4P_n(00) - P_n(000) + 1,\\ P_{n+1}(00) &= 1 + P_n(0110) - 4P_n(0) + 2P_n(010) + 4P_n(00),\\ P_{n+1}(000) &= P_{n+1}(000) + 1 + 2P_n(0110) + P_n(01110)\\ &\quad - 5P_n(0) + 3P_n(010) + 4P_n(00),\\ P_{n+1}(010) &= P_n(00100) + 2P_n(00) - 2P_n(000) - P_n(0010)\\ &\quad - P_n(00110) - P_n(0100) - P_n(01100).\end{aligned}$$

Rule 110

$$\begin{aligned}P_{n+1}(0) &= 3P_n(00) - 3P_n(0) + P_n(010) + 1,\\ P_{n+1}(00) &= 1 + P_n(0110) - 4P_n(0) + 2P_n(010) + 3P_n(00) + P_n(000),\\ P_{n+1}(000) &= P_n(0000) + 1 + 2P_n(0110) + P_n(01110)\\ &\quad - 5P_n(0) + 3P_n(010) + 4P_n(00),\\ P_{n+1}(010) &= P_n(00) - P_n(0100) - P_n(01100) - P_n(000).\end{aligned}$$

Rule 122

$$\begin{aligned}
P_{n+1}(0) &= P_n(000) + 2P_n(00) - 3P_n(0) + 2P_n(010) + 1,\\
P_{n+1}(00) &= 1 + P_n(0110) - 4P_n(0) + 2P_n(010) + 3P_n(00) + P_n(0000),\\
P_{n+1}(000) &= P_n(00000) + 1 + 2P_n(0110) + P_n(01110) - 5P_n(0) + 3P_n(010),\\
&\quad + 4P_n(00),\\
P_{n+1}(010) &= P_n(00010) + P_n(01000) + P_n(01010).
\end{aligned}$$

Rule 126

$$\begin{aligned}
P_{n+1}(0) &= P_n(000) + 2P_n(00) - 3P_n(0) + P_n(010) + 1,\\
P_{n+1}(00) &= 1 + P_n(0110) - 4P_n(0) + 2P_n(010) + 3P_n(00) + P_n(0000),\\
P_{n+1}(000) &= P_n(00000) + 1 + 2P_n(0110) + P_n(01110) - 5P_n(0)\\
&\quad + 3P_n(010) + 4P_n(00),\\
P_{n+1}(010) &= 0.
\end{aligned}$$

Rule 128

$$\begin{aligned}
P_{n+1}(0) &= 3P_n(0) - P_n(010) - 2P_n(00),\\
P_{n+1}(00) &= 2P_n(0) + P_n(0110) - P_n(00),\\
P_{n+1}(000) &= P_n(010) + P_n(0) + 2P_n(0110),\\
P_{n+1}(010) &= P_n(01110).
\end{aligned}$$

Rule 130

$$\begin{aligned}
P_{n+1}(0) &= 3P_n(0) - P_n(010) - 3P_n(00) + P_n(000),\\
P_{n+1}(00) &= 2P_n(000) + 2P_n(0) - 3P_n(00) + P_n(0110),\\
P_{n+1}(000) &= 2P_n(000) + P_n(010) - P_n(00110) + P_n(0)\\
&\quad - 2P_n(00) - P_n(0010) + 2P_n(0110),\\
P_{n+1}(010) &= -P_n(000) + P_n(01110) + P_n(00).
\end{aligned}$$

Rule 132

$$\begin{aligned}
P_{n+1}(0) &= 3P_n(0) - 2P_n(010) - 2P_n(00),\\
P_{n+1}(00) &= -2P_n(010) + 2P_n(0) + P_n(0110) - P_n(00),\\
P_{n+1}(000) &= -2P_n(010) + P_n(01010) + P_n(0) + 2P_n(0110),\\
P_{n+1}(010) &= P_n(01110) + P_n(010).
\end{aligned}$$

Rule 134

$$\begin{aligned}
P_{n+1}(0) &= 3P_n(0) - 2P_n(010) - 3P_n(00) + P_n(000),\\
P_{n+1}(00) &= P_{n+1}(0110) + 2P_n(000) + P_n(0010)\\
&\quad + 2P_n(0) - 2P_n(010) - 3P_n(00),\\
P_{n+1}(000) &= 2P_n(0110) - P_n(00110) + 2P_n(000)\\
&\quad - P_n(01000) + P_n(0010) + P_n(0)\\
&\quad - 2P_n(010) - 2P_n(00) + P_n(0100) + P_n(01010),\\
P_{n+1}(010) &= - P_n(000) + P_n(01110) - 2P_n(0010) + P_n(00) + P_n(010).
\end{aligned}$$

Rule 136

$$\begin{aligned}
P_{n+1}(0) &= - P_n(00) + 2P_n(0),\\
P_{n+1}(00) &= P_n(010) + P_n(0),\\
P_{n+1}(000) &= 2P_n(010) + P_n(00),\\
P_{n+1}(010) &= P_n(0110).
\end{aligned}$$

Rule 138

$$\begin{aligned}
P_{n+1}(0) &= -2P_n(00) + 2P_n(0) + P_n(000),\\
P_{n+1}(00) &= P_n(010) - P_n(0010) + P_n(000) - P_n(00) + P_n(0),\\
P_{n+1}(000) &= 2P_n(010) - 2P_n(0010) + P_n(000),\\
P_{n+1}(010) &= P_n(0010) + P_n(0110) - P_n(00110).
\end{aligned}$$

Rule 140

$$\begin{aligned}
P_{n+1}(0) &= - P_n(00) + 2P_n(0) - P_n(010),\\
P_{n+1}(00) &= - P_n(010) + P_n(0),\\
P_{n+1}(000) &= P_n(00) - P_n(0100),\\
P_{n+1}(010) &= P_n(010) + P_n(0110).
\end{aligned}$$

Rule 142

$$\begin{aligned}
P_{n+1}(0) &= -2P_n(00) + 2P_n(0) - P_n(010) + P_n(000),\\
P_{n+1}(00) &= P_n(000) - P_n(00) - P_n(010) + P_n(0),\\
P_{n+1}(000) &= P_n(000) - P_n(01000),\\
P_{n+1}(010) &= P_n(010) - P_n(0010) + P_n(0110) - P_n(00110).
\end{aligned}$$

Rule 146

$$\begin{aligned}
P_{n+1}(0) &= 3P_n(0) - P_n(010) - 4P_n(00) + 2P_n(000), \\
P_{n+1}(00) &= P_n(0110) + 2P_n(0) - 4P_n(00) + 2P_n(000) + P_n(0000), \\
P_{n+1}(000) &= P_n(00000) + P_n(010) - P_n(0100) + 2P_n(0110) - P_n(01100) \\
&\quad - P_n(0010) + P_n(00100) - P_n(00110) + P_n(0) - 2P_n(00) + P_n(000), \\
P_{n+1}(010) &= -2P_n(0000) + 2P_n(000) + P_n(01110).
\end{aligned}$$

Rule 150

$$\begin{aligned}
P_{n+1}(0) &= -2P_n(010) + 3P_n(0) - 4P_n(00) + 2P_n(000), \\
P_{n+1}(00) &= P_n(0110) + P_n(0010) + 2P_n(0) - 2P_n(010) - 4P_n(00) \\
&\quad + P_n(0100) + 2P_n(000) + P_n(0000), \\
P_{n+1}(000) &= P_n(00000) + 2P_n(0110) - P_n(01100) - P_n(00110) + P_n(0010) \\
&\quad + P_n(0) - 2P_n(010) - 2P_n(00) + P_n(0100) + P_n(000) + P_n(01010), \\
P_{n+1}(010) &= -P_n(00010) - 2P_n(0000) + 2P_n(000) + P_n(01110) + P_n(010) \\
&\quad - P_n(0100) - P_n(0010) + P_n(00100) - P_n(01000).
\end{aligned}$$

Rule 152

$$\begin{aligned}
P_{n+1}(0) &= -2P_n(00) + 2P_n(0) + P_n(000), \\
P_{n+1}(00) &= P_n(010) + 2P_n(000) + P_n(0) - 2P_n(00), \\
P_{n+1}(000) &= P_n(0000) + P_n(00010) - P_n(0010) + 2P_n(010) - P_n(0100), \\
P_{n+1}(010) &= P_n(0110) - P_n(000) + P_n(00).
\end{aligned}$$

Rule 154

$$\begin{aligned}
P_{n+1}(0) &= 2P_n(0) - 3P_n(00) + 2P_n(000), \\
P_{n+1}(00) &= P_n(010) - P_n(0010) + P_n(0000) + P_n(0) - 2P_n(00) + P_n(000), \\
P_{n+1}(000) &= P_n(00000) + 2P_n(010) - P_n(0100) - 2P_n(0010) + P_n(00100), \\
P_{n+1}(010) &= P_n(00010) + P_n(0110) - P_n(00110) + P_n(000) - P_n(0000).
\end{aligned}$$

Rule 156

$$\begin{aligned}
P_{n+1}(0) &= -2P_n(00) + 2P_n(0) - P_n(010) + P_n(000), \\
P_{n+1}(00) &= 2P_n(000) - P_n(010) + P_n(0) - 2P_n(00) + P_n(0100), \\
P_{n+1}(000) &= P_n(0000), \\
P_{n+1}(010) &= P_n(010) - 2P_n(0100) + P_n(0110) - P_n(000) + P_n(00).
\end{aligned}$$

Rule 160

$$
\begin{aligned}
P_{n+1}(0) &= 2P_n(0) - P_n(010) - P_n(000),\\
P_{n+1}(00) &= -2P_n(000) + 3P_n(00) + P_n(0110),\\
P_{n+1}(000) &= -P_n(000) + P_n(00100) + P_n(00110) + P_n(01100) + 2P_n(00),\\
P_{n+1}(010) &= P_n(01110) + P_n(0) - 2P_n(00) + P_n(000).
\end{aligned}
$$

Rule 162

$$
\begin{aligned}
P_{n+1}(0) &= 2P_n(0) - P_n(010) - P_n(00),\\
P_{n+1}(00) &= P_n(0110) + P_n(00),\\
P_{n+1}(000) &= P_n(01100) + P_n(000),\\
P_{n+1}(010) &= P_n(01110) + P_n(0) - P_n(00).
\end{aligned}
$$

Rule 164

$$
\begin{aligned}
P_{n+1}(0) &= 2P_n(0) - P_n(000) - 2P_n(010),\\
P_{n+1}(00) &= -2P_n(000) + 3P_n(00) - P_n(0010) - P_n(0100) + P_n(0110),\\
P_{n+1}(000) &= -P_n(000) + P_n(00110) + P_n(01100) + 2P_n(00)\\
&\quad - P_n(0010) - P_n(0100),\\
P_{n+1}(010) &= P_n(00100) + P_n(01110) + P_n(0010) + P_n(0) - 2P_n(010)\\
&\quad - 2P_n(00) + P_n(0100) + P_n(000) + P_n(01010).
\end{aligned}
$$

Rule 168

$$
\begin{aligned}
P_{n+1}(0) &= P_n(0) - P_n(000) + P_n(00),\\
P_{n+1}(00) &= P_n(0010) + 2P_n(00) - P_n(000),\\
P_{n+1}(000) &= P_n(00100) + P_n(0010) + P_n(00),\\
P_{n+1}(010) &= P_n(00110) + P_n(010) - P_n(0010).
\end{aligned}
$$

Rule 170

$$
\begin{aligned}
P_{n+1}(0) &= P_n(0),\\
P_{n+1}(00) &= P_n(00),\\
P_{n+1}(000) &= P_n(000),\\
P_{n+1}(010) &= P_n(010).
\end{aligned}
$$

Rule 172

$$\begin{aligned}
P_{n+1}(0) &= -P_n(010) + P_n(0) - P_n(000) + P_n(00), \\
P_{n+1}(00) &= 2P_n(00) - P_n(0100) - P_n(000), \\
P_{n+1}(000) &= P_n(00) - P_n(0100), \\
P_{n+1}(010) &= P_n(00100) + P_n(00110).
\end{aligned}$$

Rule 178

$$\begin{aligned}
P_{n+1}(0) &= 2P_n(0) - P_n(010) + P_n(000) - 2P_n(00), \\
P_{n+1}(00) &= P_n(0000) + P_n(0110), \\
P_{n+1}(000) &= P_n(00000), \\
P_{n+1}(010) &= -2P_n(0000) + 3P_n(000) + P_n(01110) + P_n(0) - 2P_n(00).
\end{aligned}$$

Rule 184

$$\begin{aligned}
P_{n+1}(0) &= P_n(0), \\
P_{n+1}(00) &= P_n(000) + P_n(0010), \\
P_{n+1}(000) &= P_n(0000) + P_n(00010), \\
P_{n+1}(010) &= P_n(00110) - P_n(000) + P_n(00) + P_n(010) - P_n(0010).
\end{aligned}$$

Rule 200

$$\begin{aligned}
P_{n+1}(0) &= P_n(0) + P_n(010), \\
P_{n+1}(00) &= 2P_n(010) + P_n(00), \\
P_{n+1}(000) &= P_n(0010) + P_n(0100) + P_n(01010) + P_n(000) + P_n(010), \\
P_{n+1}(010) &= 0.
\end{aligned}$$

Rule 204

$$\begin{aligned}
P_{n+1}(0) &= P_n(0), \\
P_{n+1}(00) &= P_n(00), \\
P_{n+1}(000) &= P_n(000), \\
P_{n+1}(010) &= P_n(010).
\end{aligned}$$

Rule 232

$$\begin{aligned}
P_{n+1}(0) &= 2P_n(00) - P_n(000) + P_n(010), \\
P_{n+1}(00) &= P_n(0100) + P_n(0010) + P_n(00), \\
P_{n+1}(000) &= P_n(00100) + P_n(0100) + P_n(000) + P_n(0010), \\
P_{n+1}(010) &= P_n(01010).
\end{aligned}$$

Index

Page numbers followed by "d" refer to deterministic solutions given in Appendix B, while pages followed by "p" refer to deterministic solutions of Appendix C.

H. Fukś, *Solvable Cellular Automata*, Understanding Complex Systems,
https://doi.org/10.1007/978-3-031-38700-5

The manufacturer's authorised representative in the EU is Springer Nature Customer Service Centre GmbH, Europaplatz 3, 69115 Heidelberg, Germany. If you have any concerns regarding our products, please contact ProductSafety@springernature.com

Printed and bound by CPI Group (UK) Ltd, Croydon, CR0 4YY
05/10/2026
02228122-0011